A. Azzi, K.A. Nałęcz, M.J. Nałęcz
L. Wojtczak (Eds.)

Anion Carriers of Mitochondrial Membranes

With 147 Figures

Springer-Verlag
Berlin Heidelberg New York
London Paris Tokyo

Prof. Dr. ANGELO AZZI
Institut für Biochemie
und Molekularbiologie
Universität Bern
Bühlstraße 28
CH-3012 Bern

Dr. KATARZYNA A. NAŁĘCZ
Dr. MACIEJ J. NAŁĘCZ
Dr. LECH WOJTCZAK
Department of Cellular Biochemistry
Nencki Institute of Experimental Biology
Pasteurstreet 3
PL-02-093 Warsaw

ISBN 3-540-50853-8 Springer-Verlag Berlin Heidelberg New York
ISBN 0-387-50853-8 Springer-Verlag New York Berlin Heidelberg

Library of Congress Cataloging-in-Publication Data. Anion carriers of mitochondrial membranes /
edited by Angelo Azzi ... [et al.] ; with contributions by V. Adams ... [et al.]. p. cm. Proceedings
of the International Conference on Anion Carriers of Mitochondrial Membranes, held July 5–9,
1988, in Zakopane, Poland, organized by the Nencki Institute of Experimental Biology. Includes
index. 1. Carrier proteins–Congresses. 2. Mitochondrial membranes–Congresses. I. Azzi,
A. (Angelo) II. Adams, V. III. Instytut Biologii Doświadczalnej im. M. Nenckiego. IV. International Conference on Anion Carriers of Mitochondrial Membranes (1988 : Zakopane, Poland)
QP552.C34A55 1989 574.87'342–dc 19 89-5901

Printing: Druckhaus Beltz, Hemsbach/Bergstraße
Binding: J. Schäffer GmbH & Co. KG., Grünstadt
2131/3145-543210 – Printed on acid-free paper

Preface

Almost a quarter of a century ago it became apparent that most of the important metabolites that circulate between cytosol and mitochondria do not cross the inner mitochondrial membrane freely but are transferred by specific carriers. During subsequent years, several carriers of this kind have been functionally identified. Their total number amounts, to our present knowledge, to 12-14. However, successful attempts to isolate some of them started only at the end of the last decade. Methods for isolation, purification, and functional reconstitution of mitochondrial carriers which have been developed during the last few years have opened a new and fascinating field. A few carriers have been sequenced, which enables the study of their arrangement within the membranes and elucidation of their mechanism of action. Genetic and comparative evolutionary studies also offer further possibilities. Finally, possible defects of particular carriers may help to understand the genesis of some inborn metabolic disorders.

In order to assemble the information available today on mitochondrial carriers and to set up projects for the future, an International Conference on Anion Carriers of Mitochondrial Membranes, was held on 5-9 July 1988 in Zakopane, Poland. Its program was formulated by the editors of this volume in collaboration with Pierre V. Vignais (Grenoble) and Attila Fonyó (Budapest) with the aim of bringing together practically all the scientists working in the field .

The Zakopane Conference, the first specifically devoted to mitochondrial transport proteins, was organized by the Nencki Institute of Experimental Biology in Warsaw to celebrate the 70th anniversary of its foundation. Named after Marceli Nencki (1847-1901), eminent physiologist and biochemist of Polish origin, active in Bern (Switzerland) and St. Petersburg (Russia), this institute was founded in 1918-1919, moved to Lodz after the Second World War, and in 1952 back to Warsaw as an establishment of the Polish Academy of Sciences.

One immediate result of the Conference was to demonstrate that the interest in mitochondrial carriers was not restricted to a small specialized group of scientists. Rather, due to the interactions that mitochondrial anion transport has with the rest of the cell, in catabolic and anabolic processes, and under physiological and pathological conditions, a wider echo was achieved among scientists operating in cell metabolism, organ and in vivo bioenergetics, as well as among researchers dedicated to medical problems.

The present book contains the full texts of most of the invited lectures presented at the Zakopane Conference.

Apart from the anion carriers of the inner mitochondrial membrane, the subject of the Conference was extended both to the proton-conducting protein of brown adipose tissue mitochondria, because of its structural similarity to the adenine nucleotide carrier, and also to the pore protein of the outer mitochondrial membrane, because of the methodological similarity of its isolation and its functional relationships with the inner mitochondrial membrane carriers.

Angelo Azzi   Katarzyna A. Nałęcz   Maciej J. Nałęcz   Lech Wojtczak

Contents

I. Isolation and Reconstitution of Carriers

Purification and Characterization of Three Mitochondrial Substrate Carriers:
the Phosphate, the 2-Oxoglutarate and the Dicarboxylate Carriers
F. Palmieri, G. Genchi, V. Zara, C. Indiveri and F. Bisaccia............................ 3

Purification and Reconstitution of the 2-Oxoglutarate Carrier
from Bovine Heart and Liver Mitochondria
D. Claeys, M. Müller and A. Azzi (With 4 Figures)............................ 17

Hydroxyapatite Chromatography as a Tool for the Isolation
of Anion Carriers and Other Membrane Proteins
P. Riccio (With 8 Figures)............................ 35

Purification of the Monocarboxylate Carrier by Affinity Chromatography
K.A. Nałęcz, R. Bolli, L. Wojtczak and A. Azzi (With 5 Figures)............................ 45

Recent Developments in the Extraction, Reconstitution, and
Purification of the Mitochondrial Citrate Transporter
from Normal and Diabetic Rats
R.S. Kaplan, J.A. Mayor, D.L. Oliveira and N. Johnston (With 6 Figures)............................ 59

Isolation and Functional Reconstitution of the Dicarboxylate
Carrier from Bovine Liver Mitochondria
M.J. Nałęcz, A. Szewczyk, C. Broger,
L. Wojtczak and A. Azzi (With 4 Figures)............................ 71

New Photoaffinity Derivatives of Malonate and Succinate
to Study Mitochondrial Carrier Systems
A. Szewczyk and M.J. Nałęcz (With 8 Figures)............................ 87

Reaction Mechanism of the Reconstituted Aspartate/Glutamate
Antiporter from Mitochondria. Reversible Switching to Uniport Function
T. Dierks and R. Krämer (With 7 Figures).. 99

II. Functional Evidence and Characterization of Various Carriers

Recent Studies on the Mitochondrial Phosphate Transport
Protein (PTP) and on its Relationship to the ADP/ATP Translocase (AAC)
and the Uncoupling Protein (UCP)
H. Wohlrab, C. Bukusoglu and H. DeFoe (With 4 Figures)......................... 113

Mitochondrial Phosphate Carrier: Relation of its SH Groups
to Oligomeric Organization
E. Ligeti, E. Brazda and A. Fonyó (With 7 Figures)...................................... 123

Recent Developments in the Study of the Conformational
States and the Nucleotide Binding Sites of the ADP/ATP Carrier
P.V. Vignais, G. Brandolin, F. Boulay, P. Dalbon,
M. Block and I. Gauche (With 6 Figures).. 133

Immunological and Enzymatic Approaches to the Orientation
of the Membrane Bound ADP/ATP Carrier
G.Brandolin, F. Boulay, P. Dalbon, M. Block,
I. Gauche and P.V. Vignais (With 6 Figures)... 147

The ATP/ADP Antiporter is Involved in the Uncoupling
Effect of Fatty Acids
A.Yu. Andreyev, T.O. Bondareva, V.I. Dedukhova,
E.N. Mokhova, V.P. Skulachev, L.M. Tsofina,
N.I. Volkov and T.V. Vygodina (With 7 Figures)... 159

Molecular Aspects of the Adenine Nucleotide Carrier from Mitochondria
M. Klingenberg (With 4 Figures)... 169

Kinetic Mechanisms of the Adenylic and the Oxoglutaric Carriers: a Comparison
F.E. Sluse, C.M. Sluse-Goffart and C. Duyckaerts (With 6 Figures)........................... 183

III. Porins

Porins from Mitochondrial and Bacterial Outer Membranes:
Structural and Functional Aspects
Roland Benz (With 8 Figures)... 199

Modulation of the Mitochondrial Channel VDAC by a Variety of Agents
M. Colombini, M. J. Holden and P.S. Mangan (With 3 Figures)............................... 215

Bioenergetic Consequences of the Lack of Mitochondrial Porin:
Identification of a Putative New Pore
J. Michejda, X. J. Guo and G. J.-M. Lauquin (With 5 Figures)................................ 225

Purification of Mammalian Porins
V. de Pinto, L. Gaballo, R. Benz and F. Palmieri (With 4 Figures)........................... 237

IV. Uncoupling Protein of Brown Adipose Tissue

A Molecular Biology Study of the Uncoupling Protein
of Brown Fat Mitochondria. A Contribution to the Analysis
of Genes of Mitochondrial Carriers
F. Bouillaud, S. Raimbault, L. Casteilla,
A.-M. Cassard and D. Ricquier (With 4 Figures)... 251

On the Mechanism of Transport by the Uncoupling Protein
from Brown Adipose Tissue Mitochondria
E. Rial and D.G. Nicholls (With 2 Figures)... 261

Regulation of the Amount and Activity of the Uncoupling
Protein Thermogenin in Brown Adipose Tissue
B. Cannon and J. Nedergaard (With 4 Figures)... 269

Regulation of Uncoupling Protein and Formation of Thermogenic Mitochondria
J. Houštěk and J. Kopecký (With 5 Figures)... 283

V. Carriers and their Cellular Environment

Biogenesis of Mitochondrial Proteins
M. Tropschug and W. Neupert.. 295

Insensitivity of Carbamoyl-Phosphate Synthetase Towards
Inhibition by Carbamoyl Phosphate Makes it Unlikely
that Mitochondrial Metabolite Transport Controls Ornithine Cycle Flux
A.J. Meijer (With 5 Figures).. 307

Mitochondrial Adenine Nucleotide Translocation During Fatty
Acid Metabolism in the Intact Cell
S. Soboll (With 4 Figures)... 317

Control of Oxidative Phosphorylation in Yeast Mitochondria:
The Role of Phosphate Carrier and pH
J.-P. Mazat, E. Jean-Bart, M. Rigoulet,
C. Reder and B. Guerin (With 3 Figures).. 327

The Role of Pyrophosphate and the Adenine Nucleotide Transporter
in the Regulation of the Intra-Mitochondrial Volume
A.P. Halestrap and A.M. Davidson (With 6 Figures)...................................... 337

Role of the Mitochondrial Outer Membrane in Dynamic
Compartmentation of Adenine Nucleotides
F.N. Gellerich, R. Bohnensack and W. Kunz (With 3 Figures)....................... 349

Topology of Peripheral Kinases: its Importance in Transmission
of Mitochondrial Energy
D. Brdiczka, V. Adams, M. Kottke and R. Benz (With 6 Figures)................... 361

Control of Mitochondrial Energy Production *in vivo*
D.J. Taylor (With 3 Figures).. 373

List of Contributors

You will find the addresses at the beginning of the respective contribution

Adams, V. 361
Andreyev, A.Yu. 159
Azzi, A. 17, 45, 71
Benz, R. 199, 237, 361
Bisaccia, F. 3
Block, M. 133, 147
Bohnensack, R. 349
Bolli, R. 45
Bondareva, T.O. 159
Bouillaud, F. 251
Boulay, F. 133, 147
Brandolin, G. 133, 147
Brazda, E. 123
Brdiczka, D. 361
Broger, C. 71
Bukusoglu, C. 133
Cannon, B. 269
Cassard, A.-M. 251
Casteilla, L. 251
Claeys, D. 17
Colombini, M. 215
Dalbon, P. 133, 147
Davidson, A.M. 337
de Pinto, V. 237
Dedukhova, V.I. 159
DeFoe, H. 113
Dierks, T. 99
Duyckaerts, C. 183
Fonyó, A. 123
Gaballo, L. 237
Gauche, I. 133, 147
Gellerich, F.N. 349
Genchi, G. 3
Guerin, B. 327
Guo, X.J. 225
Halestrap, A. 337
Holden, M.J. 215
Houštěk, J. 283
Indiveri, C. 3
Jean-Bart, E. 327
Johnston, N. 59
Kaplan, R.S. 59

Klingenberg, M. 169
Kopecký, J. 283
Kottke, M. 361
Krämer, R. 99
Kunz, W. 349
Lauquin, G.J.-M. 225
Ligeti, E. 123
Mangan, P.S. 215
Mayor, J.A. 59
Mazat, J.-P. 327
Meijer, A.J. 307
Michejda, J. 225
Mokhova, E.N. 159
Müller, M. 17
Nałecz, K.A. 45
Nałecz, M.J. 71, 87
Nedergaard, J. 269
Neupert, W. 295
Nicholls, D.G. 261
Oliveira, D.L. 59
Palmieri, F. 3, 237
Raimbault, S. 251
Reder, C. 327
Rial, E. 261
Riccio, P. 35
Ricquier, D. 251
Rigoulet, M. 327
Skulachev, V.P. 159
Sluse, F.E. 183
Sluse-Goffart, C.M. 183
Soboll, S. 317
Szewczyk, A. 71, 87
Taylor, D.J. 373
Tropschug, M. 295
Tsofina, L.M. 159
Vignais, P.V. 133, 147
Volkov, N.I. 159
Vygodina, T.V. 159
Wohlrab, H. 113
Wojtczak, L. 45, 71
Zara, V. 3

I. Isolation and Reconstitution of Carriers

Purification and Characterization of Three Mitochondrial Substrate Carriers: the Phosphate, the 2-Oxoglutarate and the Dicarboxylate Carriers

F. Palmieri, G. Genchi, V. Zara, C. Indiveri and F. Bisaccia

Department of Pharmaco-Biology, Laboratory of Biochemistry, University of Bari
and
CNR Unit for the Study of Mitochondria and Bioenergetics, Traversa 200 Re David 4,
I- 70125 Bari, Italy

Since several substrates of the mitochondrial enzymes must be transported from the cytosol into the matrix and several products have to leave the mitochondria, the inner mitochondrial membrane has to be equipped with transport systems which catalyze metabolic flows between the inner and the outer compartment. The existence of at least 9 transport systems for metabolites has been documented in some detail by studies performed in intact mitochondria (LaNoue & Schoolwerth, 1979; Meijer & Van Dam, 1981; Palmieri et al., 1987). With the exception of the carrier for carnitine and acylcarnitine esters, the other carriers deal with the transport of anions: substrates of oxidative phosphorylation, ADP, ATP and phosphate, fuels of the tricarboxylic acid cycle, pyruvate, β-hydroxybutyrate and acetoacetate, dicarboxylates and tricarboxylates, and substrates of amino acid metabolism. This intense traffic of anions across the mitochondrial membrane is necessary, besides for oxidative phosphorylation, the tricarboxylic acid cycle and the amino acid metabolism, for the transfer of reducing equivalents in both directions and for important metabolic pathways, whose enzymes

---

Abbreviations: GOT, glutamate-oxaloacetate transamianse; SDS, sodium dodecylsulphate; EMA, eosin-5-maleimide; PTC, phenylisothiocyanate; p-sulpho-PTC, p-sulphophenylisothiocyanate; ANPP, 4-azido-2-nitrophenyl phosphate

---

A. Azzi et al. (Eds.)
Anion Carriers of Mitochondrial Membranes
© Springer-Verlag Berlin Heidelberg 1989

are partitioned between the cytosol and the mitochondria, such as gluconeogenesis, fatty acid synthesis, ketogenesis, β-oxidation of fatty acids and, in liver, urogenesis. Many of the properties of the proposed mitochondrial anion carriers, *i.e.* the high specificity, the existence of specific inhibitors, the saturation kinetics, the different distribution in various tissues and species and the inhibition by SH reagents have pointed to the protein nature of these transport systems. Their isolation, however, has been hindered for quite a long time.

The ADP/ATP transport system has been the first mitochondrial carrier to be isolated (Klingenberg *et al.*, 1978). It was purified by solubilization of the mitochondrial membranes with Triton and chromatography on hydroxyapatite. Later, around the end of 1980, the phosphate carrier has also been purified by chromatography on hydroxyapatite (Wohlrab, 1980; Kolbe *et al.*, 1981; Palmieri *et al.*, 1981; Touraille *et al.*, 1981). Until 1984 the hydroxyapatite eluate of Triton-solubilized heart mitochondria was generally considered to contain only the ADP/ATP carrier and the phosphate carrier. However, when high resolution SDS gel electrophoresis is applied, the pass-through of hydroxyapatite of Triton-solubilized heart mitochondria can be shown to contain 5 protein bands in the molecular weight region of 30-36 kDa (Bisaccia & Palmieri, 1984). The first fraction contains four protein bands called 2-5, since band 1 appears only in later fractions. With the exception of the protein band with the lowest M_r (band 5) which is the ADP/ATP carrier, the identity of the other bands was unknown.

In the last few years we have identified three of the four remaining bands of the hydroxyapatite pass-through of Triton-solubilized heart mitochondria as carriers or pore proteins. Band 2 represents the mitochondrial porin (De Pinto *et al.*, 1985), band 3 the phosphate carrier (Bisaccia &P almieri, 1984) and band 4 the oxoglutarate carrier (Bisaccia *et al.*, 1985). This identification was based on the purification of each protein band and the use of functional reconstitution as a monitor of the transport activity during isolation. Using the same strategy we have also identified the dicarboxylate carrier and the oxoglutarate carrier in the hydroxyapatite pass-through of Triton-solubilized liver mitochondria (Bisaccia *et al.*, 1988).

The aim of this paper is to give an overview on the state of the art about identification and characterization of the mitochondrial substrate carriers we have purified from heart and/or liver mitochondria.

METHODS

The procedures for the isolation of carrier proteins have been reported previously. Essentially, the phosphate carrier was purified by chromatography on hydroxyapatite of Triton X-114 solubilized heart mitochondria supplemented with

4 mg/ml cardiolipin (Bisaccia & Palmieri, 1984). The oxoglutarate carrier was purified by chromatography on hydroxyapatite and celite of Triton X-114 solubilized heart mitochondria in the presence of 2 mg/ml cardiolipin (Bisaccia et al., 1985). Porin was purified from Triton X-100 solubilized mitochondria by chromatography on hydroxyapatite and celite in the absence of cardiolipin (De Pinto et al., 1985; De Pinto et al., 1987). The dicarboxylate carrier and the oxoglutarate carrier from rat liver mitochondria were purified by chromatography on hydroxyapatite after partial or extensive removal of Triton X-114 from the mitochondrial extract (Bisaccia et al., 1988).

Reconstitution of carrier proteins into liposomes was carried out by the freeze-thaw-sonication procedure (Kasahara & Hinkle, 1977) or by removing the detergent by chromatography on Amberlite (Krämer & Heberger, 1986; Indiveri et al., 1987a; Bisaccia et al., 1988).

Transport of labelled substrates in reconstituted liposomes was measured by the inhibitor stop method (Palmieri & Klingenberg, 1979). Polyacrylamide slab gel electrophoresis, staining and protein determination were performed as described previously (Bisaccia & Palmieri, 1984).

For immunoblotting of the oxoglutarate carrier, the proteins separated on 17.5% acrylamide slab gels were electro-blotted onto nitrocellulose. The transferred proteins were incubated with an antiserum raised against the oxoglutarate carrier purified from bovine heart mitochondria. The antibody-antigen complexes were decorated with peroxidase-conjugated antirabbit IgG. The peroxidase reaction was performed in a mixture containing 4-chloro-1-naphtol and H_2O_2.

RESULTS AND DISCUSSION

Isolation of carrier proteins from heart mitochondria

A prerequisite in the elucidation of the function of a protein is its isolation. We have indeed isolated band 2, band 3 and band 4 in a pure form, i.e. each of the protein bands present in the first fraction of the hydroxyapatite eluate of Triton-solubilized heart mitochondria. For identification we tested each isolated protein for several transport activities after their reconstitution into liposomes. In this way we found that band 3 of M_r 33 kDa is the phosphate carrier (Bisaccia & Palmieri, 1984) and band 4 of M_r 31.5 kDa is the oxoglutarate carrier (Bisaccia et al., 1985). The isolated band 2 of M_r 35 kDa did not show any transport activity for substrates of mitochondrial anion carriers. However, we thought that it could be another very hydrophobic protein which forms channels, i.e. the porin of the outer mitochondrial membrane. This was confirmed by reconstitution experiments in planar bilayer membranes in which band 2 induces a step-wise increase of the membrane conductance due to the formation of single channels or pores, which have characteristics similar to those produced by porins previously isolated from bacteria and *Neurospora crassa* (De Pinto *et al.*, 1985).

Phosphate carrier

The isolated band 3 protein, when reconstituted into liposomes, exhibits transport properties very similar to those described for the phosphate transport system in mitochondria. Thus, it catalyzes both the unidirectional transport of phosphate (uptake or efflux) and the exchange between the internal and the external phosphate. The transport of phosphate in the reconstituted system is inhibited by SH-blocking reagents. The reconstituted phosphate exchange follows a first order kinetics and is highly temperature dependent with an E_A of 64 kJ/mol in the lower temperature range. The affinity of the carrier for phosphate is low (K_m for external phosphate 2.2 mM) and the maximum exchange rate at 25°C is 71,000 μmol/min x g protein. Furthermore, the movement of phosphate in proteoliposomes is regulated by the pH gradient across the membrane, as already shown in mitochondria (Palmieri *et al.*, 1970). Based on functional reconstitution the band 3 protein, *i.e.* the phosphate carrier, has been purified 290 fold with a protein yield of 0.22%. This purification factor gives an estimation for the enrichment of the carrier protein. However, purification factors based on reconstituted activities are in general not very accurate due to possible inactivation during purification and incorporation into liposomes.

The isolated phosphate carrier consists of a single band of M_r 33 kDa upon high resolution SDS gel electrophoresis. Kolbe *et al.* (1984) published that the isolated phosphate carrier consists of two bands. We however found that the appearance of these two bands depends on the experimental conditions of how the gel electrophoresis is run. In our hands, under appropriate conditions *i.e.* in the presence of 5% 2-mercaptoethanol, we reproducebly find only one band. In collaboration with Aquila and Klingenberg we have determined more than 90% of the amino acid sequence of the phosphate carrier by Edman degradation (unpublished results). In the meantime (Runswick *et al.*, 1987) have published the entire sequence by cDNA. From the sequence data it appears that the phosphate carrier is homologous to the ADP/ATP carrier and to the uncoupling protein (Runswick *et al.*, 1987; Aquila & Klingenberg, 1987).

Although these results have greatly improved our understanding of the structure of the phosphate carrier, the mechanism of transport remains unknown. To gain more information about the mechanism, it is important to elucidate the phosphate-binding site and to find out which aminoacid residues play a key role in phosphate translocation. So far, SH groups represent the only well-defined functional groups of the phosphate carrier (Fonyo, 1978).

Since positive charges are likely to be present at the substrate-binding site we have investigated the effect of phenylisothiocyanate (PITC) and p-sulphophenyl-isothiocyanate (p-sulphoPITC) on the activity of the phosphate carrier. PITC and p-sulphoPITC were chosen as reagents for the identification of essential lysine residues, since the hydrophobic PITC is expected to modify preferably lysine residues within the core of the membrane (Sigrist & Zahler, 1978) and the polar p-sulphoPITC lysines located at the surface of the membrane. In these experiments, mitochondria were incubated with 2 mM p-sulphoPITC or 2 mM PITC at pH 7.5 and 9, respectively. After washing the mitochondria, the phosphate carrier was solubilized, purified and tested for transport activity in the reconstituted system. Both PITC and p-sulphoPITC cause a substantial decrease of the phosphate-phosphate exchange activity either at pH 7.5 or at pH 9.0. Since isothiocyanates react with nucleophiles in their unprotonated form and may therefore react also with cysteines, it was necessary to check whether the inhibition can be accounted for by an interaction of p-sulphoPITC and PITC with SH groups. This was done by including an excess of DTE in the washing medium at pH 8.5, since these conditions reverse the binding of these inhibitors with SH groups, but not with NH_2 groups. Under these conditions, therefore, the remaining inhibition can be attributed to the reaction with NH_2 groups. In the presence of p-sulphoPITC, DTE restores the activity but only partially; a significant portion of the total inhibition caused by p-sulphoPITC is still present after the treatment with DTE. In the case of PITC, on the other hand, DTE does not cause any appreciable reactivation presumably because in the lipid core of the membrane cysteines are protonated and therefore not reactive. Furthermore, the DTE-insensitive inhibition by p-sulphoPITC is pH-dependent, *i.e.* is more evident at pH 9.0 than at pH 7.5 (Genchi *et al.*, 1988), since this reagent interacts with lysine(s) located at the surface of the membrane which undergo protonation and deprotonation according the pH of the medium. In contrast, the inhibition by PITC is not affected by changing the external pH (Genchi *et al.*, 1988), since this apolar reagent interacts with the buried lysines of the hydrophobic phase of the membrane which should be independent from the bulk-phase pH. On this basis we suggest the presence of a least two types of essential lysine residues in the phosphate carrier, one type located in the hydrophobic moiety of the protein and the other at the surface of the protein. Both kinds of lysines seem to be important for the transport of phosphate because their modification leads to the inhibition of the phosphate carrier.

In order to have a tool for the characterization of the substrate-binding site of the phosphate carrier, we have investigated the reactivity of the carrier with the photoreactive substrate analogue 4-azido-2-nitrophenyl phosphate (ANPP), which was introduced by Lauquin *et al.* (1980) to label the phosphate-binding site of the F_1-ATPase. In the dark ANPP inhibits the transport of phosphate in a competitive manner. Upon illumination however the inhibition becomes irreversible indicating covalent binding of the inhibitor (Tommasino *et al.*, 1987). In contrast, the non phosphorylated analogue, 4-azido-2-nitrophenol, does not cause any inhibition of phosphate transport even after illumination. Thus the presence of the phosphate group in the ANPP molecule is an essential requirement for the inhibition of phosphate transport. This also suggests that ANPP reacts at the substrate-binding site of the phosphate carrier. This suggestion is supported by the finding that phosphate protects the carrier against inactivation by ANPP and that this effect is specific, since other anions, like sulphate, do not show any protection (Tommasino *et al.*, 1987). In other experiments we have also found that the radioactivity bound to the purified phosphate carrier is drastically reduced by the presence of phosphate during the incubation of the mitochondria with ^{32}P-ANPP. On this basis we conclude that ANPP can be used as a label of the substrate-binding site of the phosphate carrier.

2-Oxoglutarate carrier

The functional properties of the isolated band 4 protein of M_r 31.5 kDa reconstituted into liposomes closely resemble those of the oxoglutarate transport system as characterized in mitochondria (Bisaccia *et al.*, 1985). The purified and reconstituted protein catalyzes an exchange between 2-oxoglutarate and L-malate. As the transport system in mitochondria (Palmieri *et al.*, 1972), the purified protein accepts also other dicarboxylates, although with a lower affinity. It is inhibited by phthalonate and, less specifically, by some dicarboxylate analogues. It is also inhibited by certain SH reagents (mersalyl and p-chloromercuribenzoate) but not by N-ethylmaleimide. The substrate affinity for oxoglutarate is 65 μM and the maximum exchange rate at 25°C reaches 4,000-22,000 μmol/min·g protein, in dependence of the particular reconstitution conditions (Indiveri *et al.*, 1987b). The activation energy of the exchange reaction is 54 kJ/mol. The transport is independent of pH in the range between 6 and 8. The oxoglutarate carrier purified from bovine heart and reconstituted into liposomes catalyzes also some unidirectional uptake of oxoglutarate (Indiveri *et al.*, 1987b). This uptake is completely sensitive to the specific inhibitor phthalonate and is only confined to the first fraction of the oxoglutarate preparation. The main indications for an unidirectional transport of oxoglutarate are (1) the uptake of oxoglutarate into unloaded liposomes and (2) the lack of efflux of labelled oxoglutarate after addition of cold substrate to unloaded liposomes. The possibilities of a binding to the protein or a half cycle reaction, *i.e.* a single inward translocation event, are ruled out by the fact that the number of substrate molecules transported exceeds the amount of the carrier protein by 3 orders of magnitude. It is likely that the solubilized and reconstituted oxoglutarate carrier can exist in two states of conformation, which catalyze different mechanisms of ttransport. This explained, for exam'ple, the interaction with the dry hydroxyapatite/celite in the course of the adsorption chromatography, which would predominantly affect the first eluted fraction.

Apart from the inhibition by mersalyl and p-chloromercuribenzoate, nothing is known about the SH groups or other functional groups of the oxoglutarate carrier. The studies on the oxoglutarate carrier's sulphydryl groups have been hindered by their relatively low reactivity with mercurials and also by the impossibility to label the protein with radioactive N-ethylmaleimide. However, we have now found that a fluorescent maleimide, i.e. eosin-5-maleimide (EMA), strongly inhibits the oxoglutarate carrier (Zara & Palmieri, 1988). Thus half maximal inhibition is achieved with 26 μM EMA. Since EMA is impermeable and reacts covalently, we have employed this reagent for investigating the localization of the SH groups of the membrane-bound oxoglutarate carrier. This was done both in mitochondria and submitochondrial particles, which are inside-out vesicles with respect to mitochondria. In these experiments mitochondria and submitochondrial particles were incubated with 250 μM EMA for 45 min at 0°C and in the dark. After washing, the oxoglutarate carrier was purified, tested for transport activity in the reconstituted system and subjected to gel electrophoresis. As revealed by fluorography, the oxoglutarate carrier isolated from mitochondria is labelled by EMA, but the carrier isolated from submitochondrial particles is not (Zara & Palmieri, 1988). Correspondingly the activity of the reconstituted oxoglutarate carrier isolated from EMA-labelled mitochondria is inhibited by 75%, whereas that of the carrier derived from EMA-labelled submitochondrial particles is not affected. These results clearly indicate that the essential EMA-reacting SH group(s) of the oxoglutarate carrier are located at the cytosolic face of the inner mitochondrial membrane. This asymmetry of the oxoglutarate carrier seems to be preserved even after reconstitution of the protein into the liposomes, since EMA also almost completely inhibits the reconstituted carrier (Zara & Palmieri, 1988). Experiments are now in progress in our laboratory to locate these essential SH groups within the amino acid sequence of the carrier.

We have started to determine the primary structure of the oxoglutarate carrier by cDNA in collaboration with Dr. J. Walker. The results so far obtained show a high degree of homology with the ADP/ATP carrier and the phosphate carrier. It has been suggested that the ADP/ATP carrier and the phosphate carrier can be considered to be formed by three fragments of about 100 amino acids, which are homologous one to another (Runswick et al., 1987; Aquila et al., 1987). It is interesting that we have found three pieces in the oxoglutarate carrier, which are homologous to 3 pieces of the ADP/ATP carrier and the phosphate carrier, which are present in the first, the second and the third fragment of their structures. This suggests that also the oxoglutarate carrier may consist of three repeating units.

In other experiments, the reactivity of an antiserum raised against the oxoglutarate carrier from bovine heart with the same protein from other sources was investigated by immunoblot. The antiserum cross-reacts with the oxoglutarate carrier present in heart mitochondria of all the species tested (beef, pig, rat and rabbit). In contrast, there is no cross reactivity of the antiserum raised against the oxoglutarate carrier purified from bovine heart mitochondria with the same carrier of other tissues such as liver and kidney even from the same species. This indicates an organ specificity of the antigenic properties of the oxoglutarate carrier. The antiserum was also employed to estimate the amount of the oxoglutarate carrier in heart mitochondria with the enzyme-linked immunosorbant assay. In this way we have found that the oxoglutarate carrier represents about 0.2% of the total mitochondrial proteins in beef-, pig- and rat-heart. This value is much lower than the content of the adenine nucleotide carrier and close to that of the phosphate carrier in heart mitochondria.

Isolation of carrier proteins from liver mitochondria

Some mitochondrial anion-transporting systems, such as the dicarboxylate carrier and the citrate carrier, are virtually absent in heart (Sluse *et al.*, 1971; Robinson & Oei, 1975), but have high activity in liver (Palmieri *et al.*, 1971; Palmieri *et al.*, 1972). The isolation of mitochondrial substrate carriers from liver is more difficult than from heart. Until recently, only the phosphate carrier has been purified from rat liver (Kaplan *et al.*, 1986). Its isolation, however, has needed a much more elaborate procedure than the corresponding heart carrier. Similarly for the oxoglutarate carrier, the procedure developed for its isolation from heart (Bisaccia *et al.*, 1985) does not result in a pure preparation when applied to liver (unpublished data). Although our standard procedure of hydroxyapatite chromatography (Bisaccia & Palmieri, 1984) already results in a considerable enrichment of the activity of the oxoglutarate carrier and the dicarboxylate carrier, these proteins are by no means pure. In fact, the "standard" hydroxyapatite eluate of Triton-solubilized liver mitochondria contains many more bands than the corresponding eluate from heart. The first step of improvement was achieved by varying the amount of hydroxyapatite and of the concentration of Triton X-114 in the elution buffer. The breakthrough in the purification of single carriers, however, was achieved by extensive removal of the detergent by hydrophobic chromatography before application onto the hydroxyapatite columns. This led to adsorption of the carrier proteins to the hydroxyapatite and made possible the specific elution of these translocators.

Oxoglutarate carrier from liver

Once bound to the column according to this procedure, the oxoglutarate carrier can be easily eluted in pure form by application of the appropriate buffer and detergent concentration (Bisaccia et al., 1988). The isolated protein consists of one single band in SDS gel chromatography, showing a slightly higher molecular weight as compared to that of the oxoglutarate carrier isolated from heart mitochondria (Bisaccia et al., 1985). If reconstituted into liposomes the purified protein catalyzes a phthalonate-sensitive oxoglutarate/oxoglutarate exchange. Uptake of oxoglutarate is negligible when unloaded liposomes are used. Further, the purified oxoglutarate carrier transports L-malate and other dicarboxylates besides oxoglutarate and is inhibited by certain substrate analogues and by certain SH reagents, as found in mitochondria (Palmieri et al., 1972).

Dicarboxylate carrier from liver

The procedure described above for the purification of the oxoglutarate carrier from liver does not lead to purification of the dicarboxylate carrier, since this protein binds more strongly to hydroxyapatite and could be eluted from the column only in small amounts and in the presence of the oxoglutarate carrier. The experimental consequence was to optimize the concentration of detergent which should be high enough on the one hand to avoid strong interaction of the dicarboxylate carrier with the hydroxyapatite during column chromatography and low enough on the other hand to retain the contaminating proteins. By hydroxyapatite chromatography of the mitochondrial extract in which the Triton X-114 was decreased to 1.5%, we have achieved to obtain a preparation which exhibits a high dicarboxylate carrier activity after reconstitution into liposomes. On SDS gel electrophoresis this preparation consists mainly of one protein band with an M_r of 28 kDa (Bisaccia et al., 1988).

The 28 kDa protein isolated from rat liver mitochondria, when reconstituted into liposomes, shows transport properties very similar to those of the dicarboxylate carrier characterized in mitochondria. It catalyzes a very active butylmalonate-sensitive malate/phosphate exchange, and in unloaded liposomes the uptake of both malate and phosphate is negligible. As the transport system in mitochondria (Palmieri et al., 1971; Crompton et al., 1974a and 1974b), the purified and reconstituted protein transports not only dicarboxylates (but not oxoglutarate), but also phosphate and some sulphur containing compounds, i.e. sulphate and thiosulphate. It is inhibited by some dicarboxylate analogues (but very slightly by phthalonate) and by SH reagents including eosin-5-maleimide (but not by N-ethylmaleimide). In more recent experiments we have found that phosphate appears to bind to the carrier at a site different from that for dicarboxylates. The main indications are a) the inhibition of malate uptake by external phosphate is non-competitive, b) the inhibition of phosphate uptake by external malate is non-competitive and c) malonate is competitive with respect to malate and non-competitive with respect to phosphate. These results can be interpreted to show that in the presence of both malate and phosphate the carrier forms a ternary complex which is inactive. In other words although the dicarboxylate carrier can transport both dicarboxylates and phosphate, these two different classes of substrates bind to different sites.

Reconstitution of the malate/aspartate shuttle.

As reported above, our group has purified the 2-oxoglutarate carrier. Also the aspartate/glutamate carrier has been purified by Krämer *et al.* (1986). So far each substrate carrier from the inner mitochondrial membrane or from other membranes has been reconstituted and characterized separately. This is a necessary step for the identification of a transport protein and for the elucidation of its function and mechanism. However, in the case of metabolic shuttle mechanisms, usually two or even more carriers cooperate within the natural membrane to accomplish specific functions. One of the best examples for such a functional cooperation is the malate/aspartate shuttle, which includes the two carriers for the aspartate/glutamate exchange and for the oxoglutarate/malate exchange, respectively. The malate/aspartate shuttle plays an important role in the transfer of reducing equivalents from the cytosol to the mitochondria in most tissues. For the reconstituion of this shuttle it is necessary to incorporate two carriers (the oxoglutarate carrier and the aspartate/glutamate carrier) into a single liposome. Of course, to have a chance of co-reconstitution, it is essential to achieve a high protein/lipid ratio in the reconstituted proteoliposomes. This prerequisite cannot be fulfilled when using the common freeze-thaw-sonication procedure (Kasahara & Hinkle, 1977) for reconstitution, which is usually characterized by a protein/lipid ratio (w/w) of less than 10-3. Unfortunately, also the dialysis method for reconstitution (Racker, 1979), which generally leads to high protein/lipid ratios, cannot be applied, since the detergents used for solubilization and purification of these two carrier proteins are not suited for dialysis and both the oxoglutarate and the aspartate/glutamate carrier do not endure dialyzable detergents. For these reasons we have applied the new method based on the removal of detergents by chromatography on Amberlite (Krämer & Heberger, 1986). This method leads to the incorporation of sufficiently large amounts of protein into the vesicular membranes. In general 20-40% of the proteoliposomes formed by this procedure carried both carrier proteins. In order to measure the activity of the reconstituted malate/aspartate shuttle, this complex redox shuttle has to be simplified to some extent. In these experiments liposomes which contain oxaloacetate, aspartate and the transaminase GOT in the internal space, and the two reconstituted carriers within the membrane are preincubated with externally added glutamate. This leads to the formation of 2-oxoglutarate inside the liposomes by the action of the aspartate/glutamate carrier and GOT. Aspartate, which is exported in exchange for the glutamate taken up, is regenerated from oxaloacetate by the activity of GOT. This leads to further uptake of glutamate and additional formation of oxoglutarate inside. Upon adding ^{14}C-malate, the labelled substrate can now be taken up in exchange against the accumulated internal oxoglutarate by the action of the oxoglutarate carrier. Obviously, the concerted activities of the two carriers and the GOT cannot take place when glutamate is omitted in the preincubation. Thus the difference in the amount of malate taken up in the presence and in the absence of glutamate represents the activity of the reconstituted shuttle.

The results obtained with the experimental design described above show that there is a considerable activity of glutamate-induced malate uptake, *i.e.* of the reconstituted malate/aspartate shuttle (Indiveri *et al.*, 1987a). Appropriate controls have shown the absolute requirement of any of the component of the complete mixture, *i.e.* GOT, oxaloacetate, aspartate, oxoglutarate carrier or aspartate/glutamate carrier, for the observed glutamate-induced malate uptake. Even more important, no activity has been observed when the two carriers were reconstituted separately in different pools of liposomes of the same composition, which were then mixed before testing the activity. This demonstrates the absolute requirement of the co-reconstitution of the two carriers in the same liposomes. We have therefore constructed a reconstituted system which comprises the central part, *i.e.* the carrier functions, of the malate/aspartate shuttle. This system represents a satisfactory basis for studying the function and the regulation of the reconstituted malate/aspartate shuttle.

REFERENCES

Aquila H, Link TA, Klingenberg M (1987) Solute carriers involved in energy transfer of mitochondria form a homologous protein family. FEBS Lett 212:1-9

Bisaccia F, Palmieri F (1984) Specific elution from hydroxyapatite of the mitochondrial phosphate carrier by cardiolipin. Biochim Biophys Acta 766:386-394

Bisaccia F, Indiveri C, Palmieri F (1985) Purification of reconstitutively active 2-oxoglutarate carrier from pig heart mitochondria. Biochim Biophys Acta 810:362-369

Bisaccia F, Indiveri C, Palmieri F (1988) Purification and reconstitution of two anion carriers from rat liver mitochondria: the dicarboxylate and the 2-oxoglutarate carrier. Biochim Biophys Acta 933:229-240

Crompton M, Palmieri F, Capano M, Quagliariello E (1974a) The transport of sulphate and sulphite in rat liver mitochondria. Biochem J 142:127-137

Crompton M, Palmieri F, Capano M, Quagliariello E (1974b) The transport of thiosulphate in rat liver mitochondria. FEBS Lett 46:247-250

De Pinto V, Tommasino M, Benz R, Palmieri F (1985) The 35 kDa DCCD-binding protein from pig heart mitochondria is the mitochondrial porin. Biochim Biophys Acta 813:230-242

De Pinto V, Prezioso G, Palmieri F (1987) A simple and rapid method for the purification of the mitochondrial porin from mammalian tissues. Biochim Biophys Acta 905:499-502

Fonyo A (1978) SH-group reagents as tools in the study of mitochondrial anio transport. J Bioenerg Biomembr. 10:171-194

Genchi G, Petrone, G, De Palma A, Cambria A, Palmieri F (1988) Interaction of phenylisothiocyanates with the mitochondrial phosphate carrier. I. Covalent modification and inhibition of phosphate transport. Biochim Biophys Acta, in press

Indiveri C, Krämer R, Palmieri F (1987a) Reconstitution of the malate/aspartate shuttle from mitochondria. J Biol Chem 262:15979-15983

Indiveri C, Palmieri F, Bisaccia F, Krämer R (1987b) Kinetics of the reconstituted 2-oxoglutarate carrier from bovine heart mitochondria. Biochim Biophys Acta 890:310-318

Kaplan RS, Pratt RD, Pedersen PL (1986) Purification and characterization of the reconstitutively active phosphate transporter from rat liver mitochondria. J Biol Chem 261:12767-12773

Kasahara M, Hinkle, PC (1977) Reconstitution and purification of the D-glucose transporter from human erythrocytes. J Biol Chem 252:7384-7390

Klingenberg M, Palmieri F, Quagliariello E (1970) Quantitative correlation between the distribution of anions and the pH difference across the mitochondrial membrane. Eur J Biochem 17:230-238

Klingenberg M, Riccio P, Aquila H (1978) Isolation of the ADP, ATP carrier as the carboxyatractylate protein complex from mitochondria. Biochim Biophys Acta 503:193-210

Kolbe HVJ, Bottrich J, Genchi G, Palmieri F, Kadenbach B (1981) Isolation and reconstitution of the phosphate-transport system from pig heart mitochondria. FEBS Lett 124:265-269

Kolbe HVJ, Costello D, Wong A, Lu RC, Wohlrab H (1984) Mitochondrial phosphate transport: large scale isolation and characterization of the phosphate transport protein from beef heart mitochondria. J Biol Chem 259:9115-9120

Krämer R, Heberger C (1986) Functional reconstitution of carrier proteins by removal of detergent with a hydrophobic ion exchange column. Biochim Biophys Acta 863:289-296

Krämer R, Kürzinger G, Heberger C (1986) Isolation and functional reconstitution of the aspartate glutamate carrier from mitochondria. Arch Biochem Biophys 251:166-174

LaNoue KF, Schoolwerth AC (1979) Metabolite transport in mitochondria. Ann Rev Biochem 48: 871-922

Lauquin G, Pougeois R, Vignais PV (1980) 4-azido-2-nitrophenyl phosphate, a new photoaffinity derivative of inorganic phosphate. Study of its interaction with the inorganic phosphate binding site of beef heart mitochondrial adenosine triphosphatase Biochemistry 19:4620-4626

Meijer AJ, Van Dam K (1981) Mitochondrial transport. In: Boting S, De Pont J (eds) Membrane Transport. Elsevier Amsterdam, pp 235-256

Mende P, Kolbe HVJ, Kadenbach B, Stipani I, Palmieri F (1982) Reconstitution of the isolated phosphate-transport system of pig-heart mitochondria. Eur J Biochem 128:91-95

Palmieri F, Prezioso G, Quagliariello E, Klingenberg M (1971) Kinetic study of the dicarboxylate carrier in rat liver mitochondria. Eur J Biochem 22:66-74

Palmieri F, Stipani I, Quagliariello E, Klingenberg M (1972) Kinetic study of the tricarboxylate carrier in rat liver mitochondria. Eur J Biochem 29:408-416

Palmieri F, Klingenberg M (1979) Direct methods for measuring metabolite transport and distribution in mitochondria. Methods Enzymol 56:279-301

Palmieri F, Kolbe HVJ, Genchi G, Stipani I, Mende P, Kadenbach B, Quagliariello E (1981) Isolation and reconstitution of the phosphate transport system from pig heart mitochondriaIn: Palmieri F et al. (eds) Vectorial Reactions in Electron and Ion Transport in Mitochondria and Bacteria Elsevier/North Holland Biomedical, Amsterdam, pp 281-290

Palmieri F, Prezioso G, Bonvino V, Stipani I (1987) Transport of metabolites in mitochondria. In: Bertoli E, Chapman D (eds), Biomembrane and Receptor Mechanisms, vol 7, Liviana Press and Springer Verlag Heidelberg, pp 205-222

Racker E (1979) Reconstitution of membrane processes. Methods Enzymol 55:699-711

Robinson BH, Oej J (1975) Citrate transport in guinea pig heart mitochondria. Can J Biochem 53:643-647

Runswick MJ Powell SJ, Nyren P, Walker JE (1987) Sequence of the bovine mitochondrial phosphate carrier protein: structural relationship to ADP/ATP translocase and the brown fat mitochondrial uncoupling protein. EMBO J 6: 1367-1373

Sigrist H, Zahler P (1978) Characterization of phenylisothiocyanate as a hydrophobic membrane label. FEBS Lett 95:116-120

Sluse FE, Meijer AJ, Tager JM (1971) Anion translocators in rat-heart mitochondria. FEBS Lett 18:149-153

Tommasino M, Prezioso G, Palmieri F (1987) Photoaffinity labeling of the mitochondrial phosphate carrier by 4-azido-2-nitrophenyl phosphate. Biochim Biophys Acta 890:39-46

Touraille S, Briand Y, Durand R, Bonnafous JC, Marie JC (1981) Purification of a phosphate carrier in pig heart mitochondria by affinity chromatography on mersalyl ultrogel. FEBS Lett 128:142-144

Wohlrab H (1980) Purification of a reconstitutively active mitochondrial phosphate transport protein. J Biol Chem 255:8170-8173

Zara V, Palmieri F (1988) Inhibition and labelling of the mitochondrial 2-oxoglutarate carrier by eosin-5-maleimide. FEBS Lett 236:493-496

Purification and Reconstitution of the 2-Oxoglutarate Carrier from Bovine Heart and Liver Mitochondria

D. CLAEYS, M. MÜLLER AND A. AZZI

Institut für Biochemie und Molekularbiologie der Universität Bern, Bühlstrasse 28, CH-3012 Bern, Switzerland

The inner mitochondrial membrane contains a specific 2-oxoglutarate transporter which catalyzes the uptake of 2-oxoglutarate by an electroneutral counter-exchange with 2-oxoglutarate or L-malate. This antiport system accepts also other dicarboxylates as oxaloacetate, succinate and malonate (for review see LaNoue & Schoolwerth, 1979).

The 2-oxoglutarate carrier takes part in several important metabolic processes like the malate-aspartate shuttle, the gluconeogenesis from lactate, the isocitrate-2-oxoglutarate shuttle *etc.* (Meijer & Van Dam, 1981). This carrier is very active both in heart and liver mitochondria (Sluse *et al.*, 1972, Palmieri *et al.*, 1972).

The characteristics and kinetics in intact mitochondria have been investigated by several groups (Palmieri *et al.*, 1972; Sluse *et al.*, 1972; Sluse *et al.*, 1973; Sluse *et al.*, 1983). The kinetic studies with mitochondria demonstrate that the carrier has binding sites for both of its substrates on each site of the inner mitochondrial membrane, indicating that the carrier spans the membrane. Translocation is only possible when the binding sites on both sites are occupied and conformational changes may be involved during the exchange reaction.

The substrates of the carrier compete with each other when added outside the mitochondria, which suggests only one external binding site.

Abbreviations: PC, phosphatidylcholine; PE, phosphatidylethanolamine; PDG, cardiolipin; EDTA, ethylenediaminotetraacetic acid; HTP, hydroxylapatite.

A. Azzi et al. (Eds.)
Anion Carriers of Mitochondrial Membranes
© Springer-Verlag Berlin Heidelberg 1989

The uptake of 2-oxoglutarate can be strongly inhibited by several SH-reagents, but not by N-ethylmaleimide (Quagliariello & Palmieri, 1972). Some impermeable dicarboxylate analogues like phenylsuccinate and butylmalonate were demonstrated to be inhibitory as well (Palmieri *et al.*, 1972). Phtalonic acid is considered to be a specific inhibitor of the carrier (Meijer *et al.*, 1976).

Recently, the functional active 2-oxoglutarate carrier has been purified from pig and beef heart mitochondria by chromatography on hydroxylapatite and celite in the presence of cardiolipin (Bisaccia *et al.*, 1985; Indiveri *et al.*, 1987) and from rat liver mitochondria by chromatography on hydroxylapatite (Bisaccia *et al.*, 1988). SDS-polyacrylamide gel electrophoresis identified the purified carrier as a single protein with an apparent molecular weight of the 2-oxoglutarate of 31.5-32.5 kDa.

Here, we report the final purification of the 2-oxoglutarate carrier from bovine heart and liver mitochondria by a chromatography on organo-mercurial agarose. A batch procedure for the functional reconstitution of the carrier into phospholipid vesicles, by adsorption of Triton X-114 into hydrophobic Amberlite XAD-2 beads, is described. It allowed the identification of the 2-oxoglutarate carrier during the purification and some kinetic studies of the purified carrier.

MATERIALS AND METHODS

Materials

Hydroxylapatite (Bio-Gel HTP), Organo-mercurial agarose (Affigel 501; capacity 4-9 μmol Hg/ml), Dowex AG 1-X8 (100-200 Mesh) were purchased from Bio-Rad; Sephadex G-25M from Pharmacia; Celite 535 from Roth. Celite was washed several times with hot distilled water and dried at 100°C. Amberlite XAD-2 beads, Triton X-114, cholesterol, egg yolk phospholipids were obtained from Fluka. Amberlite XAD-2 beads were washed several times with isopropanol until the supernatant was clear and subsequently it was rinsed extensively with distilled water. Cardiolipin was from Sigma, phosphatidylcholine and phosphatidylethanolamine from Avanti Polar-Lipids. Asolectin from Associated Concentrates. The labeled substrates were delivered by Amersham International. All other chemicals were of analytical grade.

Preparation of mitochondria

Bovine heart mitochondria were isolated as described by Yu *et al.* (1975). Bovine liver mitochondria were prepared according to Johnson & Lardy, 1967.

Isolation of the 2-oxoglutarate carrier

Extraction of mitochondria

Beef heart mitochondria were solubilized with 3% Triton X-114, 10 mM Mops, pH 7.2, 1 mM EDTA, 50 mM NaCl, 1.8 mg/ml cardiolipin at a protein concentration of 10 mg/ml for 20 min at 0°C.

Beef liver mitochondria were first pre-extracted with 0.3% Triton X-100, 10 mM Mops, pH 7.2, 1mM EDTA, 50 mM NaCl at 15 mg protein/ml. After centrifugation at 100.000 g for 30 min, the pellet was resuspended as described for heart mitochondria.

Hydroxylapatite and celite chromatography

All operations were carried out at 4°C. Triton X-114 extract (0.5 ml) was applied without centrifugation on a column containing 500 mg cold dry Bio-Gel HTP. Elution was performed first with 1 ml 2% Triton X-114, 10 mM Mops, pH 7.2, 50 mM NaCl, 1 mM EDTA, 0.8 mg cardiolipin/ml and subsequent with the same buffer without cardiolipin until 0.75 ml hydroxylapatite eluate was collected. Several hydroxylapatite eluates were pooled. The hydroxylapatite eluate (1.2 ml) was loaded on 250 mg cold dry celite and eluted with the same buffer without cardiolipin until 1.2 ml eluate was collected. Several celite eluates were pooled.

Organo-mercurial agarose chromatography

The celite eluate (2-2.5 ml) from solubilized heart mitochondria was incubated in batch with 1 ml Affigel 501, pre-equilibrated with the buffer without cardiolipin, for 45 min at 4°C under gentle stirring. Longer incubation (2 h 30 min) was applied for the celite eluate from solubilized liver mitochondria. The 2-oxoglutarate carrier did not bind to the resin and was collected in a partially inactivated state in the supernatant. 2-Mercaptoethanol (10 mM) was added and incubated at 4°C for 30 min to reactivate the carrier.

Reconstitution of the 2-oxoglutarate carrier

Samples containing the solubilized carrier were passed through Sephadex G-25M columns, pre-equilibrated with the buffer without cardiolipin. The void volumes were collected. Liposomes were prepared by sonicating 90 mg egg yolk phospholipids (or indicated phospholipids) in 2 ml of 20 mM Mops, pH 7.2, 1mM EDTA, 50 mM NaCl for 20 min under nitrogen at 4°C, using the microtip of a Branson Sonifier B 15 in the pulsed mode at 60% duty, until the suspension was clear (20-40 min). The sonicated liposomes were centrifuged at 14,000 g for 5 min to remove metal particles. Amberlite XAD-2 beads were pre-equilibrated for 30 min at room temperature with 20 mM Mops, pH 7.2, 1 mM EDTA, 50 mM NaCl and 9 mg/ml sonicated liposomes to minimize lipid and protein adsorption during the reconstitution.

After gel-filtration on Sephadex G-25 the protein fractions were mixed with the liposomes in a 4:1 v/v ratio. 2-oxoglutarate (50 mM) or the indicated substrate was added. The transparent solutions of mixed protein-detergent-lipid micelles were kept at 4°C for 15 min. The final composition of the mixture was as follows :

50 mM substrate, 20 mM Mops pH 7.2, 1 mM EDTA, 50 mM NaCl, 2-2.5% Triton X-114, 9 mg phospholipid/ml, 0.4-8 μg protein/mg lipid. The equilibrated Amberlite XAD-2 beads were added to the mixture (500 mg/ml) and gently mixed at 4°C for 2 h 20 min. Then 500 μl proteoliposomes were removed from the beads and quickly passed through small anion exchange columns (pasteur pipettes filled with 1 ml Dowex AG 1-X8 (Cl- form), pre-equilibrated with 10 mM Mops pH 7.2, 20 mM NaCl) to remove the external substrate. From one column 500 μl of the opalescent eluate was collected and subsequently 20 mM Mops pH 7.2 were added.

Assay of the 2-oxoglutarate exchange activity

The buffered proteoliposomes were incubated at 28°C for 5 min. The exchange reaction was initiated by the addition of 0.2 mM 2-[^{14}C]-oxoglutarate (specific radioactivity of 900 dpm/nmol) and further incubated at 28°C. The exchange was stopped at indicated time intervals by passing quickly 160 μl liposomes through 350 μl Dowex AG 1-X8 (Cl- form) to remove external radioactivity. Elution occurred with 450 μl 10 mM Mops pH 7.2, 20 mM NaCl. Intraliposomal radioactivity labeled substrate was determined by liquid scintillation counting. Control samples were incubated in the presence of 0.5 mM mersalyl which was added prior to labeled substrate. Only the mersalyl sensitive 2-oxoglutarate/2-oxoglutarate exchange activity was considered.

Other methods

SDS-polyacrylamide gel electrophoresis of aceton precipitated samples was carried out as described by Bisaccia et al.(1985). Staining was performed with silver nitrate according to the Bio-Rad method. Protein was determined by the Lowry (1951) method modified for the presence of Triton X-114 and lipid by the addition of 5% sodium dodecylsulfate.

RESULTS AND DISCUSSION

Reconstitution procedure

One of the most important methodological approaches to study the structure and function of integral membrane proteins has been their reconstitution in closed phospholipid vesicles. Freeze-thaw sonication was generally applied as the reconstitution method for mitochondrial carriers, solubilized and purified with the nonionic detergents Triton X-114 and Triton X-100. However, this method had several disadvantages like the persistence of the detergent in the proteoliposomes, the partial inactivation of the reconstituted protein by sonication and the heterogeneous size of the proteoliposomes.

Here, the reconstitution of the solubilized 2-oxoglutarate carrier in a functional intact form was achieved by batch-wise adsorption of Triton X-114, used for the isolation of the carrier, by Amberlite XAD-2 beads in the presence of added phospholipids. Non-micellar Triton X-114 is efficiently adsorbed by these hydrophobic polystyrene beads (Cheetham, 1979).

Several mitochondrial carriers have recently been successfully reconstituted by this application of the Amberlite XAD-2 beads either, by using a batch-procedure (Müller et al., 1984a, Müller et al. 1984b) or, the column-procedure where the mixed micelles were passed several times through a small Amberlite column at room temperature (Bisaccia et al., 1988; Krämer & Heberger, 1986; Indiveri et al., 1987).

Krämer & Heberger (1986) analysed several important parameters for an efficient, functional reconstitution of carrier proteins in unilamellar vesicles, using Amberlite XAD-2 beads.

Besides the importance of the nature of the detergent, the most critical parameters are the phospholipid/protein ratio and detergent/phospholipid ratio in the mixed micelles, the ratio of detergent/Amberlite beads and the concentration of phospholipid during protein insertion. Some of those parameters seem to be strongly dependent on the type of carrier protein to be reconstituted. Therefore, for every carrier the conditions for reconstitution have always to be established. Particular attention in setting up a batch-procedure for the reconstitution should be given to the temperature and the amount of beads which determine the rate of detergent removal or the time necessary for the formation of proteoliposomes.

In the case of bacteriorhodopsin it has been observed that the activity of a reconstituted protein diminished when the rate of detergent removal was too high, most probably due to the scrambled orientation of the reconstituted protein or instability of formed multilamellar structures (Rigaud et al., 1988).

In view of these findings, a batch reconstitution-procedure for the 2-oxoglutarate carrier was developed. The reconstitution samples were incubated at 4°C with Amberlite XAD-2 beads using a rather low ratio of 25 mg Amberlite/mg detergent with gentle shaking.

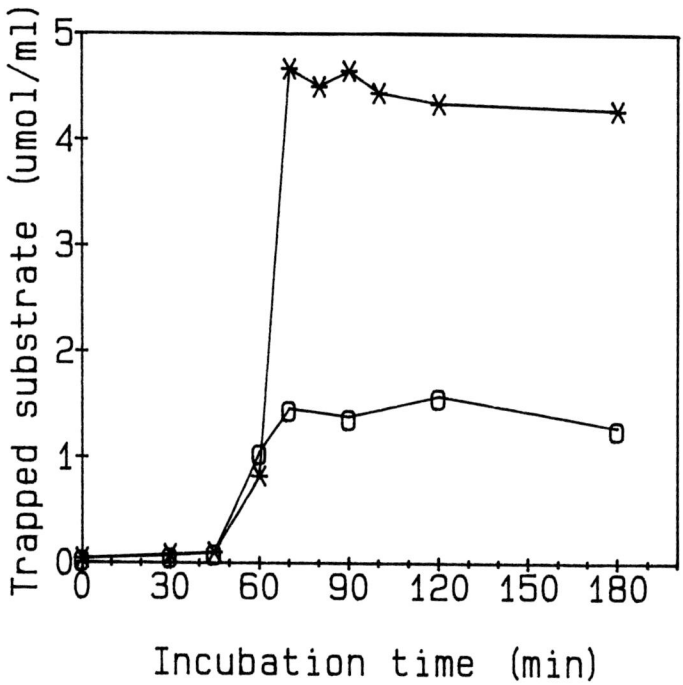

Fig. 1. Incorporation of 2-[^{14}C]-oxoglutarate into egg yolk phospholipids during incubation with Amberlite beads. Hydroxylapatite/celite eluate (3 ml) from solubilized heart mitochondria (o) and solubilization medium (x) were reconstituted in batch as described in Material and Methods, except that 50 mM 2-[^{14}C]-oxoglutarate (specific activity 8000 dpm/μmol) was present during the adsorption of Triton X-114. At the indicated time intervals 120 μl were taken from the batch solution, passed through 150 μl Dowex (Cl$^-$-form) and the eluate analysed for trapped radioactivity.

Fig. 1. shows the time course of the amount of labeled substrate trapped in liposomes and proteoliposomes formed during the incubation with the beads. The time to form closed vesicles was in both cases 1 hour. During this period no substrate could be trapped in the micelles, containing still too much Triton .

The process of enclosure of substrate in liposomes did not seem to occur gradually. Most probably in a very short time interval the vesicles sufficiently closed themselves and trapped the substrate. The amount of finally trapped substrate was 3 times higher for pure liposomes than for proteoliposomes.

The vesicles formed after 1 hour incubation were stable and not leaky for trapped 2-[^{14}C]-oxoglutarate. Only a decrease of 10% of the enclosed radioactivity was observed when the batch incubation was prolonged for 3 hours. The residual Triton X-114 was determined photometrically at 280 nm and surprisingly, there was still about 0.2% Triton X-114 (0.3 mol/mol lipid) present in the vesicular suspension after 90 min of incubation. These results are in agreement with the data from Ueno et al. (1984), demonstrating the relatively low anionic permeability of the vesicle membrane containing as much as 0.3 molecules residual detergent *pro* molecule lipid.

Further removal of detergent was progressively slow. After 135 min incubation there was still 0.1% Triton X-114 present, indicating that the complete removal of the detergent from the vesicle membrane is probably not possible by this procedure within a reasonable incubation time procedure. Although the detergent/lipid ratio was still quite high (0.15 mol/mol lipid), the vesicle membrane was sufficiently impermeable for 2-oxoglutarate to measure exchange activities over a time of 15 min.

The data of Fig. 1 do not give information about vesicle size and about the structure of the formed liposomes. The ADP/ATP carrier was reconstituted in unilamellar vesicles with a diameter of about 65 nm, using the batch procedure with Amberlite XAD-2 beads (Müller, 1983). Ueno et al. (1984) observed that the time course of detergent removal by Amberlite beads and corresponding vesiculation is a kinetic process. They studied the kinetic factors and parameters determining vesicle size of unilamellar liposomes formed in such a process. Initially formed vesicles were small due to fast removal of detergent. But, fusion and enlargement occurred spontaneously after initial vesiculation, if detergent removal became sufficiently slow, resulting in an increased, fixed vesicle size.

In view of this stabilization of liposome structure, the incubation of all reconstitution samples with the Amberlite XAD-2 beads was carried out for 2 h 20 min. The turbid proteoliposomes were passed through the Dowex column to remove external anion and to generate a substrate gradient. The pH of the eluted proteoliposomes dropped to 4.5-5 due to the anion exchanger. Therefore, 20 mM Mops pH 7.2 was added to keep the pH neutral during the exchange reaction. At low pH (5) a 40% decrease in the exchange activity was observed.

Purification of the 2-oxoglutarate carrier

The procedure used for the isolation of the 2-oxoglutarate carrier consisted of several basic steps : solubilization of mitochondria with Triton X-114 in the presence of cardiolipin, adsorption chromatography of the extract on hydroxylapatite followed by celite and finally, organo-mercurial agarose (Affigel 501) chromatography. The same procedure was used for the 2-oxoglutarate carriers from both, bovine heart and liver mitochondria, except that a pre-extraction of the liver mitochondria with 0.3% Triton X-100 was applied before solubilization with 3% Triton X-114. Pre-extraction removed low molecular weight contaminants, probably originating from enhanced proteolytic digestion in liver.

Chromatography of the mitochondrial extract on dry hydroxylapatite is generally applied as the first purification step for mitochondrial substrate carriers, whilst the bulk amount of protein is adsorbed. At least seven different substrate carriers were identified in the hydroxylapatite pass-through, depending somewhat on the concentration and kind of the detergent and phospholipid used (for review see Nałȩcz, 1986). Cardiolipin prevents elution of porin (de Pinto et al., 1987) but induces elution of the phosphate carrier (Bisaccia & Palmieri, 1984), of the dicarboxylate carrier (Bisaccia et al., 1988), of the oxoglutarate carrier (Bisaccia et al., 1985) and of the citrate carrier (Stipani et al., 1986). The requirement for this typical inner mitochondrial membrane phospholipid is reflected by a substantially increased specific transport activity of these carriers in the HTP eluate. Therefore, cardiolipin was included in the solubilization buffer at a final concentration of 1.8 mg cardiolipin/ml.

SDS-polyacrylamide gel electrophoresis analysis of the polypeptides showed that the hydroxylapatite eluate from solubilized liver mitochondria contained more distinct proteins (7-8 bands) in the apparent molecular weight range of 28-37 kDa, than the hydroxylapatite eluate from heart mitochondria (5 bands), using the same solubilization conditions (Fig. 2). This might reflect the observations that in the latter organ some carriers are missing (citrate carrier) or are only expressed to a minor extent (dicarboxylate carrier).

Chromatography of the hydroxylapatite eluates on celite resulted in a remarkable purification of the 2-oxoglutarate carrier, expressed by the 4-6 times higher specific exchange activity. The celite eluates, containing still a mixture of different proteins (Fig. 2), were processed further by chromatography on organo-mercurial agarose (Affigel 501). This last step was carried out in a batch-procedure for the appropriate incubation time as described in Material and Methods.

Only one polypeptide with an apparent molecular weight of 31.5-32 kDa was not bound to the resin as shown in Fig. 2. The same result was obtained starting either from heart or liver mitochondria. The apparent molecular weight corresponded to the one of the 2-oxoglutarate carriers purified by Bisaccia et al., 1985; Indiveri et al., 1987, Bisaccia et al., 1988. All other proteins were bound by Affigel 501 and could be eluted with 200 mM 2-mercaptoethanol (not shown).

Although, the 2-oxoglutarate carrier is strongly inhibited by SH-reagents and therefore should contain an important cysteinyl residue, it seems that this residue was not accessible to the resin due to high Triton and cardiolipin concentrations or, because its reactivity with the mercurial resin was too low. Indeed, using a high capacity Affigel

501 (9 μmol/ml) and a long batch-incubation time (2-3 hours), part of the 2-oxoglutarate carrier population was bound to the Affigel 501 and could be eluted with rather low concentrations of 2-mercaptoethanol (1 mM).

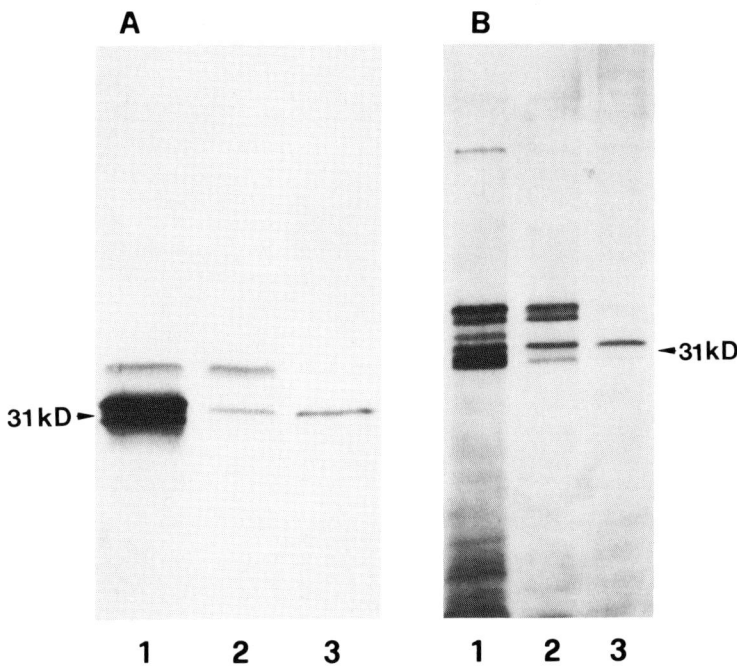

Fig. 2. Polypeptide pattern of the protein fractions, obtained after the different purification steps, of the 2-oxoglutarate carrier from (A) heart and (B) liver mitochondria. SDS-polyacrylamide gel electrophoresis and staining with silver nitrate is described in Material and Methods. Lane 1, Hydroxylapatite eluate; lane 2, celite eluate; lane 3, not bound fraction Affigel 501.

Prior to reconstitution, all fractions were passed through Sephadex G-25M columns. This step was necessary not only to standardize the composition of the samples, but also to reduce unspecific inhibitor insensitive 2-oxoglutarate transport. It was observed that the reconstituted hydroxylapatite/celite eluate of solubilization buffer had 8 times higher unspecific 2-oxoglutarate exchange activity than the same eluate passed through G-25M, before reconstitution. This inhibitor insensitive background could also be partially reduced by adding 2 mM EGTA before reconstitution, indicating the presence of interfering Ca^{2+} ions. Indeed Ca^{2+} measurements with Arsenazo III showed a elution of 2-5 mM Ca^{2+} from hydroxylapatite.

Table 1 gives the results of the purification procedure in terms of the observed specific 2-oxoglutarate/2-oxoglutarate exchange activities. Celite & Affigel 501 chromatography resulted in 20 fold increase of the specific activity with respect to the hydroxylapatite eluate. Specific activity and purification factor have to be considered apparent since, besides partial inactivation of the carrier during the whole procedure, it was observed that the efficiency of functional reconstitution increased with the purity of the preparation.

Table 1. Purification of the 2-oxoglutarate carrier.The hydroxylapatite eluates, celite eluates and not bound fractions of Affigel 501 were obtained and reconstituted as described in Material and Methods. The 2-oxoglutarate exchange reaction was measured for 1 min. Only the mersalyl sensitive activity is presented.

Purification step	Total protein (mg)	2-Oxoglutarate exchange μmol/min/g protein
Heart mitochondria		20
Hydroxylapatite	0.34	215
Celite	0.065	1310
Not bound Affigel 501	0.03	5970
Liver mitochondria		45
Hydroxylapatite	0.32	400
Celite	0.082	1730
Not bound Affigel 501	0.03	6580

Starting with 20 mg heart mitochondrial protein about 30 $\cdot \mu$g purified 2-oxoglutarate carrier was obtained. This means a protein yield of 0.15-0.2%, in contrast to 0.07% for liver mitochondria. The reconstituted purified 2-oxoglutarate carrier exhibited a high specific 2-oxoglutarate/2-oxoglutarate exchange activity (6-6.5 mmol/min/g protein), only after incubating the Affigel 501 not bound fraction with 10 mM 2-mercaptoethanol.

Although, there was no retention of the carrier to Affigel 501, it appeared that the carrier in the soluble state became reversible inactivated due to an unknown interaction with the resin. This inhibition might result from conformational changes of the carrier or disulfide-bridge formation, catalyzed by the Affigel 501 during the incubation. It cannot be excluded that the ligand (4-aminophenylmercuric-acetate), leaking from the Affigel 501 during batch incubation, caused the inhibition of the solubilized. In fact it was observed that 0.5 mM ligand, added to the hydroxylapatite eluate before reconstitution, decreased the 2-oxoglutarate exchange activity by a factor 4. In Table 2 the complete reactivation by 2-mercaptoethanol is demonstrated. The specific 2-oxoglutarate/2-oxoglutarate exchange of the reconstituted purified carrier was enhanced almost 6 times in the presence of 10 mM 2-mercaptoethanol. In contrast, adding 10 mM 2-mercaptoethanol to the hydroxylapatite eluate had a minor effect (5%

increase) on the 2-oxoglutarate exchange activity. A similar reactivation was obtained with cysteine and 1,4-dithio-DL-threitol.

Table 2. Reactivation with 2-mercaptoethanol of the 2-oxoglutarate carrier from liver mitochondria after purification with Affigel 501. Purification, reconstitution and assay conditions are described in Material and Methods. 2-mercaptoethanol was added to the not bound fraction either 30 min before Sephadex G-25 chromatography or after reconstitution, i.e. 20 min at $4^{\circ}C$ before the assay.

Reactivation	Concentration	2-Oxoglutarate exchange
	mM	mmol/8 min/g protein
Before reconstitution		
	0	5.1
	1	28.4
	10	32
After reconstitution		
	0.1	8.6
	1	29.2
	10	32.9

Thus, inhibition of the carrier in soluble state by the Affigel 501 resin was fully reversible. The fact that this could occur before or after reconstitution, indicate that the inhibitory effect had no influence on the correct reconstitution of the carrier.

PROPERTIES OF THE RECONSTITUTED 2-OXOGLUTARATE CARRIER

The time course of the mersalyl sensitive 2-oxoglutarate exchange is shown in Fig. 3. It demonstrates that the uptake was linear for about 90 seconds, reaching a saturation value of about 35 mmol/g protein after 12 min. The initial transport rate at $28^{\circ}C$ was 7.5 mmol/min/g protein. Almost complete inhibition with 0.5 mM mersalyl was observed.

The effect of intraliposomal counter-anions on the 2-oxoglutarate exchange has been investigated and the results are shown in Table 3. The indicated internal anions were present at a concentration of 50 mM during the preparation of the proteoliposomes from the purified 2-oxoglutarate carrier and egg yolk phospholipids. Similar results were obtained for the carriers from heart and liver mitochondria. The data are mean values of 4 experiments. The data of Table 3 show that beside 2-oxoglutarate, labeled 2-oxoglutarate could be exchanged with internal L-malate, malonate and succinate, which are known substrates of the carrier in mitochondria (LaNoue & Schoolwerth, 1979).

Table 3. Substrate specificity of the purified 2-oxoglutarate carrier reconstituted in liposomes. The 2-oxoglutarate carrier from heart mitochondria was purified and reconstituted as described in Material and Methods. Proteoliposomes were loaded with 50 mM of the indicated substrate. Transport was started by 0.2 mM 2-[^{14}C]-oxoglutarate or [^{14}C]-malonate. The incubation was for 2 min at 28°C. 100% exchange was 6 mmol/min/g protein.

Internal substrate (50mM)	2-[^{14}C]-Oxoglutarate exchange, %	[^{14}C]-Malonate exchange, %
2-Oxoglutarate	100	4
L-Malate	70	-
D-Malate	14	-
Malonate	25	3.5
Succinate	22	-
Glutamate	1	-
Citrate	1	-
Phosphate	1	1
Chloride	0.5	-

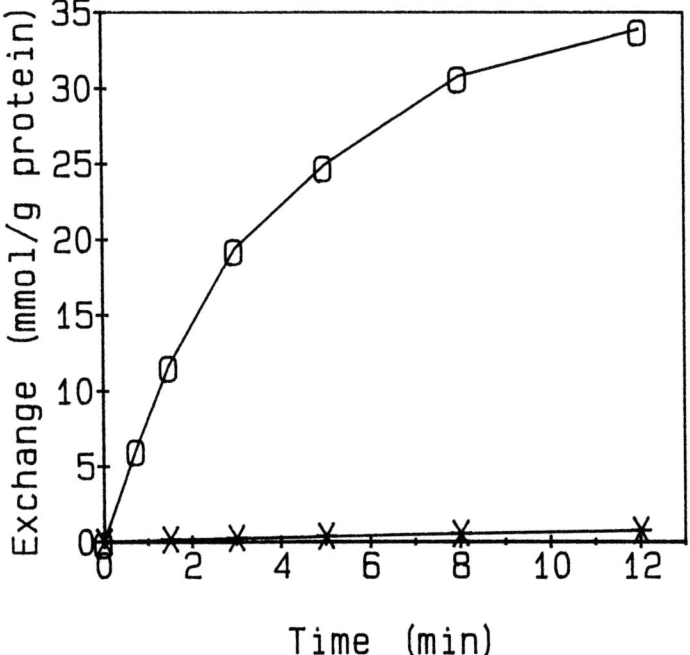

Fig. 3. Time course of 2-oxoglutarate/2-oxoglutarate exchange in proteoliposomes of the purified 2-oxoglutarate carrier from liver mitochondria in the absence (o) and in the presence (*) of mersalyl. The carrier was purified, reconstituted and assayed as discussed in Methods. 0.5 mM mersalyl was added 1 min before initiating transport.

In contrast, no significant uptake was observed when the proteoliposomes contained other substances known to be not transported by the carrier (*i.e.* phosphate, glutamate, citrate). The exchange activity in the absence of internal substrate or in the presence of only chloride was equal to the background activity of liposomes, *i.e.* vesicles without carrier protein.

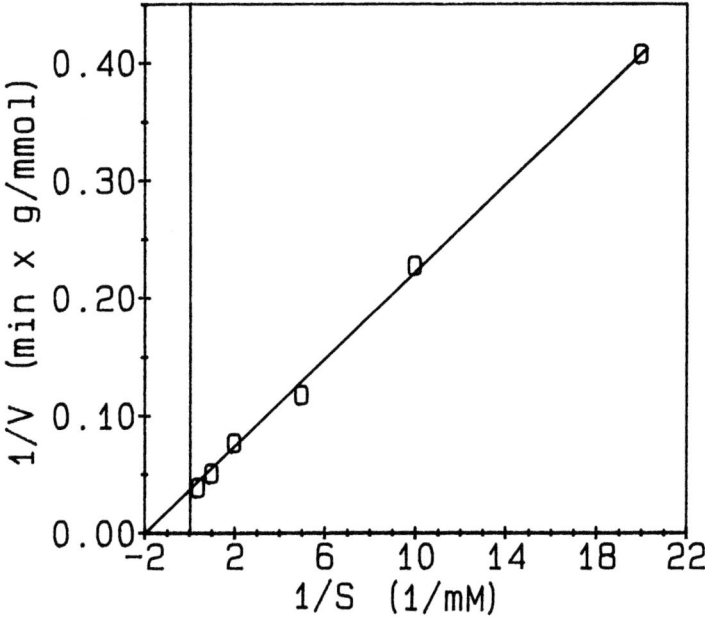

Fig. 4. Dependence of the rate of 2-oxoglutarate exchange in proteoliposomes on the external 2-[^{14}C]-oxoglutarate concentration. The 2-oxoglutarate carrier from liver mitochondria was purified and reconstituted as described in Material and Methods. The internal 2-oxoglutarate concentration was 50 mM. The externally added 2-oxoglutarate concentration was varied between 0.05 and 2.5 mM. Incubation time was 90 s at 28°C.

Thus, unidirectional uptake of 2-[^{14}C]-oxoglutarate was virtually absent in proteoliposomes of the purified carrier. Further, the purified carrier did not catalyze malonate/phosphate exchange, the characteristic of the dicarboxylate carrier. It is concluded that upon reconstitution into phospholipid vesicles the purified carrier displayed a similar substrate specificity as the carrier in mitochondria.

The dependence of the rate of 2-oxoglutarate/2-oxoglutarate exchange in proteoliposomes, loaded with 50 mM 2-oxoglutarate, on the external 2-[^{14}C]-oxoglutarate concentration is presented in Fig. 4 as a Line-Weaver Burk plot. The calculated K_m and V_{max} values of the exchange reaction were 0.5 ± 0.05 mM and 28.5 ± 0.2 mmol/min/g protein, respectively. The obtained K_m value was different from that found for the purified 2-oxoglutarate carrier from bovine heart mitochondria

(65 μM) by Indiveri *et al* (1987), reflecting most probably the different reconstitution and assay conditions. A K_m value for 2-oxoglutarate of 50 μM in rat liver mitochondria was reported (LaNoue & Schoolwerth, 1979).

Table 4. Inhibitor sensitivity of the 2-oxoglutarate/2-oxoglutarate exchange activity of the purified carrier from heart mitochondria. Purification, reconstitution and assay conditions are as described in Material and Methods. The inhibitors were added to the buffered proteoliposomes at the indicated time and temperature, before the labeled substrate (0.2 mM 2-[^{14}C]-oxoglutarate). The exchange activity of pure liposomes represents 100% inhibition. The exchange incubation was 10 min at 28°C.

Inhibitor	Preincubation	Inhibition,%
None		0
Phtalonic acid	1 mM 2 min, 26°C, pH 7	70
Mersalyl	0.5 mM 2 min, 26°C, pH 7	97
p-Hydroxymercuribenzoate	1 mM 2 min, 26°C, pH 7	94
2-Cyano-4-hydroxycinnamate	1 mM 2 min, 26°C, pH 7	65
Phenylsuccinate	1 mM 2 min, 26°C, pH 7	40
Eosin-5-maleimide	10 μm 10 min, 4°C, pH 7	90
	25 μm 10 min, 4°C, pH 7	98
	80 μm 10 min, 4°C, pH 7	99.5
N-Ethylmaleimide	100 μm 10 min, 4°C, pH 7	20
HgCl$_2$	10 μm 10 min, 4°C, pH 7	60
SITS	25 μm 10 min, 26°C, pH 7.5	50
	100 μm 10 min, 26°C, pH 7.5	75
p-Hydroxyphenylglyoxal	2 mM 25 min, 26°C, pH 7.5	55

Table 4 reports the sensitivity of the reconstituted 2-oxoglutarate carrier towards several inhibitors. The exchange was strongly inhibited by sulphydryl blocking reagents like mersalyl, p-hydroxymercuribenzoate and HgCl$_2$ which are known inhibitors of the carrier in mitochondria. Interestingly, eosin-5-maleimide, an SH-reagent which at low concentrations (less than 300 μM, Houštek & Pedersen, 1985) does not permeate the membrane, was the most potent inhibitor. In contrast, N-ethylmaleimide had only a small effect. These data indicate that the reconstituted carrier contains at least one sulphydryl group that might be involved in transport of 2-oxoglutarate. Also 2-cyano-4-hydroxycinnamate, a known inhibitor of the pyruvate carrier (Halestrap, 1975), gave 65% inhibition of the 2-oxoglutarate exchange activity.

Reagents that are relatively specific for arginine and lysine residues were also tested. The concentration of p-hydroxyphenylglyoxal, necessary for half maximum inhibition of the 2-oxoglutarate exchange in proteoliposomes, was found to be 2 mM. This corresponded to the inhibitory effect of p-hydroxyphenylglyoxal on the reconstituted citrate carrier (Stipani *et al.*, 1986). SITS (4-acetamino-4'-isothiocyanatostilbene-2,2'-disulfonate), a lysine reagent known to inhibit the anion translocator in the erythrocyte membrane was also an effective inhibitor. These results may indicate the presence of essential arginine and lysine residues.

Phtalonic acid, a specific inhibitor of the 2-oxoglutarate carrier, inhibited the exchange 70% at a concentration of 1 mM.

The exchange activity of the purified reconstituted 2-oxoglutarate carrier was also influenced by the phospholipid composition of the liposomes used for reconstitution. The highest 2-oxoglutarate exchange activity was obtained using proteoliposomes prepared from egg yolk phospholipids. The carrier activity decreased 35% when the reconstitution occurred with asolectin as the source of liposomes.

Table 5. Influence of different lipid composition of proteoliposomes on the 2-oxoglutarate/2-oxoglutarate exchange activity. The 2-oxoglutarate carrier was purified from heart mitochondria, reconstituted and assayed as described in Material and Methods. Preparation of the liposomes: the different lipids were dissolved in the indicated ratios in diethylether. The solvent was evaporated and further dried under vacuum. Liposomes were prepared by sonication at a total lipid concentration of 45 mg/ml as described in Material and Methods.

Lipids (weight ratio)	Exchange rate (mmol/6 min/g protein)
PC	2.2
PC/PE (60/40)	4.6
PC/PE/PDG (45/35/20)	4.7
PC/PE/Cholesterol (57/38/5)	10.7

Table 5 compares the different 2-oxoglutarate exchange activities of the purified 2-oxoglutarate carrier reconstituted either in pure phosphatidylcholine liposomes prepared from the indicated lipid mixtures. Addition of phosphatidylethanolamine to pure phosphatidylcholine increased the transport activity 2 fold. The same effect was achieved using a mixture of PC/PE/CL at a ratio comparable to the one in the inner mitochondrial membrane. However, when 5% cholesterol was supplemented to the PC/PE mixture, an almost 5 fold enhancement of the exchange activity was observed, resulting in the same high activity obtained with egg yolk phospholipid proteoliposomes. Cholesterol was observed to activate also the nucleotide carrier (Krämer, 1982).

The activation of the 2-oxoglutarate exchange reaction in proteoliposomes by cholesterol, possible consequent to a more rigid bilayer structure or less leaking vesicle membrane could be responsible for the effect. Alternatively, the changed membrane fluidity could have affected conformational changes of the reconstituted 2-oxoglutarate carrier, activating the exchange reaction. The presence of cholesterol in the liposomes could also have favored the extent or efficiency of incorporation of functional active carrier molecules. Finally, this modulation of the exchange activity by cholesterol does not exclude direct or specific interactions of the reconstituted carrier protein with regions of the bulky cholesterol molecule, resulting in activation of transport. Cholesterol, abundant in the plasma membrane, is a major determinant of membrane fluidity, but is almost not present in the inner mitochondrial membrane. However, the rigidifying role of cholesterol in the inner mitochondrial membrane can be obtained by other components such as proteins.

CONCLUSIONS

The results presented show that the 2-oxoglutarate carrier could be purified from bovine heart and liver mitochondria using the same isolation procedure. Hydroxylapatite, celite and Affigel 501 chromatography of the Triton X-114 extracted mitochondria resulted in the isolation of a single polypeptide band with an apparent molecular weight of 31.5-32 kDa. This purified protein was functionally identified as the 2-oxoglutarate carrier. Reconstituted into phospholipid vesicles, it catalyzed an efficient and specific transport of 2-oxoglutarate by a strict counter exchange with 2-oxoglutarate.

Removal of Triton, in the presence of added phospholipids, was achieved by adsorption into Amberlite XAD-2 beads. The proteoliposomes obtained by this method are more suitable for transport measurements and kinetic studies than freeze-thaw-sonication vesicles, where the detergent remains in the vesicles. Application of the Amberlite beads in a batch-procedure resulted in satisfactorily removal of Triton X-114, producing tight proteoliposomes.

This technique allowed monitoring the purification procedure by measuring the 2-oxoglutarate exchange and the characterization of substrate specificity and inhibitor sensitivity of the isolated carrier. The reconstituted purified carrier exhibited specific transport properties which were similar to those found in intact mitochondria. The rate of 2-oxoglutarate exchange in proteoliposomes was strongly dependent on the type of phospholipids used for the reconstitution. The presence of cholesterol in the liposomes significantly enhanced the exchange rate.

Finally, no functional differences between heart and liver carrier could be observed. However, only 50% carrier could be isolated from liver compared to heart mitochondria.

REFERENCES

Bisaccia F, Palmieri F (1984) Specific elution from hydroxylapatite of the mitochondrial phosphate carrier by cardiolipin. Biochim Biophys Acta 766:386-394

Bisaccia F, Indiveri C, Palmieri F (1985) Purification of reconstitutively active 2-oxoglutarate carrier from pig heart mitochondria. Biochim Biophys Acta 810:362-369

Bisaccia F, Indiveri C, Palmieri F (1988) Purification and reconstitution of two anion carriers from rat liver mitochondria: the dicarboxylate and the 2-oxoglutarate carrier. Biochim Biophys Acta 933:229-240

Cheetham PSJ (1979) Removal of Triton X-100 from aqueous solution using Amberlite XAD-2. Anal Biochem 92:447-452

de Pinto V, Prezioso G, Palmieri F (1987) A simple and rapid method for the purification of the mitochondrial porin from mammalian tissues. Biochim Biophys Acta 905:499-502

Halestrap AP (1975) The mitochondrial pyruvate carrier. Kinetics and specificity for substrates and inhibitors. Biochem J 148:85-96

Houštek J, Pedersen PL (1985) Adenine nucleotide and phosphate transport systems of mitochondria. Relative location of sulfhydryl groups based on the use of the novel fluorescent probe eosin-5-maleimide. J Biol Chem 260:6288-6295

Indiveri C, Krämer R, Palmieri F (1987) Reconstitution of the malate/aspartate shuttle from mitochondria. J Biol Chem 262:15979-15983

Indiveri C, Palmieri F, Bisaccia F, Krämer R (1987) Kinetics of the reconstituted 2-oxoglutarate carrier from bovine heart mitochondria. Biochim Biophys Acta 890:310-318

Johnson D, Lardy H (1967) Isolation of liver or kidney mitochondria. Methods Enzymol 10:94-96

Krämer R (1982) Cholesterol as activator of ADP-ATP exchange in reconstituted liposomes and in mitochondria. Biochim Biophys Acta 693:296-304

Krämer R, Heberger C (1986) Functional reconstitution of carrier proteins by removal of detergent with a hydrophobic ion exchange column. Biochim Biophys Acta 863:289-296

LaNoue KF, Schoolwerth AC (1979) Metabolite transport in mitochondria. Ann Rev Biochem 48:871-922

Lowry OH, Rosenbrough NJ, Farr AL, Randall RJ (1951) Protein measurement with the Folin Phenol reagent. J Biol Chem 193:265-275

Meijer AJ, Van Woerkom GM, Eggelte TA (1976) Phthalonic acid; an inhibitor of 2-oxoglutarate transport in mitochondria. Biochim Biophys Acta 430:53-61

Meijer AJ, Van Dam K (1981) Mitochondrial transport: In: Boting S, De Pont J (eds) Membrane Transport. Elsevier Amsterdam, pp 235-256

Müller M (1983) Lipid-protein and protein-protein interactions in the inner mitochondrial membrane. PhD thesis ETH nr 7392

Müller M, Cheneval D, Carafoli E (1984) Doxorubicin inhibits the phosphate-transport protein reconstituted in liposomes. A study on the mechanism of the inhibition. Eur J Biochem 140:447-452

Müller M, Krebs JJR, Cherry RJ, Kawato S (1984) Rotational diffusion of the ADP/ATP translocator in the inner membrane of mitochondria and in proteoliposomes. J Biol Chem 259:3037-3043

Nałecz MJ (1986) Metabolite transport systems of inner mitochondrial membrane: characteristics and isolation attempts. In: Kuczera J, Przestalski S (eds) Biophysics of Membrane Transport, vol.2. Publ. Dept. of the Agricultural University of Wroclaw, pp 13-45

Palmieri F, Quagliariello E, Klingenberg M (1972) Kinetics and specificity of the oxoglutarate carrier in rat-liver mitochondria. Eur J Biochem 29:408-416

Quagliariello E, Palmieri F (1972) Kinetics of substrate uptake by mitochondria. Identification of carrier sites for substrates and inhibitors. In: Azzone GF, Carafoli E, Lehninger AL, Quagliariello E, Siliprandi N (eds) Biochemistry and Biophysics of Mitochondrial Membranes. Academic Press New York and London, pp 659-680

Rigaud J, Paternostre M, Bluzat A (1988) Mechanisms of membrane protein insertion into liposomes during reconstitution procedures involving the use of detergents. 2. Incorporation of the light-driven proton pump bacteriorhodopsin. Biochemistry 27:2677-2688

Sluse FE, Ranson M, Liebecq C (1972) Mechanism of the exchanges catalysed by the oxo-glutarate translocator of rat-heart mitochondria. Eur J Biochem 25:207-217

Sluse FE, Goffart G, Liebecq C (1973) Mechanism of the exchanges catalysed by the oxoglutarate translocator of rat-heart mitochondria. Kinetics of the external-product inhibition. Eur J Biochem 32:283-291

Sluse-Goffart CM, Sluse FE, Duyckaerts C, Richard M, Hengesch P, Liebecq C (1983) Conformational changes and possible structure of the oxoglutarate translocator of rat-heart mitochondria revealed by the kinetic study of malate and oxoglutarate uptake. Eur J Biochem 134:397-406

Stipani I, Zara V, Zaki L, Prezioso G, Palmieri F (1986) Inhibition of the mitochondrial tricarboxylate carrier by arginine-specific reagents. FEBS Lett 205:282-286

Ueno M, Tanford C, Reynolds A (1984) Phospholipid vesicle formation using nonionic detergents with low monomer solubility. Kinetic factors determine vesicle size and permeability. Biochemistry 23:3070-3076

Yu ChA, Yu L, King TE (1975) Studies on cytochrome oxidase. Interactions of the cytochrome oxidase protein with phospholipids and cytochrome c. J Biol Chem 350:1383-1392

Hydroxyapatite Chromatography as a Tool for the Isolation of Anion Carriers and Other Membrane Proteins

P. RICCIO

Dipartimento di Biochimica e Biologia Molecolare and Centro di Studio sui Mitocondri e Metabolismo Energetico, C.N.R., Università di Bari, Via Amendola 165/A, 70126 Bari, Italy

In recent years it has become increasingly clear that ionic adsorption chromatography on hydroxyapatite is extremely effective in fractionation of membrane proteins, a result which is not to be found in the case of water-soluble proteins. The aim of this paper is to suggest reasons for this difference on the basis of the possible roles of detergents and lipids, thereby giving indications for the use of hydroxyapatite in isolating membrane proteins.

Hydroxyapatite (not hydroxylapatite, as recommended by Bernardi, 1971), (HA) $Ca_{10} (PO_4)_6 (OH)_2$, is a crystalline form of calcium phosphate which can be used for the fractionation of proteins as well as nucleic acids and viruses (Bernardi, 1971; Spencer et al., 1978). The process of interaction leading to the fractionation of proteins is called adsorption, is of an ionic nature and involves a different degree of protein surface affinity to the resin.

Abbreviations: HTP (Hydroxyapatite Tiselius Powder) is the commonly used name for the commercial product of Bio-Rad Lab.; EDTA, ethylenediaminotetraacetate; EGTA, ethyleneglycolbis-(2-aminoethyl)-tetraacetate; octyl-POE, n-octylpentaoxyethylene; POE, pentaoxyethylene; PLP, proteolipid; MBP, myelin basic protein

---

A. Azzi et al. (Eds.)
Anion Carriers of Mitochondrial Membranes
© Springer-Verlag Berlin Heidelberg 1989

36

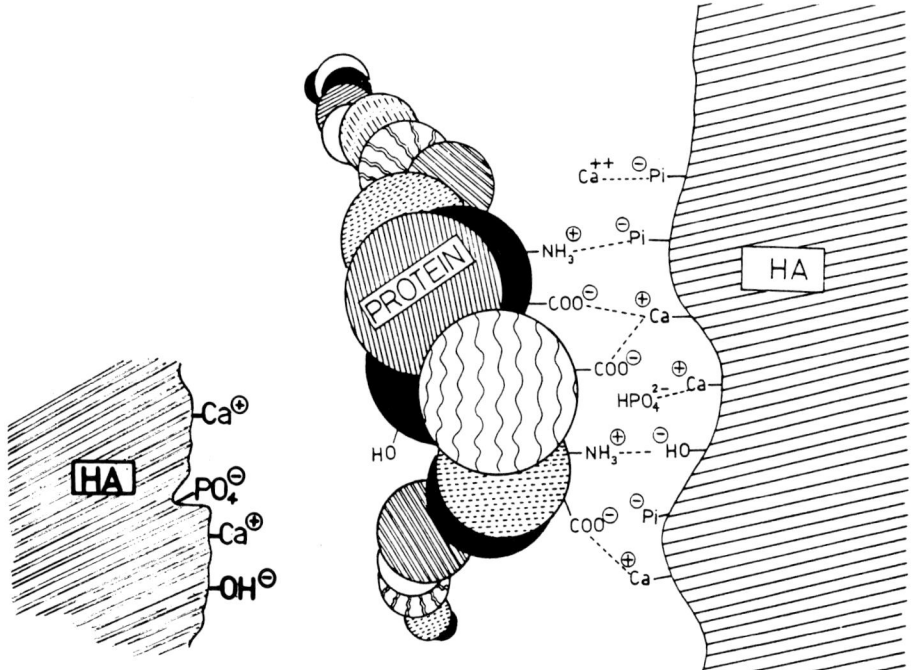

Fig. 1. The adsorbing sites of hydroxyapatite.
Fig. 2. Schematic representation of possible ionic interaction at protein and hydroxyapatite surfaces.

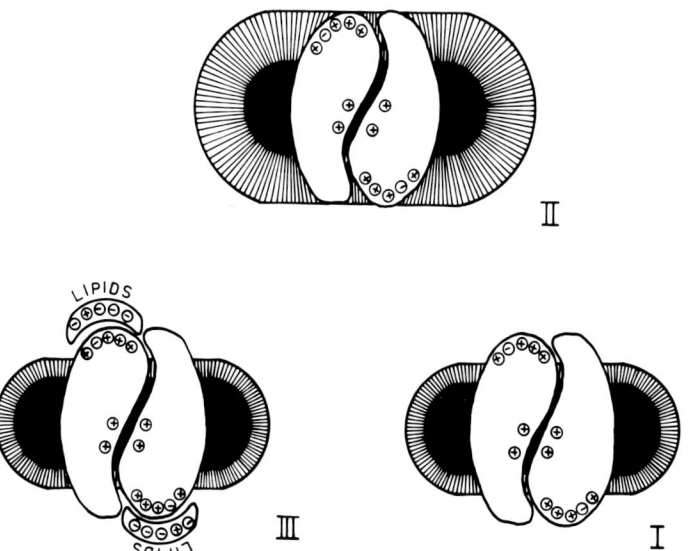

Fig. 3. Mixed protein-detergent micelles. Protein is represented as a hypothetical dimeric anion carrier. Detergent hydrophilic domain is in black and white.

Adsorption chromatography on hydroxyapatite columns was introduced in the 50's by Tiselius and coworkers (Swingle & Tiselius, 1951; Tiselius *et al.*, 1956) with the development of a procedure for the preparation of the inactive brushite $CaHPO_4$ $2H_2O$ and its conversion into the active hydroxyapatite, useful for column chromatography.

Thus, the use of hydroxyapatite was introduced before ion-exchange chromatography (Peterson & Sober, 1956) and gel filtration (Porath & Floid, 1959). Nevertheless adsorption chromatography has to date been largely neglected in the case of water-soluble proteins when compared with the latter techniques. On the contrary, subsequent to isolation of the adenine nucleotide carrier in Martin Klingenberg's laboratory (Riccio *et al.*, 1975a, 1975b; Klingenberg *et al.*, 1978, 1979), hydroxyapatite chromatography has become a fundamental tool for isolation of membrane proteins: anion and electron carriers, pore-forming and lipid-binding proteins, for which the chromatographic method appears much more suitable than for water-soluble proteins.

In order to explain this different behaviour we must first know how this resin works. On the basis of studies carried out with water-soluble proteins by Bernardi (1971) and by Gorbunoff (1984), and with membrane proteins by Riccio (1983) the following brief account can be given on the type of interaction between proteins and hydroxyapatite.

Hydroxyapatite can be considered as a mixed-bed ion-exchange containing calcium, phosphate and hydroxyl ions as adsorbing sites (Fig. 1).

The ionic adsorption process of proteins by hydroxyapatite involves multiple interaction between more or less widely distributed charge clusters on the two counterpart surfaces as schematically represented in Fig. 2. Binding strength depends on surface affinity increasing with the extension of the surface available for interaction, and with the complementary density and distribution of charges on both surfaces. Single charge binding is not important.

Although in a way similar to ion-exchange chromatography because of dependence on pH, isolelectric point of proteins, and type of counterions in the equilibrating medium, hydroxyapatite is on the other hand completely different from that when considering both the large surface and multiplicity of charges required for binding. In addition, it is possible in this case to have different degrees of ionic interaction strength, both in adsorption and elution steps, since devalent cations and anions can form complexes and thus act much more powerfully than they would do by simply following the Debye-Huckel law.

Thus, as shown in Table 1, acidic proteins will bind by their carboxyl groups to the calcium ions of hydroxyapatite; these proteins will require for elution phosphate (30-200 mM) or other calcium-complexing anions such as citrate (5-25 mM), EDTA or

EGTA (1-5 mM). On the other hand, basic proteins will bind to hydroxyapatite primarily by general electrostatic interactions to the phosphate and hydroxyl groups of the resin, and will be specifically eluted by 150-600 mM NaCl or by 1-7 mM $CaCl_2$. In any case, phosphate up to 500 mM can be used as a general eluant for all types of proteins but, unlike potassium salt, sodium phosphate can be used in the cold only up to 250 mM. It should be pointed out that although the use of citrate or EGTA as eluants is here proposed on the basis of our previous work (Riccio et al., 1977) as well as current personal experience, in their application the dissolving and destabilizing action of these anions on hydroxyapatite must be taken into account.

TABLE 1

PROTEINS	HA BINDING COUNTERPART	ELUANTS	ELUTING POWER OF SPECIFIC ELUANTS
ACIDIC	Calcium	NaP_i:0.030-0.200 M Na-Citrate:0.005-0.025M EGTA:0.001-0.005M (no NaCl or $CaCl_2$ up to 3M)	EGTA > Citrate > > Phosphate
NEUTRAL	Calcium, Phosphate	NaP_i, NaCl ($MgCl_2$ better than $CaCl_2$)	Phosphate > > Chloride
BASIC	Phosphate	NaCl: 0.150-0.600M NaP_i: 0.200-0.500M $CaCl_2$:0.001-0.007M	$Ca^{++}, Mg^{++} > > Na^+, K^+$

Although membrane proteins may be expected to interact with hydroxyapatite in the same way as hydrophilic proteins do, as mentioned above, their fractionation on hydroxyapatite may be by comparison remarkably easy. What makes the difference is all that which is present in a membrane extract and which is not needed in water-soluble proteins: detergent and lipid molecules. These amphiphilic molecules, the former required for membrane disruption and solubilization and the latter present as membrane components (Helenius & Simons, 1975), constitute additional parameters which can change protein chromatographic behaviour by masking or forming charged groups on the protein surface which are necessary for binding to the chromatographic resin. How this may occur is more readily understandable when we consider that, regardless of the model chosen to explain the association of detergents to membrane proteins, whether insertion of protein into preformed detergent micelles (Tanford & Reynolds, 1976), or coating of protein hydrophobic domain by a detergent film (Clarke, 1975), intrinsic membrane proteins will exist in solution as mixed, possibly lipid containing, protein-detergent complexes organized in the form of micelles. This is represented in Fig. 3 showing three possible different types of complexes of a protein with nonionic detergents and lipids. Type I and type II protein-detergent complexes differ in terms of the different hydrophilic portion length of the detergent, which in type II is sufficiently extended to mask protein charges and inhibit interaction with the resin. In the type III micelle possible ionic binding of lipids to the hydrophilic portion of the protein is taken into consideration. In this case as well protein binding to the resin could be diminished. The three different situations could be assembled as in Fig. 4 in a single representation accounting for the three types of micelles cited above.

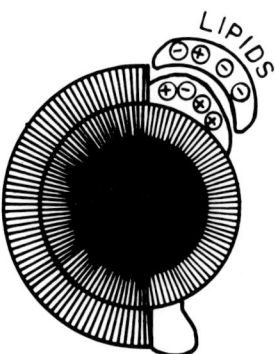

Fig. 4. Single representation of different possible assemblages of a protein (in white) with detergents and lipids.

Detergent hydrophilic domain may in this case appear overly conspicuous: in fact, mixed protein-detergent micelles are usually represented with the hydrophilic part of detergents being in some way not relevant, as is the polar head of most membrane lipids. However, the actual size of the water-soluble portion of a detergent may be very consistent. An example is shown in Fig. 5 where four different types of the family of the nonionic detergent Triton, with different hydrophilic portion lengths are represented. By comparison, the molecule of another nonionic detergent, n-octylpentaoxyethylene (octyl-POE) is also shown.It appears clear here that the hydrophilic portion of a detergent should not be neglected: it may well be large enough to embed the hydrophilic part of a protein and shield it from interaction with hydroxyapatite. As a result, considering that nonionic detergents do not react with the resin, the protein will not be adsorbed and will pass through the column (type II complex in Fig. 3). Detergents can thus be chosen with different hydrophilic portion lengths so that the hydroxyapatite-protein interaction is consequently modified, *i.e.* the latter diminishes as the former increases. This relationship remains sound when there is no change in protein conformation with the change of detergent.

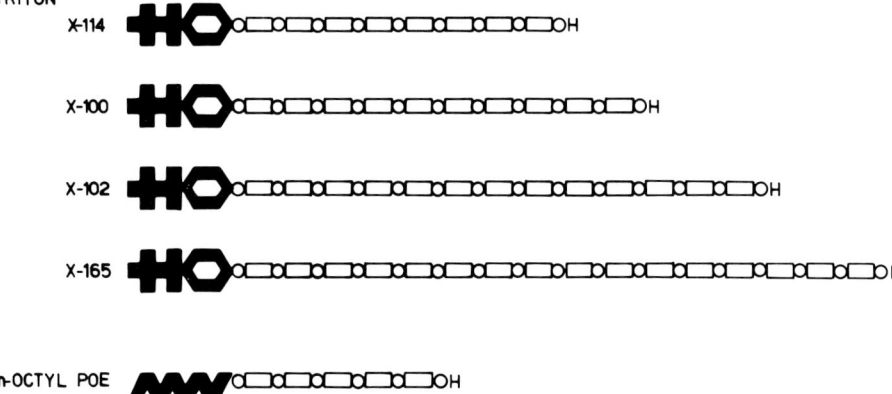

Fig. 5. Some derivatives of the Triton, p-tert.-octylphenylpolyoxyethylene, series with different numbers of oxyethylene groups (   ). POE is an abbreviation of pentaoxyethylene.

This effect of detergents could explain why most carriers or pore-proteins, which cross the membrane, are very hydrophobic and probably do not possess a notably extended hydrophilic domain, are not adsorbed when in Triton X-100.

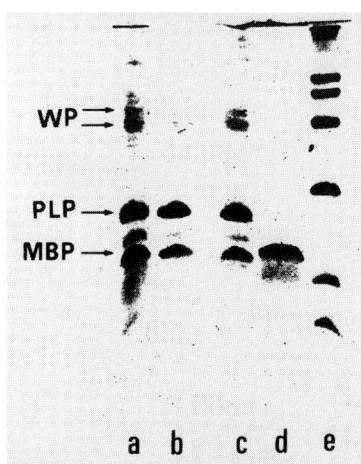

Fig. 6. Electropherogram of a SDS slab gel exhibiting myelin proteins stained with Coomassie blue. From left to right, the samples are:lanes a-d: as in the text, lane e: standard proteins (77, 66, 45, 30, 17, 12 kDa). WP, Wolfgram proteins; PLP, proteolipid protein; MBP, myelin basic protein.

In fact, when in Triton X-100, many anion carriers (ATP/ADP, phosphate, tricarboxylate, dicarboxylate, monocarboxylate, 2-oxoglutarate) and other membrane proteins are recovered in the void volume of an even dry hydroxyapatite column with a high adsorbing capacity, *i.e.* in the absence of eluants such as phosphate (Riccio *et al.*, 1975b; Lin & Klingenberg, 1980; Wohlrab, 1980; Kolbe *et al.*, 1981; Saint Macary & Foucher, 1985; De Pinto *et al.*, 1987).

cytosolic side

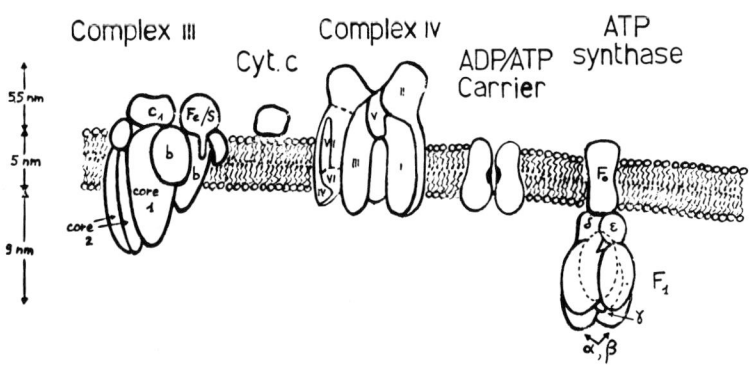

Fig. 7   matrix side

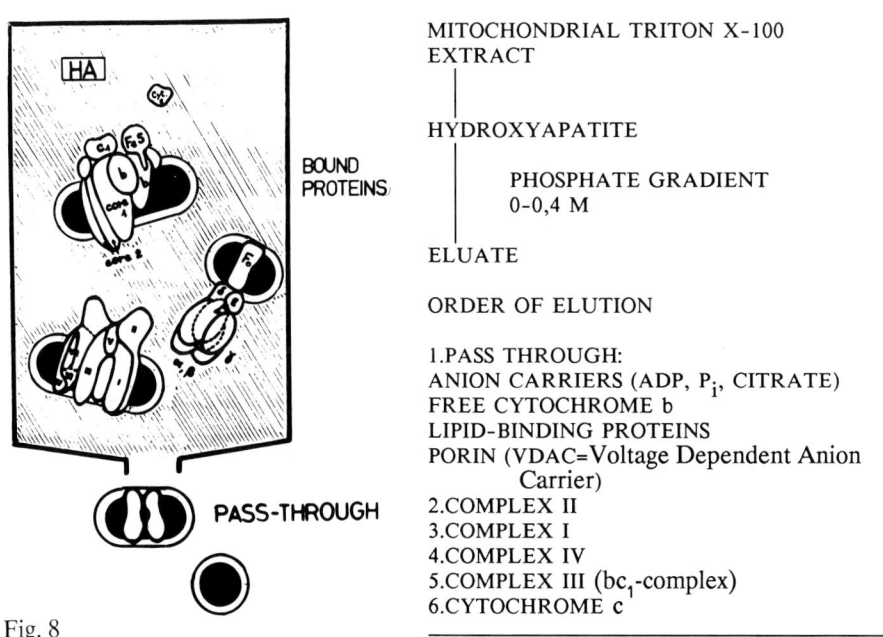

Fig. 8

TABLE 2

MITOCHONDRIAL TRITON X-100
EXTRACT

HYDROXYAPATITE

PHOSPHATE GRADIENT
0-0,4 M

ELUATE

ORDER OF ELUTION

1.PASS THROUGH:
ANION CARRIERS (ADP, P_i, CITRATE)
FREE CYTOCHROME b
LIPID-BINDING PROTEINS
PORIN (VDAC=Voltage Dependent Anion
        Carrier)
2.COMPLEX II
3.COMPLEX I
4.COMPLEX IV
5.COMPLEX III (bc_1-complex)
6.CYTOCHROME c

On the other hand, in the presence of Triton X-114, which has a shorter hydrophilic head, some of these proteins remain bound to hydroxyapatite (Bisaccia &

Palmieri, 1984) or require the addition of cardiolipin for elution (Bisaccia *et al.*, 1985; Kaplan & Pedersen, 1985; Nałęcz *et al.*, 1986; Szewczyk *et al.*, 1987). A possible demonstration for the proposed role of nonionic detergents is given in Fig. 6 showing the SDS electrophoretic pattern of brain myelin proteins extracted respectively by either Triton X-100 (lane a) or by n-octyl POE (lane c). The corresponding pass-through fractions of a hydroxyapatite eluate are on lane b (Triton X-100) and lane d (octyl-POE). It is evident that the proteolipid (PLP), a transmembrane protein of myelin sheath surrounding nerve axons, is eluted in the void volume only when associated to Triton X-100 (Riccio *et al.*, 1984a).

Although this result cannot be conclusive, because proteolipid properties are not known in either detergent, further evidence collected by De Pinto *et al.* (1988) with mitochondrial porine and other detergents should confirm the proposed role of detergents. Besides binding to a nonionic detergent with a sufficiently extended hydrophilic domain, other factors may facilitate passage of proteins through hydroxyapatite columns: a high protein/HA ratio; the presence of salts with eluting power; protein conformation, native in the case of membrane proteins, unfolded in the case of water-soluble proteins; a pH close to the isoelectric point of the protein to be eluted; and finally the ionic binding of lipids or other specific ligands to the charged surface of the protein. This would be the case of the type III micelle in Fig. 3, perhaps explaining the above mentioned requirement of cardiolipin for elution of most mitochondrial carrier when in Triton X-114. Cardiolipin requirement for the activity of some of these carriers (Kadenbach *et al.*, 1982) is not in contrast with cardiolipin role during chromatography. The chromatographic behaviour on hydroxyapatite of myelin basic protein (MBP) may be cited as an example for the proposed role of membrane lipids. When in the lipid-free and water-soluble unfolded state, this cationic protein binds to HA and can be eluted by NaCl, but when extracted by oxtyl-POE it retains its native conformation binding to all myelin lipids and is not adsorbed (Riccio *et al.*, 1984b). It should be mentioned that in the lipid-bound form the protein is no longer cationic. In conclusion, protein ability to bind amphiphilic lipids either before or after interaction with HA may strongly influence protein chromatographic behaviour in such a way as to cause them to be eluted.

Finally, on the basis of what has so far been discussed, with attention focused on mitochondrial membrane proteins alone (Fig. 7), the following chromatographic behaviour could be suggested as depicted in Fig. 8 for the proteins dissolved in Triton X-100. When eluted with phosphate the possible order of elution could be as in Table 2. At this point, it is worth noting that two cationic proteins such as ADP/ATP carrier (intrinsic) and cytochrome c (extrinsic) are separated because the translocase binds to Triton X-100 whereas cytochrome c does not.

It should be mentioned here that detergent concentration in the eluants is a further parameter to be considered. Although the elution order cited in Table 2 can be obtained in the presence of 0.5% Triton X-100, in the presence of 0.05% Triton X-100 the elution order of complexes III and IV is mutually changed. In general, decreased detergent concentration may lead to an increase in the salt concentration needed for elution of certain proteins.

ACKNOWLEDGEMENTS

I would like to point out that my observations on fractionation of membrane proteins on hydroxyapatite is based on experience gained by working with the following extremely valid colleagues: Martin Klingenberg (ADP/ATP carrier), Ferdinando Palmieri (citrate carrier), Gebhard von Jagow (bc_1-complex), Ernesto Quagliariello (complex III subunits, brain myelin proteins).

In addition, I would like to thank Dr. Antonella Bobba for her help in the preparation of the drawings, Mr Nicola Cataldo for the photographic work, Mr Michele Cinquepalmi for typing the manuscript and Prof. John Credico for revision of the final English version.

As in my first report on adsorption chromatography (Riccio, 1983), I would like to dedicate this paper to the memory of my dear parents Lina and Leonardo Riccio.

REFERENCES

Bernardi G (1971) Chromatography of proteins on hydroxyapatite. In: Jakoby WB (ed) Methods in Enzymology, vol 22. Academic Press, New York, p 325

Bisaccia F, Palmieri F (1984) Specific elution from hydroxyapatite of the mitochondrial phosphate carrier by cardiolipin. Biochim Biophys Acta 766:386-394

Bisaccia F, Indiveri C, Palmieri F (1985) Purification of reconstitutively active 2-oxoglutarate carrier from pig heart mitochondria. Biochim Biophys Acta 810:362-369

Clarke S (1975) The size and detergent binding of membrane proteins. J Biol Chem 250:5459-5469

De Pinto V, Prezioso G, Palmieri F (1987) A simple and rapid method for the purification of the mitochondrial porin from mammalian tissues. Biochim Biophys Acta 905:499-502

De Pinto V, Gaballo L, Benz R, Palmieri F (1989) Purification of mammalian porins. In: Azzi A, Nałęcz KA, Nałęcz MJ, Wojtczak L (eds) The Anion Carriers of the Mitochondrial Membranes. Springer Verlag Heidelberg, pp 237-248

Gorbunoff MJ (1984) The interaction of proteins with hydroxapatite. Anal Biochem 136:425-445

Helenius A, Simons K (1975) Solubilization of membranes by detergents. Biochim Biophys Acta 415:29-79

Kadenbach B, Mende P, Kolbe HVJ, Stipani I, Palmieri F (1982) The mitochondrial phosphate carrier has an essential requirement for cardiolipin. FEBS Lett 139:109-112

Kaplan RS, Pedersen PC (1985) Isolation and reconstitution of the n-Butylmalonate-sensitive dicarboxylate transporter from rat liver mitochondria. J Biol Chem 260:10293-10298

Klingenberg M, Riccio P, Aquila H (1978) Isolation of the ADP, ATP carrier as the carboxyatractylate protein complex from mitochondria. Biochim Biophys Acta 503:193-210

Klingenberg M, Aquila H, Riccio P (1979) Isolation of functional membrane proteins related to or identical with the ADP, ATP carrier of mitochondria. In: Fleischer S, Packer L (eds) Methods in Enzymology, vol 56. Academic Press, New York, p 407

Kolbe HVJ, Bottrich J, Genchi G, Palmieri F, Kadenbach B (1981) Isolation and reconstitution of the phosphate-transport system from pig heart mitochondria. FEBS Lett 124:265-269

Lin CS, Klingenberg M (1980) Isolation of the uncoupling protein from brown adipose tissue mitochondria. FEBS Lett 113:299-303

Nałęcz MJ, Nałęcz KA, Broger C, Bolli R, Wojtczak L, Azzi A (1986) Extraction, partial purification and functional reconstitution of two mitochondrial carriers transporting keto acids: 2-oxoglutarate and pyruvate. FEBS Lett 196:331-336

Peterson EA, Sober HA (1956) Chromatography of proteins: I. Cellulose ion-exchange adsorbent. J Amer Chem Soc 78:751-755

Porath J, Flodin P (1959) Gel Filtration: A method for the salting and group separation. Nature 183:1657-1659

Riccio P, Aquila H, Klingenberg M (1975a) Solubilization of the carboxyatractylate-binding protein from mitochondria. FEBS Lett. 56:129-132.

Riccio P, Aquila H, Klingenberg M (1975b) Purification of the carboxyatractylate-binding protein from mitochondria. FEBS Lett 56:133-138

Riccio P, Schagger H, Engel WD, von Jagow G (1977) bc$_1$-complex from beef heart: one step purification by hydroxyapatite chromatography in Triton X-100, polypeptide pattern and respiratory chain characteristics. Biochim Biophys Acta 459:250-262

Riccio P (1983) Adsorption chromatography of proteins in nonionic detergents. In: Frigerio A (ed) Chromatography in Biochemistry, Medicine and Environmental Research, vol 1. Elsevier, Amsterdam, p 177

Riccio P, De Santis A, Bobba A, Simone SM, Quagliariello E (1984a) Fractionation of myelin proteins. Ital J Biochem 33:216A-217A

Riccio P, Rosenbusch JP, Quagliariello E (1984b) A new procedure for the isolation of the brain myelin basic protein in a lipid-bound form. FEBS Lett 177:236-240

Saint Macary M, Foucher B (1985) Comparative partial purification of the active dicarboxylate transport system of rat liver, kidney and heart mitochondria. Biochem Biophys Res Commun 133:498-504

Spencer M, Neave EJ, Webb NL (1978) Hydroxyapatite for chromatography: III. Cation and pH effects on fractionation of tRNA for crystallization. J Chromatogr 166:447-454

Swingle SM, Tiselius A (1951) Tricalcium phosphate as an adsorbent in the chromatography of proteins. Biochem J 48:171-174

Szewczyk A, Nałęcz MJ, Broger C, Wojtczak L, Azzi A (1987) Purification by affinity chromatography of the dicarboxylate carrier from bovine heart mitochondria. Biochim Biophys Acta 894:252-260

Tanford C, Reynolds JA (1976) Characterization of membrane proteins in detergent solutions. Biochim Biophys Acta 457:133-170

Tiselius A, Hjerten S, Levin O (1956) Protein chromatography on calcium phosphate columns. Arch Biochem Biophys 65:132-155

Wohlrab H (1980) Purification of a reconstitutively active mitochondrial phosphate transport protein. J Biol Chem 255:8170-8173

Purification of the Monocarboxylate Carrier by Affinity Chromatography

KATARZYNA A. NAŁĘCZ , REINHARD BOLLI, LECH WOJTCZAK
AND ANGELO AZZI

Department of Cellular Biochemistry, Nencki Institute of Experimental Biology,
Pasteur str. 3, 02-093 Warsaw, Poland
and
Institut für Biochemie und Molekularbiologie der Universität Bern, Bühlstrasse 28,
CH-3012 Bern, Switzerland

The existence of a system transporting pyruvic acid through the inner mitochondrial membrane has been for a long time a matter of debate. In 1971, when such a system was described for the first time (Paradies et al., 1971) and since then it has been characterized in terms of its substrate specificity. Both other 2-oxo acids and their derivatives substituted in position 2 with halogens turned out to be good and competitive substrates in the reactions of either net pyruvate uptake according to the transmembrane pH difference or in the exchange reaction (Halestrap, 1975; Paradies & Papa, 1977). Halestrap described a series of chemical compounds which exhibited an inhibitory effect on the pyruvate transport measured in intact mitochondria, among them the most widely applied being α-cyanocinnamate and its 4-hydroxy derivative (Halestrap, 1975).

A simple method of hydroxylapatite chromatography used in case of proteins solubilized from the inner mitochondrial membrane allows to separate a group of

---

Abbreviations: DTE, dithioerithritol; DTT, dithiotreitol; Mops, 4-morpholinepropane-sulphonic acid; pCMB, p-chloromercuribenzoic acid; SDS sodium dodecylsulphate

---

A. Azzi et al. (Eds.)
Anion Carriers of Mitochondrial Membranes
© Springer-Verlag Berlin Heidelberg 1989

carrier proteins; among them the pyruvate carrier activity was detected as well (Nałęcz et al., 1986). Depending on the applied detergents and lipids some of the carriers were obtained in a pure form (Krämer & Klingenberg, 1977; Kolbe et al., 1981; Kolbe et al., 1984; Bisaccia et al., 1984; Bisaccia et al., 1985).

This study deals with several attempts to separate the pyruvate carrier, and describes the application of affinity chromatography using an α-cyano-4(OH)cinnamate - Sepharose column as the method of choice to purify it.

MATERIALS AND METHODS

Materials

Pyruvate (sodium salt) was from Boehringer, Triton X-100, Triton X-114 and egg yolk lecithin were from Fluka. $[1-^{14}C]$Pyruvate was delivered by Amersham International. All other chemicals were purchased by the firms given by Nałęcz et al. (1986).

Bovine heart mitochondria were prepared according to a standard procedure, as described by Yu et al. (1975). Submitochondrial particles were obtained as described elsewhere (Nałęcz et al., 1986).

Experimental procedure

Solubilization of the submitochondrial particles was performed in the presence of 3% Triton X-114 (if not stated otherwise) and 10 mg/ml asolectin, as given by Nałęcz et al. (1986). The procedure of hydroxylapatite chromatography followed that described by Nałęcz et al. (1986), in case of hydroxylapatite/celite column both gels were mixed in 1:1 w/w ratio and fractions of 350 μl were collected. Gel filtration experiments were done with Sephadex G-50 column (50 x 0.8 cm) equilibrated with 50 mM NaCl, 0.2% Triton X-100, 200 μM α-cyano-4(OH)cinnamate, 10 mM Mops, pH 7.2 and run with the same buffer with the velocity of 2 ml/h.

Synthesis of the affinity resin was started by coupling a six carbon spacer to Sepharose 4B according to the procedure given by Szewczyk et al. (1987), what was followed by an elongation of the spacer performed by addition of succinic anhydride (2 mmol/ml gel) in the presence of N-ethyl-N-3(dimethylaminopropyl)carbodiimide hydrochloride at pH 4.8. Nitrobenzoylazide attachment, reduction and diazotation were performed as described by Cuatrecasas and Anfinsen (1971), except that coupling of α-cyano-4(OH)cinnamate as a ligand was done at pH 8.1.

α-cyano-4(OH)cinnamate present in the fractions obtained from both gel filtration experiments and the affinity column had to be removed and the salt concentration had to be reduced to the same value in all samples. Therefore these fractions were incubated for 10 min with DTE (at concentration 5 times higher than that of α-cyano-4(OH)cinnamate) and passed through small Sephadex G-25 (coarse) columns equilibrated with 2% Triton X-100, 50 mM KCl and 20 mM Mops at indicated pH value. Samples corresponding to the void volume were subjected to reconstitution either with the use of freeze-thaw-sonication procedure (Nałęcz et al., 1986) or using Amberlite XAD-2 beads. In the case of the latter technique, preformed egg yolk lecithin vesicles (45 mg lipid/ml in 50 mM KCl and 20 mM Mops at indicated pH) were used. Such a suspension of liposomes was diluted 5 times with a sample containing protein, 2% Triton X-100, 50 mM KCl, 20 mM Mops and subsequently added to moist Amberlite XAD-2 beads, previously equilibrated with the lipids. The following proportions for reconstitution were chosen: 2.8 mg Triton X-100 per mg of egg yolk lecithin, 27 mg Triton X-100 per g "moist beads" and 7 μg protein per mg of phospholipid. For the exchange experiments media used for reconstitution were supplemented with 50 mM substrate. Samples prepared for reconstitution were incubated for 1.5-2 h at 25°C and the process was stopped by removing Amberlite and the external medium by passing through Dowex-Cl columns equilibrated with 170 mM sucrose. pH value of the effluent was adjusted to 7.2 for exchange experiments and to 8 for uptake measurements.

The assay of 2-oxoglutarate exchange (50 mM inside, 0.2 mM outside) followed the procedure described by Bisaccia et al. (1985). Measurements of exchange activity of the pyruvate carrier were performed at 15°C as described by Nałęcz et al. (1986). The concentration of 2-oxo acid entrapped inside proteoliposomes was 50 mM, that of added labeled pyruvate was 0.5 mM. The measurements of pyruvate uptake were started by simultaneous addition of the substrate and HCl to obtain pH value of 6 outside the vesicles. The reaction was stopped as in the case of exchange measurements.

For gel electrophoresis protein was precipitated with acetone as described by Nałęcz et al. (1986) and gel electrophoresis was performed according to De Pinto et al. (1985) under conditions given by Nałęcz et al. (1986).

Protein concentration was determined according to Lowry et al. (1951) with modifications described elsewhere (Nałęcz et al., 1986).

RESULTS AND DISCUSSION

The technique of hydroxylapatite chromatography of solubilized inner mitochondrial membrane proteins was used initially to purify adenine nucleotide translocator (Krämer & Klingenberg, 1977), but subsequently it was generally applied for the purification of other carriers (Klingenberg et al., 1979).

Depending on the type of detergent used for solubilization and on the phospholipid content of the medium, the protein pattern of the hydroxylapatite eluate was found to be different when analysed by polyacrylamide gel electrophoresis (Bisaccia

Table 1. Purification of the pyruvate carrier from bovine heart mitochondria by hydroxylapatite/celite chromatography. Solubilization, hydroxylapatite(HTP)/celite chromatography and reconstitution with the use of freeze-thaw-sonication procedure were performed as described in Materials and Methods. Pyruvate/pyruvate exchange was measured at 15°C, only the α-cyano-4(OH)cinnamate - sensitive reaction was taken into account.

Sample	Protein (μg)	Activity μmol/mg/min	Purification degree
Triton extract	3600	0.007	1
HTP/celite eluate			
-fraction 1	5	5.0	710
-fraction 2	6	3.0	430
-fraction 3	8	0.5	70
-fractions 4-7	5-10	0	-

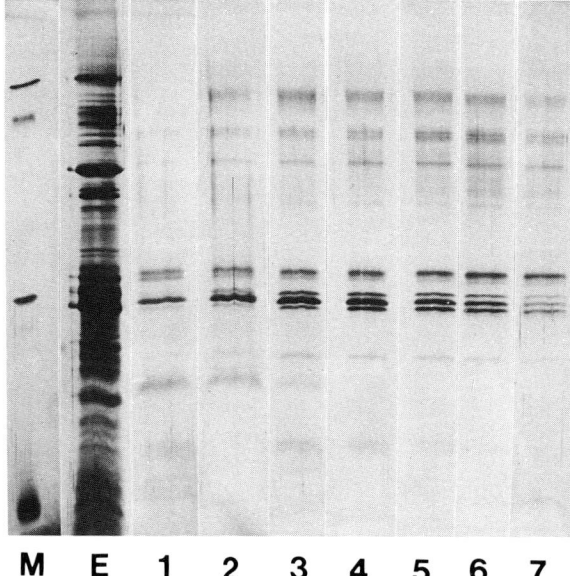

M E 1 2 3 4 5 6 7

Fig. 1. Polypeptide pattern of the Triton extract and the fractions after hydroxylapatite /celite chromatography. 300 μl of the fractions were precipitated with acetone, centrifuged, solubilized with SDS buffer and applied to SDS-polyacrylamide gel electrophoresis as in Materials & Methods. The Triton extract (E) is presented on the first lane, the numbers 1-7 represent corresponding fractions from hydroxylapatite/celite column. The following molecular weight markers (M) were used: phosphorylase b (94 kDa), bovine serum albumin (67 kDa), carbonic anhydrase (30 kDa) and cytochrome c (12 kDa).

& Palmieri, 1984; Kolbe *et al.*, 1984; Nałęcz *et al.*, 1986), in some cases resulting in preparation of pure carrier protein (Krämer & Klingenberg, 1977; Kolbe *et al.*, 1981; Kolbe *et al.*, 1984; De Pinto *et al.*, 1985; Bisaccia *et al.*, 1985). As it was reported elsewhere (Nałęcz *et al.*, 1986), also the transport activity of pyruvate homo-exchange was detected in the hydroxylapatite eluate.

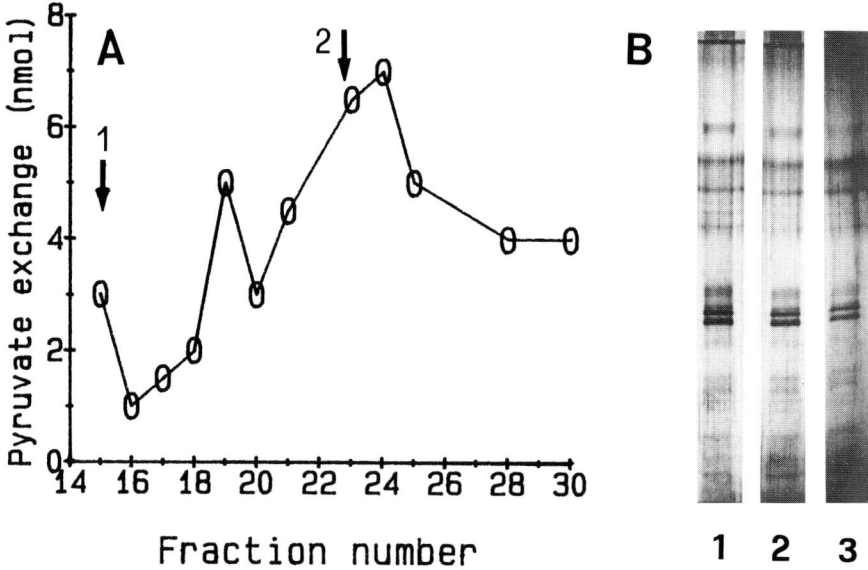

Fig. 2. Gel filtration of hydroxylapatite eluate. A. Activity profile of hydroxylapatite eluate of Triton X-114 extract. 100 μg of protein was loaded on the Sephadex G-50 coarse column. Elution, reconstitution with Amberlite XAD-2 and assay conditions are described in Materials and Methods. The arrows indicate the elution volume of blue dextran (1) and cytochrome *c* (2); 1 ml fractions were collected. B. SDS-PAGE of the fractions 15 (lane 1), 19 (lane 2), 22 (lane 3) was performed as described in the legend to Fig. 1.

Solubilization and purification of monocarboxylate carrier performed in media of different phospholipid content indicated that, in order to obtain high activity of the purified and reconstituted pyruvate transporting protein, a high cardiolipin concentration was necessary during solubilization of the membranes with Triton X-114. Hydroxylapatite chromatography could, however, be run with low-lipid-containing media (Nałęcz *et al.*, 1986).

As reported by Bisaccia *et al.* (1985), solubilization of mitochondria from pig heart with Triton X-114 and chromatography on hydroxylapatite and celite in the presence of cardiolipin resulted in a purification of the 2-oxoglutarate carrier.

Therefore, in the present work the hydroxylapatite/celite chromatography was used as well in an attempt to purify the pyruvate carrier from bovine heart mitochondria. According to previous observations concerning the activity of monocarboxylate carrier, cardiolipin was substituted by asolectin in all media used for purification. Table 1 presents the results of such procedure. The activity of pyruvate exchange was found mainly in the first two fractions, before the main peak of protein (fraction 3). This resulted in a very high degree of purification when compared with hydroxylapatite chromatography alone (Nałęcz et al., 1986). The polypeptide content of the fractions obtained from hydroxylapatite/celite column was analysed by SDS-PAGE (Fig. 1).

Table 2. The effect of deoxycholate (DOC) on the pyruvate/pyruvate exchange activity of the reconstituted carrier. Hydroxylapatite (HTP) eluate was passed through Sephadex G-25 columns equilibrated with varying buffers as indicated. The collected void volume fractions were subjected to Amberlite XAD-2 reconstitution and the pyruvate exchange was measured as described in Materials and Methods.

Sample	Sephadex G-25 running buffer	Activity nmol/mg protein	%
HTP eluate	-	145	100
HTP eluate after Sephadex	2% Triton X-100, 10 mM Mops, pH 7.4	233	160
	2% Triton X-100, 0.7 % DOC, 10 mM, Mops, pH 7.4	75	52

Several bands were visible in all fractions (lanes 1-4), what is in agreement with observations of Bisaccia et al. (1985) in case when the celite chromatography of the pass-through of hydroxylapatite column was run without cardiolipin. It has to be emphasized, however, that the PAGE patterns of the first two fractions were enriched in two proteins, namely of M_r of 34 kDa and of about 17 kDa.

Thomas and Halestrap (1981) reported that, among proteins of rat liver and heart mitochondria which were labeled with [^{3}H]N-phenylmaleimide, polypeptide(s) of low molecular weight of 12-15 kDa were protected against this labeling by incubation with α-cyanocinnamate. This could have suggested that the 17 kDa protein was a more likely candidate for the pyruvate carrier. An attempt was made to separate proteins of M_r of about 30 kDa from the smaller ones. A gel filtration technique was applied with use of a Sephadex G-50 column. In order to stabilize the activity of pyruvate carrier, the chromatography was run in the presence of inhibitor. Samples of the fractions were afterwards reactivated with DTE, known to reverse the inhibition by α-cyanocinnamate (Paradies, 1988). The activity of pyruvate exchange was detected in void volume of the Sephadex G-50 column (Fig. 2A), what could suggest the presence of some aggregates.

Table 3. Uptake and exchange activities of the reconstituted pyruvate carrier measured at different pH. Hydroxylapatite eluate after passing through Sephadex G-25 column was reconstituted with Amberlite XAD-2 beads, as described in Materials and Methods, at indicated pH. For the uptake experiment the pH gradient was created by simultaneous addition of HCl and radioactive pyruvate. 1 mM α-cyano-4(OH)cinnamate was applied as the inhibitor. Liposomes were formed from egg yolk lecithin.

Assay	Activity (nmol/min per mg protein)					
	Proteoliposomes			Liposomes		
	Total	+ Inhibitor		Total	+ Inhibitor	
Uptake pH 8	98	73	25	88	63	25
Uptake pH_{in} 8 pH_{out} 6	375	88	287	150	100	50
Exchange pH 8	88	75	13	88	73	15
pH 7.2	70	39	31	32	26	6

The main peak of the activity was localized, however, in the region in which cytochrome c was eluted. This could suggest elution of a protein of small molecular weight. SDS-PAGE analysis of the active fractions revealed, however, the presence of at least one band in the 30 kDa region (Fig. 2B). This observed retardation could be explained as a general phenomenon described for various asymmetric membrane proteins subjected to gel filtration (Nozaki et al., 1976).

Some of the carriers from the inner mitochondrial membrane were obtained in a pure form due to the application of affinity chromatography (Brandolin et al., 1974; De Pinto et al., 1982; Szewczyk et al., 1987). Following the procedure used for purification of the dicarboxylate carrier from bovine heart mitochondria (Szewczyk et al., 1987), an affinity resin was synthesized. α-cyano-4(OH)cinnamate was used as the ligand and the length of the spacer was doubled. As in the case of the dicarboxylate carrier (Szewczyk et al., 1987), the medium used for the affinity chromatography had to be supplemented with 0.7% deoxycholate; otherwise none of the proteins present in the pass-through of hydroxylapatite or hydroxylapatite/celite columns were bound to the affinity resin. A similar effect of an ionic detergent was observed by Robinson et al. (1980) for anionic dye affinity chromatography. The presence of deoxycholate in the medium made it difficult to measure the activity of pyruvate exchange when the freeze-thaw-sonication reconstitution method was used. Therefore, a method of detergent removal with Amberlite XAD-2 beads was applied. Moreover, the major part of deoxycholate was removed before reconstitution by passing the samples through a small Sephadex G-25 column. Table 2 demonstrates the effect of DOC removal. In comparison with control sample of hydroxylapatite eluate, its filtration through Sephadex G-25 column increased the total activity, whilst in the presence of deoxycholate the activity was decreased. Therefore this detergent was omitted in all media after the fractions from the affinity column were collected.

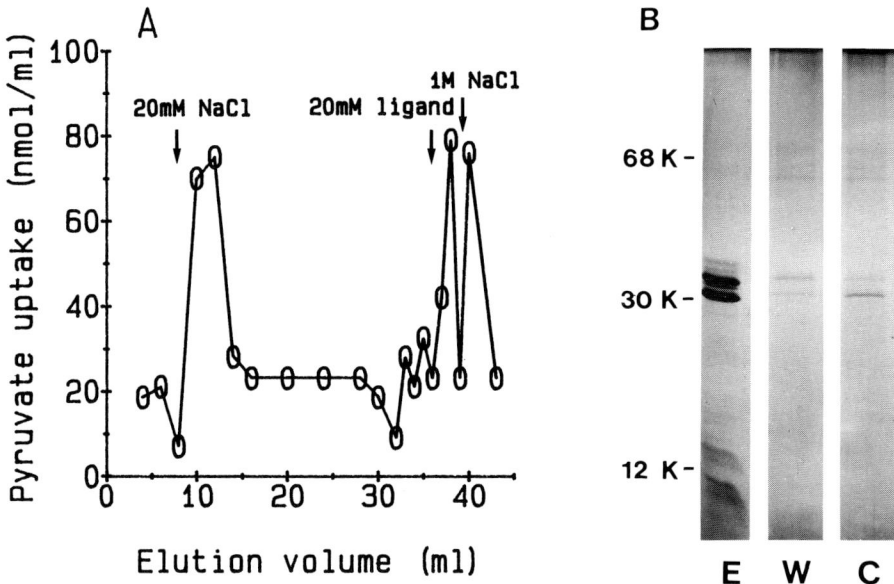

Fig. 3. Affinity chromatography of hydroxylapatite eluate performed at pH 7.2. A. Elution profile. 800 μl of hydroxylapatite eluate supplemented with 0.7% deoxycholate was loaded on 2 ml of the affinity resin equilibrated with 2% Triton X-100, 0.7% deoxycholate, 50 mM NaCl, 10 mM Mops, pH 7.2. Elution was performed with the same buffer with the additions, as indicated. 20 mM α-cyano-4(OH)cinnamate was used as the ligand. 200 μl of each fraction was taken for the reconstitution and uptake measurement, as described in Materials and Methods. Only the inhibitor-sensitive reaction was taken into account. B. SDS-PAGE pattern of the polypeptides loaded on the affinity resin (E) and present in the fractions eluted with 20 mM NaCl (W), 20 mM ligand (C). Treatment of the samples and conditions of electrophoresis as indicated in the legend to Fig. 1.

The possibility of a complete removal of detergents with the use of Amberlite beads gave the opportunity to create a pH gradient in proteoliposomes. Both exchange and uptake activities of the pyruvate translocating system were measured. From the data presented in Table 3 one can see that both reactions do not occur at pH 8, values obtained with liposomes and proteoliposomes being practically the same. The activity of pyruvate exchange measured at pH 7.2 was not very high but values obtained with liposomes were significantly lower (50%). In case of artificially created   pH by acidification of the external medium the activity of pyruvate uptake was about 10-fold higher when compared with pyruvate exchange at pH 7.2. Therefore the measurements of pyruvate uptake were chosen for monitoring the pyruvate carrier at various purification steps during further procedure.

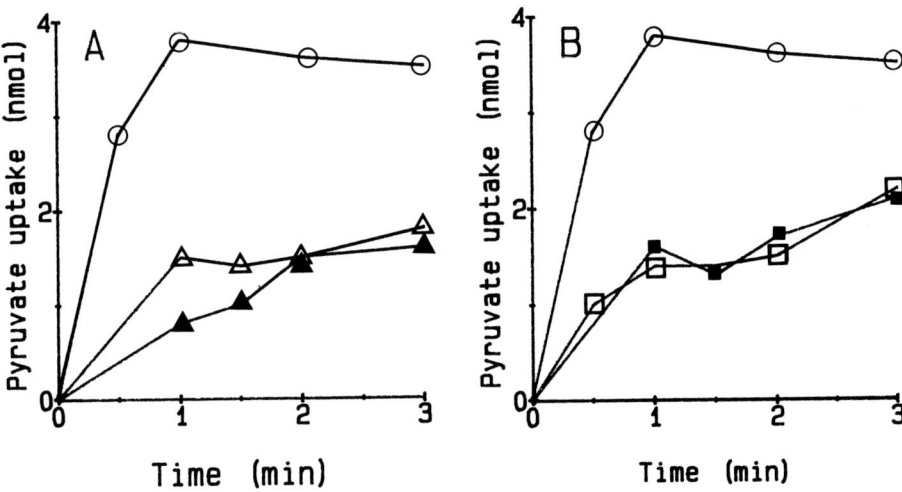

Fig. 4. Pyruvate uptake activity of the fractions eluted with α-cyano-4-(OH)cinnamate at pH 7.2. The fractions eluted with α-cyano-4(OH)cinnamate from the affinity resin were pooled and the inhibitor was removed by passing through Sephadex G-25. Reconstitution with Amberlite XAD-2 and pyruvate uptake were done as described in Materials and Methods in the absence (o) or in the presence of inhibitors (A) and other 2-oxo acids (B). The following concentrations were used: 0.2 mM pCMB (△), 1 mM α-cyano-4(OH)cinnamate (▲), 5 mM α-oxoisocaproate (■), 5 mM phenylpyruvate (□).

The affinity chromatography experiments were run with eluates from hydroxylapatite and hydroxylapatite/celite columns. The material was loaded and the chromatography was run at pH 7.2. Elution was performed with the medium supplemented first with 20 mM NaCl, and then with 20 mM ligand, and the final wash was done with 1 M NaCl. As can be seen in Fig. 3A, the activity of pyruvate uptake could be detected not only in fractions eluted with the ligand, but already a low salt concentration released a part of the carrier.

This would point out the ionic character of the protein-ligand interaction. This is supported by experiments in which α-cyano-4(OH)cinnamate was substituted by either β-mercaptoethanol or DTT. Pyruvate uptake was not detected in fractions eluted with these compounds (not shown).

Table 4. Transport activity of the pure proteins separated with the cinnamate affinity resin. Hydroxylapatite eluate after addition of deoxycholate (0.7%) and adjusting pH to 6.4 was loaded on the affinity resin equilibrated with 2% Triton X-100, 0.7% deoxycholate, 10 mM Mops, pH 6.4. The 34 kDa protein was eluted upon addition of 0.2 M NaCl at pH 6.4; the 31.5 kDa protein after increasing pH to 8 and a further addition of 20 mM α-cyano-4(OH)cinnamate. The activities of pyruvate uptake and oxoglutarate exchange were measured as described in Materials and Methods.

Molecular weight of protein	Activity	
	Pyruvate uptake nmol/mg prot./min	2-Oxoglutarate exchange nmol/mg prot./12 min
34 kDa	7000	29
31.5 kDa	0	2200

From the polypeptide pattern of different fractions (Fig. 3B) it can be concluded that many proteins are bound to the column irreversibly, since many bands disappear. This can be due to the fact that α-cyanocinnamate and its derivatives can react with -SH groups. In fractions eluted with α-cyano-4(OH)cinnamate two bands of 31.5 and 34 kDa were well visible, both being present in fractions obtained after washing the column with low salt concentration. The activity of pyruvate uptake measured in fractions eluted with the ligand could be ascribed to the pyruvate carrier. It was inhibited by α-cyano-4(OH)cinnamate and the -SH group reagent pCMB (Fig. 4A), what is in agreement with the effects observed in mitochondria (Halestrap, 1975, 1978) and in the reconstituted system (Nałęcz et al., 1986).

Some other α-oxo acids known to inhibit pyruvate transport (Halestrap, 1975; Paradies & Papa, 1977) decreased the uptake of pyruvate to the same extent as inhibitors. The polypeptide of 31.5 kDa was bound stronger to the resin even in the presence of low salt concentration. Different variations of pH and salt concentrations were tested, in order to separate these two bands. When proteins obtained in the hydroxylapatite eluate were loaded on the affinity resin at pH 6.5 and elution was run at this pH with 0.2 M NaCl, a pure preparation of the 34 kDa protein was obtained . Upon increasing pH to the value of 8 the protein of 31.5 kDa molecular weight was eluted. Bisaccia et al. (1985) defined 2-oxoglutarate carrier as a protein of 31.5 kDa. Therefore, pyruvate uptake and oxoglutarate exchange were measured in fractions containing pure proteins. From the data presented in Table 4 it can be concluded that the 34 kDa protein catalyses pyruvate uptake in the reconstituted system, whilst the protein 31.5 kDa is capable to catalyse 2-oxoglutarate exchange only.

The uptake of pyruvate measured with the purified protein reconstituted into egg yolk vesicles was inhibited by α-cyano-4(OH)cinnamate as well as by another 2-oxo acid α-oxoisocaproate (Fig. 5).

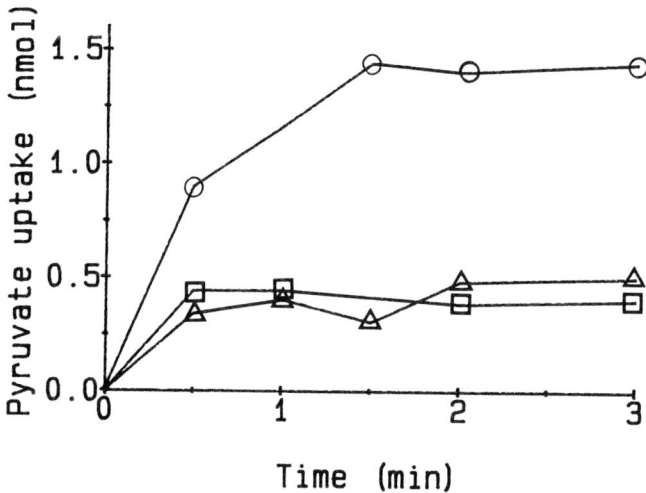

Fig. 5. Pyruvate uptake activity of the 34 kDa protein. The 34 kDa protein, purified as described in the legend to Table 4, was reconstituted and the uptake was measured as given in Materials and Methods in the absence (o) or presence of either 1 mM α-cyano-4(OH)cinnamate (▵) or 5 mM α-oxoisocaproate (□)

Table 5. Pyruvate exchange with other 2-oxo acids catalysed by the 34 kDa protein. The 34 kDa protein, purified as described in the legend to Table 4, was reconstituted with Amberlite XAD-2 as described in Materials and Methods. The concentration of the oxo acid inside proteoliposomes was 50 mM and the exchange with 0.5 mM pyruvate (outside) was measured.

2-Oxo acid inside	Additions	Exchange nmol/mg protein/5 min
Pyruvate	-	114
Pyruvate	α-cyano-4(OH) cinnamate	-
		73
Oxoisocaproate	-	146
Oxoglutarate	-	52

Moreover, the same protein in its reconstituted form catalysed the exchange reaction of pyruvate added from outside with either pyruvate or α-oxoisocaproate present inside, whilst the exchange with 2-oxoglutarate did not occur. The exchange reaction catalysed by the 34 kDa protein was inhibited by α-cyano-4(OH)cinnamate (Table 5). These results coincide with the described characteristics of the mitochondrial pyruvate translocator and indicate that the 34 kDa protein is the pyruvate (monocarboxylate) carrier.

CONCLUSIONS

The transporting system translocating pyruvate through the inner mitochondrial membrane was purified with the aid of an affinity resin using α-cyano-4(OH)cinnamate as ligand. Several proteins from hydroxylapatite eluate were bound to this resin.

At neutral pH only two of those polypeptides (31.5 and 34 kDa) were eluted specifically with the ligand. By varying the separation conditions, elution of these proteins was possible. The 31.5 kDa protein catalysed the exchange of 2-oxoglutarate, whilst the 34 kDa protein was capable to carry out the uptake and exchange of pyruvate.

ACKNOWLEDGEMENTS

This research was supported by the Swiss National Research Council and the Polish Central Program for Research and Development (CPBR 3.13.4.4.2).

REFERENCES

Bisaccia F, Palmieri F (1984) Specific elution from hydroxylapatite of the mitochondrial phosphate carrier by cardiolipin. Biochim Biophys Acta 766:386-394

Bisaccia F, Indivieri C, Palmieri F (1985) Purification of reconstitutively active α-oxoglutarate carrier from pig heart mitochondria. Biochim Biophys Acta 810:362-369

Brandolin G, Meyer C, Defaye G, Vignais PM, Vignais PV (1974) Partial purification of an ATR-binding protein from mitochondria. FEBS Lett 46:149-153

Cuatrecasas P, Anfinsen CB (1971) Affinity chromatography. Methods Enzymol 22:345-378

De Pinto V, Tommasino M, Palmieri F, Kadenbach B (1982) Purification of the active mitochondrial phosphate carrier by affinity chromatography with an organomercurial agarose column. FEBS Lett 148:103-106

De Pinto V, Tommasino M, Benz R, Palmieri F (1985) The 35 kDa DCCD-binding protein from pig heart mitochondria is the mitochondrial porin. Biochim Biophys Acta 813:230-242

Halestrap AP (1975) The mitochondrial pyruvate carrier. Kinetics and specificity for substrates and inhibitors. Biochem J 148:85-96

Halestrap AP (1978) Pyruvate and ketone-body transport across the mitochondrial membrane. Exchange properties, pH-dependence and mechanism of the carrier. Biochem J 172:377-387

Klingenberg M, Aquila H, Riccio P (1979) Isolation of functional membrane proteins related to or identical with the ADP, ATP carrier of mitochondria. Methods Enzymol .46:407-414

Kolbe HVJ, Böttrich J, Genchi G, Palmieri F, Kadenbach B (1981) Isolation and reconstitution of the phosphate-transport system from pig heart mitochondria. FEBS Lett 124:265-269

Kolbe HVJ, Costello D, Wong A, Lu RC, Wohlrab H 1984) Mitochondrial phosphate transport. Large scale isolation and characterization of the phosphate transport protein from beef heart mitochondria. J Biol Chem 259:9115-9120

Krämer R, Klingenberg M (1977) Reconstitution of adenine nucleotide transporter with purified ADP, ATP-carrier. FEBS Lett 82:363-367

Lowry OH, Rosenbrough NJ, Farr AL, Randall RJ (1951) Protein measurement with the Folin phenol reagent. J Biol Chem 193:265-275

Nałęcz KA, Bolli R, Wojtczak L, Azzi A (1986) The monocarboxylate carrier from bovine heart mitochondria: partial purification and its substrate-transporting properties in a reconstituted system. Biochim Biophys Acta 851:29-37

Nozaki Y, Schechter NM, Reynolds JA, Tanford C (1976) Use of gel chromatography for the determination of the Stokes radii of proteins in the presence and absence of detergents. A reexamination. Biochemistry 15:3884-3890

Papa S, Francavilla A, Paradies G, Meduri B (1971) The transport of pyruvate in rat liver mitochondria. FEBS Lett 12:285-288

Paradies G (1988) The effect of phenylglyoxal on the translocation of pyruvate in rat-heart mitochondria. Biochim Biophys Acta 932:1-7

Paradies G, Papa S (1977) On the kinetics and substrate specificity of the pyruvate translocator in rat liver mitochondria. Biochim Biophys Acta 462:333-346

Robinson JB, Strottmann JM, Wick DG, Stellwagen E (1980) Affinity chromatography in nonionic detergent solutions. Proc Natl Acad Sci (USA) 77:5847-5851

Szewczyk A, Nałęcz MJ, Broger C, Wojtczak L, Azzi A (1987) Purification by affinity chromatography of the dicarboxylate carrier from bovine heart mitochondria. Biochim Biophys Acta 894:252-260

Thomas AP, Halestrap AP (1981) Identification of the protein responsible for pyruvate transport in rat liver and heart mitochondria by specific labelling with [^{3}H]N-phenylmaleimide. Biochem J 196:471-479

Yu ChA, Yu L, King TE (1975) Studies on cytochrome oxidase. Interactions of the cytochrome oxidase protein with phospholipids and cytochrome c. J Biol Chem 350:1383-1392

Recent Developments in the Extraction, Reconstitution, and Purification of the Mitochondrial Citrate Transporter from Normal and Diabetic Rats

R.S. Kaplan, J.A. Mayor, D.L. Oliveira and N. Johnston

Department of Pharmacology, University of South Alabama, College of Medicine, Mobile, AL 36688, U.S.A.

The mitochondrial tricarboxylate (*i.e.* citrate) transporter catalyzes an electroneutral exchange across the mitochondrial inner membrane of a tricarboxylate (*e.g.* citrate, threo-D_s-isocitrate, cis-aconitate) for either another tricarboxylate, a dicarboxylate (*e.g.* malate or succinate), or phosphoenolpyruvate (Robinson *et al.*, 1971, Palmieri *et al.*, 1972). This transporter plays a particularly important role in hepatic intermediary metabolism because citrate efflux from mitochondria provides the cytoplasm with: a) a carbon source for both triacylglycerol and sterol biosyntheses (Spencer & Lowenstein, 1962; Srere & Bhaduri, 1962; Greville, 1969); and b) a source of NAD^+ and NADPH (*via* the concerted action of malate dehydrogenase and malic enzyme) which can be utilized by the glycolytic and lipid biosynthetic pathways respectively.

Due to the importance of the citrate transporter in cellular metabolism, the properties of this transporter have been extensively characterized in isolated mitochondria from normal (Robinson *et al.*, 1971; Palmieri *et al.*, 1972) and diseased (Cheema-Dhadli & Halperin, 1973; Kaplan *et al.*, 1982) tissues. Furthermore, during the last several years studies have been carried out to partially purify this transporter in reconstitutively active form (Stipani & Palmieri, 1983; Palmieri *et al.*, 1986).

This work was supported by NIH grant 1 R29 GM38785-01 and an American Diabetes Feasibility Grant to R.S.K.

A. Azzi et al. (Eds.)
Anion Carriers of Mitochondrial Membranes
© Springer-Verlag Berlin Heidelberg 1989

In this paper we describe procedures which result in: a) a substantially optimized extraction and reconstitution of the functional citrate transporter; and b) a highly purified preparation of the citrate transporter. Furthermore, for the first time, studies with the partially purified transporter have been extended to mitochondria that were obtained from streptozotocin-induced diabetic rats.

EXPERIMENTAL PROCEDURES

Isolation of the citrate transport protein

Frozen mitoplasts (approximately 75 mg of protein) were thawed and diluted to a final concentration of 30 mg protein/ml with ice-cold Buffer A (final concentration of 50 mM NaCl, 20 mM Hepes, 1 mM EDTA, pH 7.2, 5 mM citrate). All subsequent steps were carried out at 0-4°C. The transporter was then extracted by adding an equal volume of 6% Triton X-114 + 6 mg cardiolipin/ml, in Buffer A. Following a 20 min incubation, the suspension was centrifuged at 138,000 x g (average) for 45 min (including acceleration time). The supernatant (*i.e.* the Triton X-114 extracted mitoplasts) was removed and 500 μl aliquots were applied to separate hydroxylapatite (HA) columns prepared essentially as described previously (Kaplan *et al.*, 1986). The columns were eluted with Buffer A + 1% Triton X-114 and 4 ml of eluate were collected. The eluates from several HA columns were pooled and 1 mg cardiolipin (in Buffer B)/ml was added. This material was then added to 15 ml of a slurry of DEAE-Sepharose CL-6B. This slurry had been previously washed twice with Buffer B (20 mM NaCl, 10 mM Hepes, 1 mM EDTA, 5 mM citrate, pH 7.2, 1% Triton X-114) and then equilibrated with Buffer C (*i.e.* Buffer B + 1 mg cardiolipin/ml). Following a 90 min incubation of the slurry with the combined eluate (employing gentle rotation) the slurry was pelleted *via* centrifugation at 200 x g for 40 s (including acceleration time). After removal of the supernatant, the slurry (which contained resin-bound citrate transporter) was subjected to sequential washes (30 min gentle rotation/wash) with 14 ml of Buffer C followed by 7 ml of Buffer C + 100 mM NaCl. The transporter was then eluted *via* a 45 min incubation with 10 ml of Buffer C + 200 mM NaCl.

Preparation of citrate-loaded proteoliposomes

Asolectin vesicles were prepared by bath sonication of dried asolectin (233.2 mg) in 2.1 ml of Buffer D (120 mM Hepes, 50 mM NaCl, 1 mM EDTA, pH 7.4) as described previously (Kaplan & Pedersen, 1985a). Aliquots of freshly isolated protein fractions (0.25-0.35 ml) were added to 0.525 ml asolectin vesicles in the presence of 50 mM citrate. This mixture was vortexed and then rapidly frozen in liquid nitrogen.

Immediately prior to assay, the samples were thawed for 10 min in a water bath at room temperature and then sonicated with a probe sonicator (Branson Sonifier 250; output control = 1.6; 70% duty cycle; 30 bursts; total sonication time = 21 s; on ice). The proteoliposomes were then placed on a Dowex-1 resin (Sigma; Cl⁻ form; dry mesh 100-200) in a Pasteur pipette which had been pre-equilibrated with Buffer D. The liposomes were eluted with Buffer D and the opalescent fraction (approximately 1.25 ml) was collected and assayed for transport.

Measurement of 1,2,3-benzenetricarboxylate-sensitive citrate uptake

All transport incubations were carried out for 4 min at 30°C unless indicated otherwise. Aliquots (50 μl) of proteoliposomes were preincubated for 1 min with either 4 μl of 200 mM 1,2,3-benzenetricarboxylate (BTC) (final concentration of 0.01 M) (CONTROL) or with 4 μl of the buffer that the BTC was dissolved in (EXPERIMENTAL). Transport was then initiated by the addition of 25 μl of 4.26 mM [1,5-^{14}C]citrate (Amersham Corp.; reaction mix extraliposomal citrate: specific radioactivity approximately 2.3×10^4 dpm/nmol; 1.35 mM). Four min later, transport was quenched by the addition of 4 μl of 200 mM BTC (final concentration 0.01 M) to the EXPERIMENTAL incubation. The CONTROL received 4 μl of buffer. Aliquots (60 μl) of each reaction mix were then added to Dowex resin in separate Pasteur pipettes. The liposomes were eluted with Buffer D. Eluates (4 ml) were collected and 2 ml aliquots from each eluate were then mixed with 18 ml of Budget Solve (Research Products International Corp.). Intraliposomal [^{14}C]citrate radioactivity was then quantified utilizing an LS-7800 Beckman scintillation counter employing an open window. The BTC-sensitive citrate uptake rate was then calculated by subtracting the CONTROL value from the EXPERIMENTAL value. Employing the above procedure, control studies (*i.e.* liposomes without added protein) indicated negligible BTC-sensitive citrate uptake.

Animals

Male CD rats (retired breeders from Charles River Breeding Laboratories) were utilized for the isolation of citrate transporter from normal mitochondria. For the diabetic (and their respective control) studies, male Wistar rats (250-300 g) from Charles River were utilized. A diabetic state was induced *via* i.p. injection of 70 mg streptozotocin/kg dissolved in 31 mM citrate, 39 mM NaP$_i$, pH 4.0-4.5. Control rats were injected with buffer alone. Blood glucose levels were measured using the glucose oxidase method (Glucose Reagent Set, American Scientific Products) on a Beckman Glucose Analyzer 2.

Miscellaneous procedures

SDS-PAGE, Coomassie staining, and the mitoplast preparations were carried out essentially as described (Kaplan & Pedersen, 1985a) except, in the latter procedure, we performed all homogenization steps at 950 rpm. Protein content of the isolated transporter fractions was determined as previously detailed (Kaplan & Pedersen, 1985b).

RESULTS AND DISCUSSION

Optimization of the extraction and reconstitution of the functional citrate transporter

Initial experiments focused on rigorously defining conditions which result in an optimal extraction and reconstitution of the functional citrate transporter. Our previous studies (Kaplan & Pedersen, 1985a; Kaplan *et al.*, 1986) as well as those from other laboratories (Szewczyk *et al.*, 1987; Bisaccia *et al.*, 1988) have shown that the nonionic detergent Triton X-114 effectively extracts several different functional mitochondrial anion transport proteins.

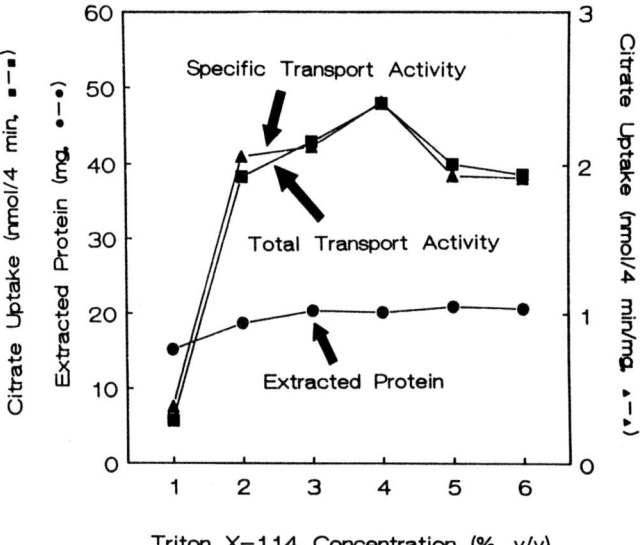

Fig. 1. Extraction and reconstitution of the functional citrate transporter with Triton X-114. Rat liver mitoplasts (28 mg protein) were extracted with 1-6% Triton X-114. The extracts were then incorporated into liposomes and assayed for transport as described under Experimental Procedures. Transport data represent means obtained from 6 incubations.

Accordingly, in the present study, we attempted to solubilize the functional citrate transporter with this detergent. Fig. 1 shows that whereas little citrate transport activity was extracted with 1% Triton X-114, maximal amounts of total and specific transport activities were obtained when the detergent concentration was increased to 3-4%. It should be noted that these extractions were performed in the presence of 3 mg exogenous cardiolipin/ml and 5 mM citrate.

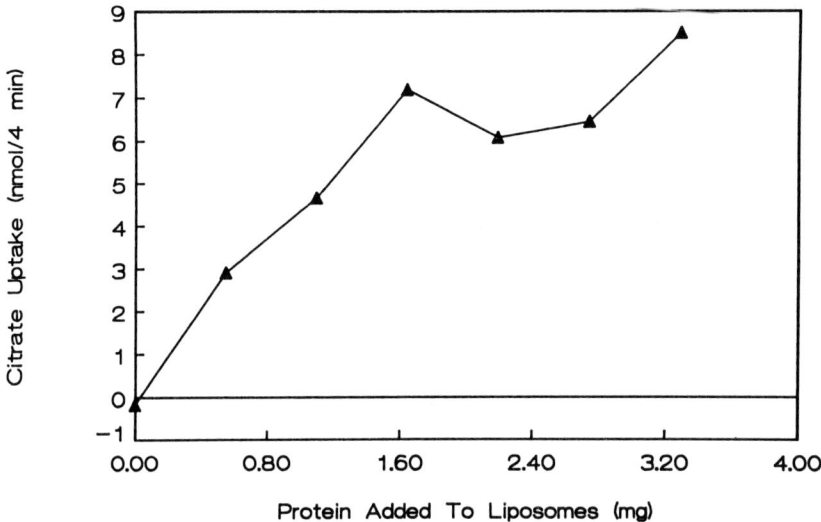

Fig. 2. Rate of 1,2,3-benzenetricarboxylate-sensitive citrate uptake into proteoliposomes as a function of the amount of added Triton X-114 extract. Varying amounts of Triton X-114 extract were added to preformed asolectin vesicles in the presence of 50 mM citrate. The volume of the resulting proteoliposomes was maintained at 0.9 ml by adding varying volumes of the extraction buffer (*i.e.* 50 mM NaCl, 20 mM Hepes, 1 mM EDTA, pH 7.2, 5 mM citrate, 3% Triton X-114, 3 mg cardiolipin/ml). Other conditions were as described under Experimental Procedures. Each datum point represents the mean of 5 incubations.

Experiments were then conducted to examine the dependence of the observed BTC-sensitive citrate transport rate on the amount of Triton X-114 extract which was added to liposomes. As depicted in Fig. 2, in the absence of added protein, no BTC-sensitive uptake of [^{14}C]citrate was observed. As increasing quantities of protein were added, the transport rate increased in an approximately linear manner up to 1.6 mg of added protein. As greater amounts of extract were added, a further non-linear increase in the transport rate was observed.

Fig. 3 shows that as expected for a transporter which catalyzes an obligatory exchange reaction, the solubilized citrate carrier demonstrated a strict requirement for intraliposomal substrate (*i.e.* citrate). Thus at the lowest internal citrate concentration tested (*i.e.* 1.4 mM citrate which originated from the extraction buffer) little uptake of external [^{14}C]citrate was observed. The rate of citrate uptake increased in response to increasing intraliposomal citrate concentrations, with maximal uptake rates observed in the presence of 50-60 mM intraliposomal citrate.

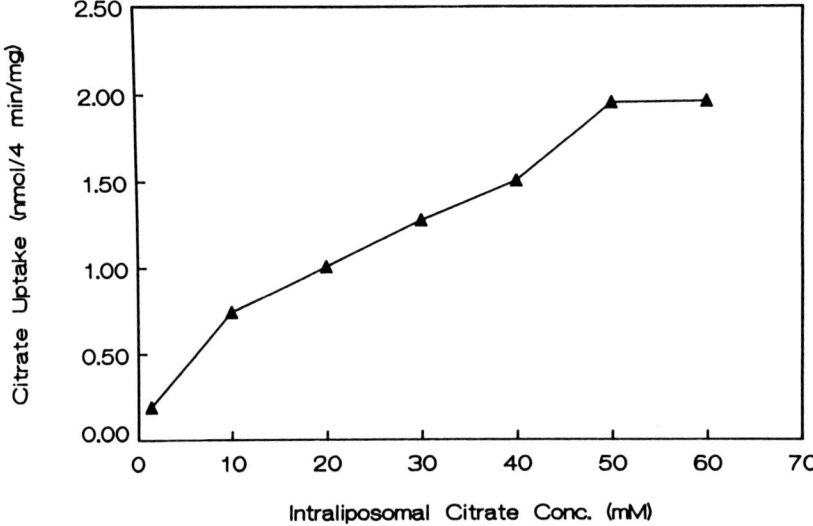

Fig. 3. Rate of 1,2,3-benzenetricarboxylate-sensitive citrate uptake into proteolipo-somes as a function of the intraliposomal citrate concentration. The Triton X-114 extract was incorporated into asolectin vesicles in the presence of 1.4-60 mM citrate. Transport incubations were carried out as described under Experimental Procedures. Each datum point represents the mean of incubations repeated a total of 6 times.

Finally, experiments were conducted in order to examine the effect of probe sonication for varying times on the reconstituted citrate transport activity. We observed (data not shown) that sonication for 21 s on ice (30 x 0.7 s bursts) yielded a 2-3 fold increase in the observed transport activity. Sonication for up to 43 s yielded no further increase.

Having optimized the reconstitution procedure with respect to the parameters described above, we then proceeded to develop a purification scheme for the functional transporter.

Partial purification of the reconstitutively active citrate transporter

Our procedure for the partial purification of the reconstitutively active citrate transporter from rat liver mitoplasts consisted of three basic steps. First, mitoplasts were extracted with Triton X-114 in the presence of cardiolipin and citrate as described above. Second, following a high speed centrifugation, the extract was chromatographed on hydroxylapatite (HA). HA chromatography has proven to be a remarkably effective purification step with several other mitochondrial transporters (Klingenberg *et al.*, 1978; Kolbe *et al.*, 1984; Kaplan & Pedersen, 1985a; Szewczyk *et al.*, 1987; Bisaccia *et al.*, 1988;) as well as with the citrate carrier (Stipani & Palmieri, 1983). It is important to note however, that in contrast to the latter study, we have performed the initial extraction and the subsequent chromatographic steps in the presence of citrate (a transport substrate) rather than BTC (a relatively high affinity transport inhibitor). We believe that our conditions permit a more reliable estimate of the transport rate since they eliminate the possibility that residual BTC might be affecting the transport process. The elution profile of the functional transporter from HA is depicted in Fig. 4.

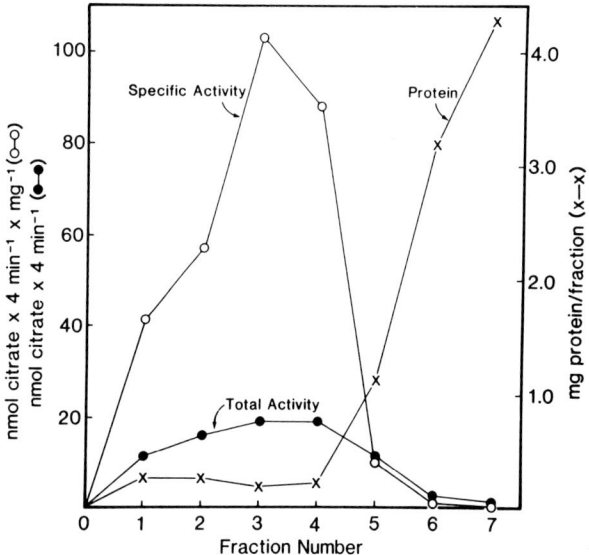

Fig. 4. Elution profile of the functional citrate transporter from hydroxylapatite. Successive 1 ml fractions of the HA eluate were incorporated into liposomes in the presence of 30 mM citrate. Data calculations were normalized to 34.5 mg starting mitoplast protein. Each transport datum point represents the mean of 4 incubations.

Clearly, most of the citrate transport activity appears in the first 4 ml of the eluate. Furthermore, an excellent separation of active transporter from the other extracted proteins (which elute in later fractions) is achieved, resulting in a 23-fold increase in the specific activity of this material (*i.e.* combined fractions 1-4) relative to the initial extract (Table 1).

The extensive purification achieved by HA chromatography is verified by the high resolution SDS-polyacrylamide gradient gels depicted in Fig. 5. Thus while many protein bands are present in the initial Triton extract (Fig. 5, Panel 1, lane B), there is a substantial reduction in the number of protein bands present in the initial fractions of the HA eluate (Fig. 5, Panel 1, lanes C-F). Most of these are located in the 30-36 kDa region.

Table 1. Purification of the citrate transport protein from rat liver mitoplasts. Frozen rat liver mitoplasts (82 mg protein) were employed as the starting material. Purified protein fractions were incorporated into preformed phospholipid vesicles in the presence of 50 mM citrate. Transport reactions were carried out as described under Experimental Procedures.

Fraction	1,2,3-Benzenetri-carboxylate-sensitive citrate uptake (nmol/4 min/mg protein)	Specific activity enhancement (fold)
Triton X-114 extracted mitoplasts	4.1	—
Hydroxylapatite eluate (fractions 1-4)	96.0	23
DEAE-Sepharose eluate	155.8	38

The final step in the purification procedure involved chromatography on DEAE-Sepharose CL-6B. This step enabled the removal of most of an approximately 60 kDa protein band and a 30-31 kDa band as depicted in Fig. 5, Panel 2 (lanes C and D), thus yielding a substantially purified citrate transporter preparation which, based on SDS-PAGE analysis, consisted of 4 main protein bands plus several minor bands. The specific transport activity of this fraction was increased by an additional 62% relative to the HA eluate (Table 1), thereby confirming that the 60 kDa and the 30-31 kDa proteins were in fact contaminants within the preparation. The transport activity of the final DEAE-Sepharose purified transporter preparation is enhanced 38-fold relative to the initial detergent extract.

Extraction and partial purification of the mitochondrial citrate transporter from diabetic animals

Two important alterations in intermediary metabolism which occur in Type I diabetes are an increase in hepatic fatty acid oxidation resulting in substantial ketone body formation, and a decrease in hepatic fatty acid synthesis (for review see Foster, 1983). Since citrate efflux from mitochondria, *via* the citrate carrier, provides the cytoplasm with a potential carbon source to support fatty acid biosynthesis, it was of interest to examine the properties of this transporter when isolated from animals that were in a diabetic state. This is especially true, since an earlier study showed that the properties of this transporter were altered (*i.e.* increased K_m) in intact mitochondria isolated from diabetic animals relative to their normal counterparts (Cheema-Dhadli & Halperin, 1973).

Table 2. Extractable citrate transport activity from normal and diabetic rats. Rat liver mitoplasts (78 mg protein) were employed as the starting material. Diabetic animals received 70 mg streptozotocin (in citrate/phosphate buffer)/kg i.p. 8 weeks prior to the mitoplast preparation. Control animals received the buffer alone. Shortly before the mitoplast isolation, blood glucose levels of 487 mg/dl and 118 mg/dl were determined with the diabetic and control animals, respectively. Mitoplasts were prepared from 3 animals in each group. Each datum point represents the mean value obtained from 4 transporter preparations.

Mitochondrial source	1,2,3-Benzenetricarboxylate-sensitive citrate uptake			
	Total transport activity (nmol/4 min)	Decrease (%)	Specific transport activity (nmol · mg · protein /4min)	Decrease (%)
Triton X-114 extracted mitoplasts				
Normal	224.8	—	4.3	—
Diabetic	149.2	34	2.2	49
Hydroxylapatite eluate				
Normal	175.5	—	129.4	—
Diabetic	82.4	53	49.9	61

Table 2 provides the first comparative information on the amount of functional citrate carrier that can be extracted from normal *versus* diabetic animals. Both the total and specific activities of the Triton X-114 solubilized transporter from diabetic animals were decreased by 34% and 49% respectively when compared with control animals. Furthermore, following partial purification of the transporter by HA chromatography, the observed decreases in the total and specific transport activities became even more pronounced (*i.e.* 53% and 61% respectively). These fractions were analyzed *via* SDS-PAGE in order to determine whether the measured activity decreases correlated with an observable alteration in the amount and/or the gross molecular properties of a protein in the 30-36 kDa range (*i.e.* the most likely molecular mass range of the citrate transporter based on a comparison with the known molecular masses of several other mitochondrial transporters). However, no major decreases in the Coomassie staining

68

intensity (diabetic *versus* normal) or alterations in protein mobility were apparent in this molecular mass range (Fig. 6).

Panel 1

Panel 2

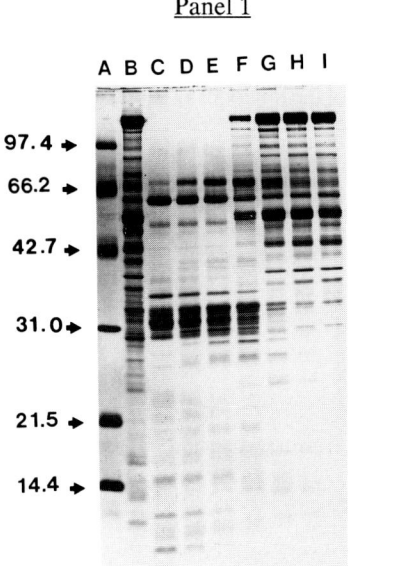

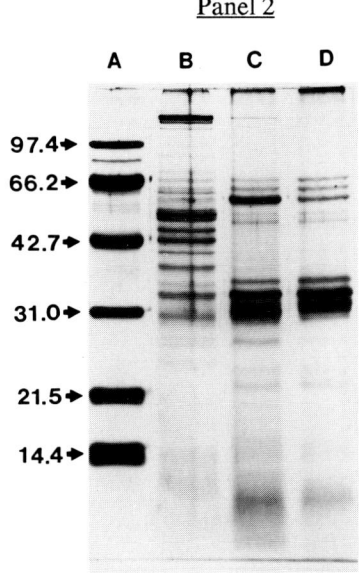

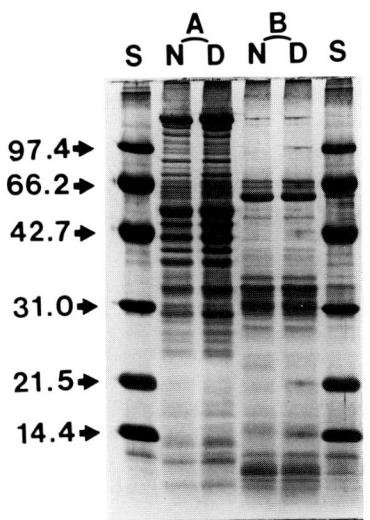

Fig. 5. Coomassie-stained SDS-polyacrylamide gradient gel electrophoretic profile of sequential steps in the purification of the functional citrate transporter. Proteins were run in a 4.5% polyacrylamide stacking gel followed by a highly resolving 14-20% gradient gel essentially as described previously (Kaplan & Pedersen, 1985a). Panel 1: lane A, 3 μg of each Bio-Rad SDS-PAGE low molecular weight standard protein: phosphorylase b (97,400), bovine serum albumin (66,200), ovalbumin (42,699), carbonic anhydrase (31,000), soybean trypsin inhibitor (21,500), and lysozyme (14, 400); lane B, 40 μg of the Triton X-114 mitoplast extract; lanes C-I, approximately 17-25 of hydroxylapatite eluate fractions 1-7 from Fig. 4. Panel 2: lane A, 3 μg of each Bio-Rad SDS-PAGE low molecular weight standard protein; lane B, 34 μg of the Triton X-114 mitoplast extract; lane C, 19 μg of the hydroxylapatite eluate; lane D, approximately 10-12 of the DEAE-Sepharose CL-6B purified citrate transporter.

Fig. 6. SDS-PAGE of the partially purified citrate transporter obtained from normal and diabetic rats. SDS-PAGE was out carried as described for Fig. 5. Lanes in S, 3 μg of each Bio-Rad SDS-PAGE standard protein. Lanes in A, 40 μg of the Triton X-114 extract utilizing mitoplasts obtained from either normal (N) or diabetic (D) rats. Lanes in B, 18 μg of the HA eluate prepared from normal (N) or diabetic (D) extracts.

Further work is currently in progress in this laboratory to purify the citrate transporter to essential homogeneity from both normal and diabetic animals. This will permit an elucidation of the mechanism of action and the regulation of this anion transporter at the molecular level in both the normal and diabetic states.

REFERENCES

Bisaccia F, Indiveri C, Palmieri F (1988) Purification and reconstitution of two anion carriers from rat liver mitochondria: the dicarboxylate and the 2-oxoglutarate carrier. Biochim Biophys Acta 933:229-240
Cheema-Dhadli S, Halperin ML (1973) The role of the mitochondrial citrate transporter in the regulation of fatty acid synthesis: effect of fasting and diabetes. Can J Biochem 51:1542-1544
Foster DW (1983) Diabetes mellitus. In: Stanbury JB, Wyngaarden JB, Frederickson DS, Goldstein JL, Brown MS (eds) The Metabolic Basis of Inherited Disease. McGraw-Hill, New York London, pp 99-117
Greville GD (1969) Intracellular compartmentation and the citric acid cycle. In: Lowenstein JM (ed) Citric Acid Cycle Control and Compartmentation. Marcel Dekker, New York London, pp 1-136
Kaplan RS, Morris HP, Coleman PS (1982) Kinetic characteristics of citrate influx and efflux with mitochondria from Morris hepatomas 3924A and 16. Cancer Res 42:4399-4407
Kaplan RS, Pedersen PL (1985a) Isolation and reconstitution of the n-butylmalonate-sensitive dicarboxylate transporter from rat liver mitochondria. J Biol Chem 260:10293-10298
Kaplan RS, Pedersen PL (1985b) Determination of microgram quantities of protein in the presence of milligram levels of lipid with Amido Black 10B. Anal Biochem 150:97-104
Kaplan RS, Pratt RD, Pedersen PL (1986) Purification and characterization of the reconstitutively active phosphate transporter from rat liver mitochondria. J Biol Chem 261:12767-12773
Klingenberg M, Riccio P, Aquila H (1978) Isolation of the ADP, ATP carrier as the carboxyatractylate-protein complex from mitochondria. Biochim Biophys Acta 503:193-210
Kolbe HVJ, Costello D, Wong A, Lu RC, Wohlrab H (1984) Mitochondrial phosphate transport: large scale isolation and characterization of the phosphate transport protein from beef heart mitochondria. J Biol Chem 259:9115-9120
Palmieri F, Stipani I, Quagliariello E, Klingenberg M (1972) Kinetic study of the tricarboxylate carrier in rat liver mitochondria. Eur J Biochem 29:408-416
Palmieri F, Stipani I, Prezioso G, Krämer R (1986) Partial purification and reconstitution of the tricarboxylate carrier from rat liver mitochondria. Methods Enzymol 125:692-696
Robinson BH, Williams GR, Halperin ML, Leznoff CC (1971) Factors affecting the kinetics and equilibrium of exchange reaction of the citrate-transporting system of rat liver mitochondria. J Biol Chem 246:5280-5286
Spencer AF, Lowenstein JM (1962) The supply of precursors for the synthesis of fatty acids. J Biol Chem 237:3640-3648
Srere PA, Bhaduri A (1962) Incorporation of radioactive citrate into fatty acids. Biochim Biophys Acta 59:487-489
Stipani I, Palmieri F (1983) Purification of the active mitochondrial tricarboxylate carrier by hydroxylapatite chromatography. FEBS Lett 161:269-274
Szewczyk A, Nałecz MJ, Broger C, Wojtczak L, Azzi A (1987) Purification by affinity chromatography of the dicarboxylate carrier from bovine heart mitochondria. Biochim Biophys Acta 894:252-260

Isolation and Functional Reconstitution of the Dicarboxylate Carrier from Bovine Liver Mitochondria

MACIEJ J. NAŁĘCZ, ADAM SZEWCZYK, CLEMENS BROGER, LECH WOJTCZAK AND ANGELO AZZI

Department of Cellular Biochemistry, Nencki Institute of Experimental Biology, Pasteur str. 3, 02-093 Warsaw, Poland
and
Institut für Biochemie und Molekularbiologie der Universität Bern, CH-3012 Bern Bühlstrasse 28, Switzerland

In order to understand the structure and function relationship of mitochondrial transporters, recent efforts have been focused on isolation of the individual carriers and their functional reconstitution into proteoliposomes (for review see Nałęcz , 1986). In line with these studies we reported a successful purification by affinity chromatography of the dicarboxylate carrier from bovine heart mitochondria (Szewczyk et al., 1987). In general, heart mitochondria are considered optimal for studies on purification of the inner membrane proteins. This comes from the fact that the cristae/matrix ratio of these organelles is exceptionally high (more membranes per unit of weight of the mitochondrial preparation) and, in addition, from that the activity of proteolytic enzymes in heart homogenates is relatively low. On the other hand, however, it is known that the mitochondrial dicarboxylate carrier activity is much higher in liver than in heart (Sluse et al., 1971). Although it has not been clarified whether this is due to

Abbreviations: HTP, hydroxyapatite; MOPS, 3-(N-morpholino)propanesulfonic acid; SDS, sodium dodecylsulfate; SMP, submitochondrial particles.

A. Azzi et al. (Eds.)
Anion Carriers of Mitochondrial Membranes
© Springer-Verlag Berlin Heidelberg 1989

difference in total amount of the carrier protein in mitochondria, we found that only very little of the translocator protein could actually be isolated from heart (Szewczyk *et al.*, 1987). Thus it seemed logical to try to isolate the dicarboxylate carrier from liver mitochondria.

MATERIALS AND METHODS

Biological materials

Bovine liver mitochondria and submitochondrial particles (inner membrane inside-out particles) were obtained by standard procedures. Submitochondrial particles were stored in 240 mM sucrose plus 10 mM MOPS (pH 7.2) at -70°C. Bovine heart mitochondria were prepared and stored as described elsewhere (Nałęcz *et al.*, 1986a).

Extraction and hydroxyapatite chromatography

The purification procedure started from extraction of the membranes with Triton X-100 followed by hydroxyapatite chromatography and various subsequent steps: affinity chromatography, anion exchange chromatography, immobilized-metal chromatography or dye affinity chromatography. To 0.6 ml of SMP (approximately 15 mg protein) 1.2 ml of the solution containing 4% Triton X-100, 50 mM NaCl, 20 mM MOPS (pH 7.2) and 1 mM EDTA was added and the mixture was allowed to stay on ice for 20 min. 0.6 ml of this suspension was then loaded onto 600 mg of dry hydroxyapatite and eluted with 1.8 ml of the solubilization buffer (as above). Only the pass-through of this chromatography was collected (approximately 0.6 ml) and used for further studies. Details of subsequent purification procedures are described under Results and Discussion.

Reconstitution

The material obtained at various steps of the purification procedure was reconstituted into liposomes by five different procedures:

Procedure A was a freeze-thaw-sonication performed exactly as described previously (Nałęcz et al., 1986a, 1986b).

Procedure B was based on the removal of the detergent by Amberlite XAD-2 and was performed as follows: malate-containing liposomes were prepared by sonicating 180 mg asolectin (soy bean phospholipids, Associated Concentrates) in 4 ml of solution composed of 250 mM malate, 50 mM NaCl and 20 mM MOPS (pH 7.2). Amberlite XAD-2 (uncharged polystyrene resin for hydrophobic interactions, Fluka AG) was pre-equilibrated with the suspension of these liposomes (5 times diluted with water) for 30 min at room temperature, at the ratio of 1 ml of diluted liposomal suspension per 1 g of Amberlite. The sample for reconstitution (2 ml final volume) was prepared by mixing 400 μl of malate-containing liposomes (final concentration of malate 50 mM), 0.5 ml of the carrier-containing sample (HTP eluate or any other material from further purification steps), 0.5 ml of the solubilization buffer (as above, containing 4% Triton) and 0.6 ml water. The final sample was clear and transparent and its absorption spectrum was measured at 200-300 nm in order to check the Triton content. To the Amberlite pre-equilibrated with liposomes the reconstitution sample was added at the ratio of 1 ml of the sample per 1 g of Amberlite. The mixture, placed in a small flask, was then incubated with shaking for 1 h at 30°C (water bath). After this time the reconstituted proteoliposomes were collected and their absorption spectrum was measured to confirm the removal of Triton. At this stage the sample was turbid and opalescent. In order to remove external malate from proteoliposomal preparation the sample was passed through Dowex (Fluka AG, 100-200 mesh, chloride form, pre-equilibrated with 170 mM sucrose): 1 ml of proteoliposomes was loaded onto 500 mg of Dowex (packed into Pasteur pipette) and eluted with 900 μl of 170 mM sucrose. Only the turbid lipid-containing fractions were collected. MOPS buffer (pH 7.2) was then added to final concentration of 100 mM. In some experiments other substrates at 50 mM final concentration were also enclosed inside liposomes: phosphate, malonate, sulfate, citrate or glutamate.

Procedure C was carried out as Procedure B except for higher pH kept throughout inside and outside the vesicles; MOPS buffer of pH 7.2 was replaced by Tris-HCl buffer of pH 8.7 at respective concentrations.

Procedure D was performed as Procedure B or C except that the radioactive substrate was trapped inside proteoliposomes. In this case the final concentration of the substrate inside was 5 mM and not 50 mM, plus 90 mM NaCl to compensate for the same ionic strength of the internal medium.

Procedure E was identical as Procedure D, *i.e.* proteoliposomes reconstituted by this method contained 5 mM malate + 90 mM NaCl inside, but no radioactivity was added.

Assay of the activity

The dicarboxylate carrier activity was assayed at 30°C after reconstitution of the protein material at different steps of the purification procedure. Three methods were used:

Assay A: This was performed at pH 7.2 for 20 min and concerned, depending on the ion enclosed in the liposomes, the malate/malate, malate/phosphate, malate-

/malonate or malate/sulfate exchange reactions. The malate/glutamate exchange system was taken as a control for unspecific malate accumulation since no such reaction is catalysed in mitochondria. 300 μl of proteoliposomes was supplemented with 30 μl of 2 mM radioactive malate (0.2 mM final concentration) + mersalyl or another inhibitor of the dicarboxylate carrier (*e.g.* butylmalonate, phenylsuccinate, bathophenathroline etc.) to verify the inhibitor specificity of the measured transport reaction. After the incubation 250 μl of the sample was loaded onto 100 mg of Dowex (packed in a Pasteur pipette) in order to remove external radioactive malate and the sample was eluted with 450 μl of 170 mM sucrose. Radioactivity accumulated in the proteoliposomes was measured in every sample (of approximately 700 μl total volume) after addition of 5 ml of Packard Emulsifier-Safe-TM solution. In some experiments this assay was performed at pH 8.7 using proteoliposomes reconstituted by Procedure C.

In order to measure the loss of accumulated radioactivity from proteoliposomes two other assays were used:

Assay B: Proteoliposomes loaded with radioactive malate during reconstitution Procedure D were used. Loss of internal radioactivity was measureded in absence or presence of externally added substrates and/or inhibitors of the dicarboxylate carrier.

Assay C: Loading of radioactive malate into proteoliposomes was obtained after reconstitution, *via* the incorporated carrier. The experiment was performed in the following way: Proteoliposomes reconstituted according to Procedure E (about 6 ml final volume) were passed through a 500 mg Dowex column into 100 mM MOPS (pH 7.2) or 100 mM Tris-HCl (pH 8.7) and external radioactive malate was added to final concentration of 2 mM. The sample was incubated for 20 min at 30°C in order to load with radioactive substrate *via* the malate/malate exchange that fraction of proteoliposomes which contained the active carrier. Subsequently, the sample was passed through 500 mg Dowex (into 100 mM buffer, as above) in order to remove the remaining external radioactivity, and divided into 1.5 ml portions. Some of the samples were supplemented with substrates and/or inhibitors of the dicarboxylate carrier and incubated. 250 μl aliquots were collected at various times to measure the leak of accumulated radioactive malate. They were passed through a 100 mg Dowex column and the radioactivity retained by the proteoliposomes was measured.

RESULTS AND DISCUSSION

Pre-purification of the dicarboxylate carrier from liver mitochondria and studies on its functional reconstitution

As mentioned in the Introduction, it has been supposed that liver rather than heart mitochondria are better starting material for isolation of the dicarboxylate carrier. Table 1 presents a comparison of the results of the identical pre-purification procedure employed with mitochondria from these two tissues. It has been observed that more protein is extracted from liver and that a higher total amount of protein is recovered in the HTP eluate. Nevertheless, much higher specific activity of the dicarboxylate carrier has also been found in the HTP eluate obtained from liver mitochondria (Table 1). All this suggests that, besides higher total amount of extracted protein, the liver material is specifically enriched in the dicarboxylate carrier. Such a conclusion strongly supports a further use of liver extracts.

Table 1. Purification by hydroxyapatite chromatography of the dicarboxylate carrier from heart and liver submitochondrial particles. 400 mg protein was taken as the starting material for both heart and liver. The activity is presented as butylmalonate-sensitive accumulation of radioactive malate (malate/phosphate exchange) measured by Assay A in proteoliposomes reconstituted by Procedure A.

Tissue	Protein (mg)	Activity total (nmol/mg)	specific (nmol/min/mg)
Heart			
Triton extract	16.2	7.4	10.46
HTPeluate	0.36	6.35	17.64
Liver			
Triton extract	28.1	16.9	10.60
HTP eluate	0.62	29.4	47.42

Before describing results of further purification of the dicarboxylate carrier, however, another important problem should be discussed, *i.e.* the reconstitution. Optimal reconstitution system as well as a high carrier activity in proteoliposomes are vital for functional studies. One of the usual problems with reconstitution is the leakiness of liposomes and the rapid loss of entrapped substrates. The phenomenon is often protein-dependent since the presence of protein influences lipid packing, membrane fluidity and other parameters important for diffusion of charged molecules across the lipid bilayer.

Thus control experiments performed with lipid vesicles not containing protein should be considered not conclusive.

Moreover, controls which involve enclosure of radioactive substrate into proteoliposomes during the reconstitution procedure may not give a true answer either. If the majority of vesicles do not contain reconstituted protein, which is most likely the case, a possible leak of the radioactivity from proteoliposomes may be obscured by a relative stability of vesicles consisting of lipid only.

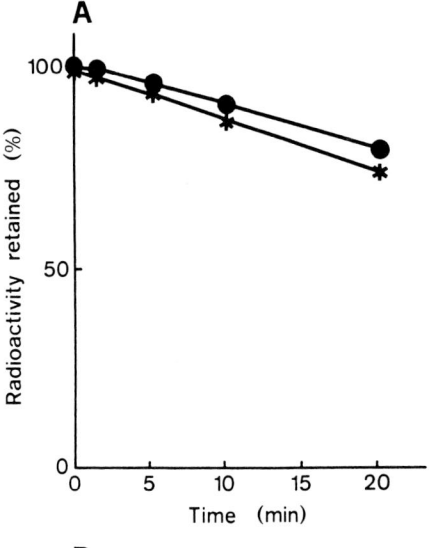

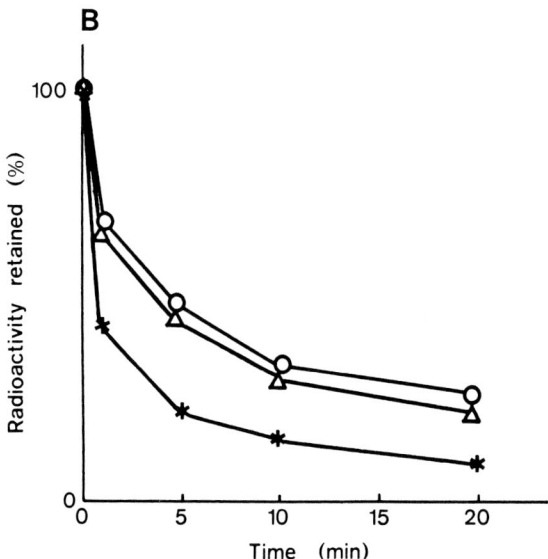

Fig. 1. Leak of radioactive malate from proteoliposomes incubated at 30°C. The pH was 7.2 throughout the experiment. A: Proteoliposomes loaded with radioactivity during the reconstitution (Procedure D). Assay B was used to monitor the leak of malate, either with no additions (closed circles) or in the presence of 10 mM phosphate outside the vesicles (stars). B: Proteoliposomes were reconstituted by Procedure E and loaded with radioactivity *via* the carrier (Assay C). The loss of internal malate was measured with no additions (open circles), in the presence of 10 mM phosphate outside the vesicles (stars) or in the presence of 10 mM phosphate and 1 mM bathophenathroline (triangles).

In fact, when radioactive malate was enclosed into proteoliposomes during reconstitution of HTP eluate (Procedure D, Assay B), only 20% loss of radioactivity was observed during 20 min of incubation at 30°C in the medium containing MOPS buffer and NaCl (Fig. 1A, open circles). External phosphate should be expected to produce a loss of the radioactivity due to its exchange with internal malate on the dicarboxylate carrier. However, only a small increase of leak was found (Fig. 1A, stars). Such observation cannot be explained by the absence of the carrier since its functional reconstitution has already been documented by other reconstitution and assay systems

(Table 1). All this can be understood, however, when one assumes that we were looking mainly at the fractions of the vesicles which did not contain protein.

On the other hand, when a similar experiment was performed under conditions described as Assay C, a rapid leak of radioactive substrate from protein-containing vesicles was observed (Fig. 1B, closed circles). The malate efflux was more pronounced upon addition of external phosphate (Fig. 1B, stars) and the effect of phosphate was found sensitive to bathophenanthroline (Fig. 1B, triangles). This points to the fact that the portion of malate efflux which is phosphate-dependent and mersalyl-sensitive can be interpreted as being due to the activity of the reconstituted carrier. Nevertheless, it amounted to no more than 15% of the total efflux, suggesting a profound unspecific leak of the entrapped substrate under these conditions (Procedure E, Assay C).

Therefore it appears that proteoliposomes reconstituted as described above lose their internally enclosed substrate very rapidly. This, of course, must result in the underestimation of the carrier activity measured by usual methods (*e.g.* Assay A). What comes out is that most of the activity data reported so far for the reconstituted dicarboxylate carrier may not really reflect the true translocating capacity of the transport protein.

The simplest explanation of the leak is that the reconstitution procedure did not remove all of the detergent and that some Triton, possibly bound to protein, was still present in proteoliposomes. However all attempts to improve removal of Triton (*e.g.* lowering its concentration prior to the reconstitution, increasing the amount of Amberlite, changing 2-3 portions of Amberlite during the reconstitution procedure, increasing the time of reconstitution *etc.*) failed to decrease the leak of malate from proteoliposomes.

Other explanations of the leak were also considered: 1. that it is due to the presence of mitochondrial porin; 2. that it is facilitated by the physical state of membrane lipids (*e.g.* high fluidity); 3. that it is induced by surface-charge phenomena. It is not worthy to describe here all control experiments performed to verify these hypotheses since, finally, a simple way to diminish the observed leak was found, namely the reconstitution at high pH. Fig. 2A shows that when Procedure E and Assay C were used at pH 8.7, the loss of radioactive malate amounted to no more that 15% after 20 min incubation (Fig. 2A, closed circles). On the other hand, clearly visible and bathophenanthroline-sensitive phosphate effect (Fig. 2A, open circles and stars) pointed to the presence of functionally reconstituted dicarboxylate carrier.

Specific activity of the HTP-eluate (malate/phosphate exchange) measured under these conditions was much higher than observed with other reconstitution systems and the total accumulation of radioactive malate in proteoliposomes amounted to much higher values as well (see Fig. 2B).

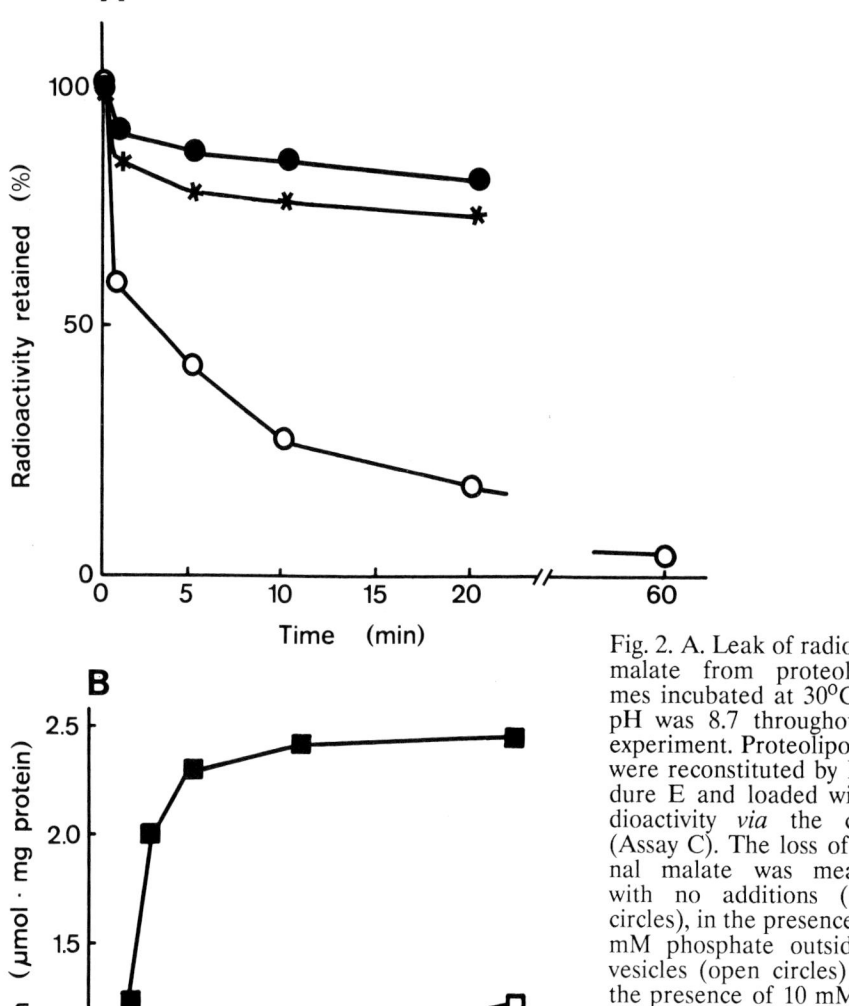

Fig. 2. A. Leak of radioactive malate from proteoliposomes incubated at 30°C. The pH was 8.7 throughout the experiment. Proteoliposomes were reconstituted by Procedure E and loaded with radioactivity *via* the carrier (Assay C). The loss of internal malate was measured with no additions (closed circles), in the presence of 10 mM phosphate outside the vesicles (open circles) or in the presence of 10 mM phosphate and 1 mM bathophenatroline (stars). B. Time-course of radioactive malate accumulation in proteoliposomes during 20 min incubation at 30°C. Stars, proteoliposomes reconstituted by Procedure A, Assay A; open squares, proteoliposomes reconstituted by Procedure B (pH 7.2), Assay A; closed squares, proteoliposomes reconstituted by Procedure C (pH 8.7), Assay B.

It is difficult to decide what is the actual mechanism diminishing the leak of malate at pH 8.7. It could be that the mitochondrial porin does not reconstitute or becomes inactivated under such conditions. Other possibilities are unspecific decrease of the negative value of the surface charge (for detailed description of surface charge effects on membrane transport see Wojtczak & Nałęcz , 1985) and specific deprotonation of some essential amino groups of proteins and/or lipids. Whatever the reason, it became clear that the mitochondrial dicarboxylate carrier can be efficiently reconstituted at high pH and that the unspecific loss of entrapped substrates is almost abolished under these conditions. This way of reconstitution was considered optimal and has been used in further studies.

Further purification steps

Untill now three different groups have reported attempts to purify mitochondrial dicarboxylate carrier from liver (Saint-Macary & Foucher, 1983, 1985; Kaplan & Pedersen, 1985; Bisaccia et al., 1988). Each of them utilized "classical" isolation procedures, i.e. Triton extraction of the membranes and hydroxyapatite chromatography performed under various conditions. In all cases, however, the final preparation consisted of a multipeptide mixture or, in one case, of a single but inactive peptide of the putative dicarboxylate carrier (Bisaccia et al., 1988). Interestingly, purification procedures described for heart mitochondria seemed not to be satisfactory for liver and, furthermore, liver mitochondria appeared generally more difficult to use for isolation of carrier proteins. In our hands the affinity chromatography column developed for purification of the dicarboxylate carrier from heart mitochondria (Szewczyk et al., 1987) failed to bind the carrier protein from liver (Szewczyk & Nałęcz , unpublished). As pointed out by the group of Palmieri (Bisaccia et al., 1988), these apparently strange observations may have at least three explanations: 1. intrinsic difference between the proteins from different sources; 2. different protein/lipid and membrane protein/soluble protein ratios in mitochondria from different tissues; 3. specific lipid composition of liver mitochondria which may influence protein-lipid-detergent interactions thus affecting chromatographic behaviour of proteins. In addition, high activity of proteolytic enzymes in liver homogenates may also have a negative influence on the final result of the isolation procedure. The fact that liver mitochondria appeared particularly difficult to work with on the isolation of carrier proteins suggests that some new methodical approach is needed to overcome this difficulty. Below we mention some of our attempts to find such a methodical breakthrough.

Ion exchange chromatography.
    Anion exchange chromatography on DEAE-Cellulose (Whatman) and DEAE-Sepharose (Bio-Rad) was checked as a potentially powerful method for further

purification of the dicarboxylate carrier. Unfortunately, weak retention and poor resolution of proteins during elution procedure were observed.

The use of sulfo-propyl Sephadex (SP-Sephadex C-50, LOBA Chemie, Wien) seemed more promising due to the presence of a hydrophobic derivative of the sulfonyl group being structurally similar to one of the substrates of the dicarboxylate carrier, namely sufate. Nevertheless, no carrier activity was retained by the column under various conditions applied.

Immobilized metal chromatography.

The fact that bathophenathroline, a copper chelator, is a strong inhibitor of the mitochondrial dicarboxylate carrier led to proposals that a metal ion may be located in the active site of the carrier (for review see LaNoue & Schoolwerth, 1979). On the other hand, it has been observed in the present study that the addition of various metal salts to the suspension of proteoliposomes containing reconstituted carrier leads to inhibition of transport activity. The actual experiment was performed by adding 0.5 mM (final concentration) of chloride salts of Zn^{2+}, Co^{2+}, Cu^{2+}, Mn^{2+} and Fe^{3+} to proteoliposomes containing phosphate (reconstituted by the Procedure B), prior to the addition of radioactive malate (malate/phosphate exchange). It was found that only copper was strongly inhibitory, reaching 100% inhibition already after 5 min of incubation. This pointed to copper as a possible ligand in metal affinity chromatography (for the principles of this method see Porath, 1985).

Copper ions were immobilized on chelating Sepharose 6B (Pharmacia) as described in the column leaflet supplied by the producer. HTP eluate from bovine liver mitochondria was loaded onto the column (1 ml of the eluate per 2 ml of the resin) which was then eluted with increasing concentrations of sulfate, imidazol, histidine, cysteine or EDTA (Porath, 1985). It has been observed that proteins present in the HTP-eluate can be resolved electrophoretically into separate groups of single, double or triple bands, thus suggesting that the procedure is promising in purification of many mitochondrial carriers (not shown). The dicarboxylate carrier, however, could not be identified in any of the fractions. The failure of this method was due to a complete disappearance of the carrier activity, most likely produced by a strong and irreversible inhibitory effect of copper during the purification procedure.

Dye affinity chromatography.

Finally, we employed the technique of immobilized dye affinity chromatography (for review see Lowe & Pearson, 1984). In particular, Cibacron Blue F3G-A seemed suitable for isolation of the mitochondrial dicarboxylate carrier. As known from the literature (Lowe & Pearson, 1984), Cibacron Blue F3G-A appears especially effective for purification of pyridine nucleotide-dependent enzymes, phosphokinases, coenzyme

A-dependent enzymes, hydrolases, acetyl-, phosphoribosyl- and aminotransferases. Molecular models show a rough resemblance between Blue F3G-A and NAD, but the most important similarities concern the planar ring structure and the location of negatively charged groups (Fig. 3).

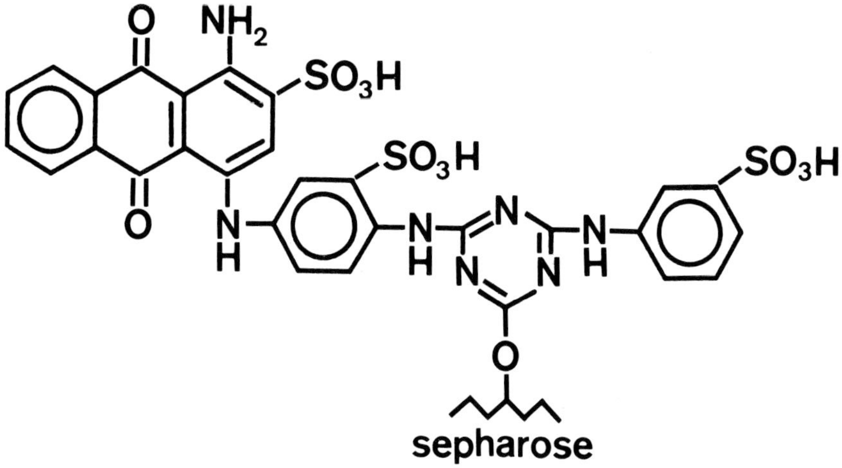

Fig. 3. Chemical structure of Cibacron Blue F3G-A covalently attached to Sepharose.

In respect to the planar structure, Cibacron Blue can also be considered similar to bathophenathroline and therefore was chosen for the purification of the dicarboxylate carrier. Additionally, Blue F3G-A also contains three sulfonic groups, possibly able to interact with the substrate binding site of the carrier. Nevertheless, the concept that Cibacron Blue and related dyes can interact only as specific ligands for the substrate and/or effector binding domain in protein should be viewed with caution. Bovine serum albumin is the most notable example of the contrary. Some other proteins such as interferon or serum lipoproteins were also shown to interact with several triazine dyes (Lowe & Pearson, 1984).

Purification of the dicarboxylate carrier from liver mitochondria was performed with Cibacron Blue F3G-A covalently attached to Sepharose CL-6B (Reactive Blue 2-Sepharose CL-6B, Sigma). Before use the column was washed with 5 M urea in 0.5 M NaOH in order to remove any loosely bound dye. Subsequently, it was equilibrated with the medium composed of 2% Triton X-100, 0.7% deoxycholate, 20 mM MOPS and 50 mM NaCl. As discussed by Robinson *et al.* (1980) the use of nonionic detergents in affinity chromatography sometimes leads to unsatisfactory results due to the masking of the ligand by noncharged detergent micelles, thus preventing, at least partially, the formation of a specific ligand protein complex.

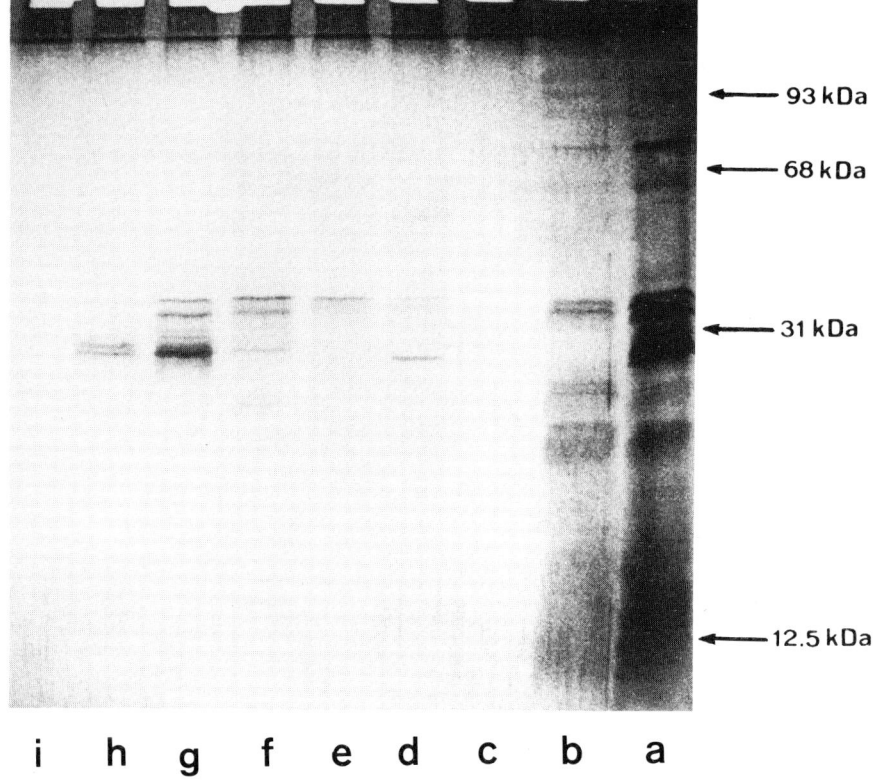

Fig. 4. SDS-polyacrylamide gel electrophoretical pattern of the bovine liver mitochondrial material collected during the purification experiment on Cibacron Blue F3G-A column: a, HTP eluate; b, non-retarded material of the pass-through from the column; c, material recovered at the end of washing with the equilibration medium; d-i, material eluted with increasing concentrations of NaCl: d, 100 mM; e, 200 mM; f, 400 mM; g, 800 mM; h, 1000 mM ; i, 2000 mM.

It has been suggested that the addition of an anionic detergent, *e.g.* SDS or deoxycholate, might resolve this difficulty (Robinson *et al.*, 1980). The presence of salts in the affinity chromatography eluent has also been considered useful in abolishing unspecific interaction between proteins and the charged resin. For purification of the dicarboxylate carrier from heart mitochondria (Szewczyk *et al.*, 1987) we used the medium supplemented with deoxycholate and NaCl as described above. Deoxycholate was also added to the HTP eluate, up to the final concentration of 0.7%, and the eluate was loaded onto the Cibacron Blue column (1 ml of the HTP eluate per 500 μl of the affinity resin).

Table 2. Functional properties of the dicarboxylate carrier from beef liver purified by dye-affinity chromatography on Cibacron Blue F3GA-Sepharose CL-6B column. The carrier was reconstituted by Procedure C and measured by Assay B.

Material	Assay system (50 mM in / 0.2 mM out)	Specific activity (nmol/min/mg)
HTP eluate (+0.7% deoxy-cholate)	phosphate/malate	124
Cibacron-Blue eluate in 100 mM NaCl	phosphate/malate	941
	sulfate/malate	816
	malonate/malate	1001
	2-oxoglutarate/malate	14
	citrate/malate	9
	glutamate/malate	21

Non-retarted proteins were collected in the pass-through from the column and adsorbed proteins were subsequently eluted. In the first set of experiments only one step elution with 3 M NaCl was applied. It was observed that Cibacron Blue Sepharose displays a high yield of protein binding (the binding capacity of the column was estimated to be about 1 mg of HTP eluate protein per 1 ml of the gel) and that mainly polypeptides of apparent molecular weight of about 30 kDa were bound to the resin (see Fig. 4, lanes d-h). Non-retarded pass-through fractions, instead, contained all other proteins present in the HTP-eluate (Fig. 4, compare lanes a and b). It is interesting to note that the gel region of M_r of about 30 kDa is exactly the one in which most of the already purified or identified mitochondrial carriers are located (Nałęcz, 1986). This suggests that the Cibacron Blue Sepharose column may be suitable for isolation of not only the dicarboxylate carrier but also of many other mitochondrial transport systems.

Subsequently, a step elution with increasing concentrations of NaCl was applied to the Cibacron Blue column (Fig. 4). It was observed that a reasonable resolution of bound proteins could be obtained in this way. Furthermore, when all fractions collected from the column were reconstituted (Procedure C), the dicarboxylate carrier activity (Assay A) was found almost exclusively in the fraction eluted with 100 mM NaCl. When applied to SDS-polyacrylamide gel electrophoresis, this fraction was found to contain mainly one polypeptide of M_r of 28 kDa (Fig. 4, lane d). Exactly the same molecular weight has been reported for the putative (inactive) dicarboxylate carrier from rat liver mitochondria (Bisaccia et al., 1988).

This strongly points to the possibility that the polypeptide recovered in the 100 mM NaCl fraction from Cibacron Blue column may represent the mitochondrial

dicarboxylate carrier. In order to verify this hypothesis, further functional characterization of the eluted polypeptide was performed.

It was found that the activity of the reconstituted fraction containing 28 kDa
polypeptide showed substrate specificity typical for the mitochondrial dicarboxylate
carrier (Table 2). It was also observed that the specific activity of the phosphate/malate
exchange reaction measured for the Cibacron Blue eluted material was increased
approximately 10 times compared to the HTP eluate (Table 2). All this suggests that
the 28 kDa band showed in Fig. 4, lane d, indeed represents isolated and functionally
active dicarboxylate carrier from bovine liver mitochondria.

CONCLUSIONS

The scope of the present paper was to isolate the dicarboxylate carrier from
bovine liver mitochondria and to find optimal conditions for its functional
reconstitution. The reconstitution method based on detergent removal with Amberlite
XAD-2 at high pH (8.7) diminishes the leakiness of proteoliposomes and prevents an
unspecific loss of entrapped substrates. This, in turn, leads to a high activity of the
carrier after its reconstitution. This method was considered optimal and has been used
throughout this study.

The final isolation of the carrier was achieved by extraction of the membranes
with Triton X-100 followed by hydroxyapatite chromatography and subsequent affinity
chromatography on an immobilized dye. For the latter procedure a resin composed of
Cibacron-Blue F3G-A covalently attached to Sepharose CL-6B was used. The carrier
was eluted from such column with 100 mM NaCl. The fraction was found
electrophoretically to contain one polypeptide band of apparent molecular weight of 28
kDa, postulated to represent the active mitochondrial dicarboxylate carrier.

ACKNOWLEDGEMENTS

This work was supported by Swiss National Science Foundation Grant N. 3.525-
086 and the Polish Central Program for Research and Development (CPBR 3.13.4.4.2).

REFERENCES

Bisaccia F, Indiveri C, Palmieri F (1988) Purification and reconstitution of two anion
    carriers from rat liver mitochondria: the dicarboxylate and the 2-oxoglutarate
    carrier. Biochim Biophys Acta 933:229-240
De Pinto V, Tommasino M, Benz R, Palmieri F (1985) The 35 kDa DCCD-binding
    protein from pig heart mitochondria is the mitochondrial porin. Biochim Biophys
    Acta 813:230-242

Kaplan R, Pedersen PL (1985) Isolation and reconstitution of the n-butylmalonate-sensitive dicarboxylate transporter from rat liver mitochondria. J Biol Chem 260:10293-10298

LaNoue KF, Schoolwerth AC (1979) Metabolite transport in mitochondria. Ann Rev Biochem 48:871-922

Lowe CR, Pearson JC (1984) Affinity chromatography on immobilized dyes. Methods Enzymol 104:97-113

Nałęcz MJ (1986) Metabolite transport systems of inner mitochondrial membrane: Characteristics and isolation attempts. In: Kuczera J, Przestalski S (eds) Biophysics of Membrane Transport, vol 2. Publ. Dept. of the Agricultural University of Wroclaw, pp 13-45

Nałęcz MJ, Nałęcz KA, Azzi A (1986a) Extraction, partial purification and reconstitution of a mixture of carriers from the inner mitochondrial membrane. In: Azzi A, Masotti L, Vecli A (eds) Membrane Proteins, Springer-Verlag, Berlin Heidelberg, pp 67-75

Nałęcz MJ, Nałęcz KA, Broger C, Bolli R, Wojtczak L, Azzi A (1986b) Extraction, partial purification and functional reconstitution of two mitochondrial carriers transporting ketoacids: 2-oxoglutarate and pyruvate. FEBS Lett 196:331-336

Porath J (1985) Immobilized metal ion affinity chromatography - A powerful method for protein purification. In: Tschesche H (ed) Modern Methods in Protein Chemistry, vol 2. Walter de Gruyter & Co, Berlin New York, pp 85-95

Robinson JP Jr, Strottmann JM, Wick DG, Stellwagen E (1980) Affinity chromatography in nonionic detergent solutions. Proc Natl Acad Sci (USA) 10:5847-5851

Saint-Macary M, Foucher B (1983) Reconstitution of the dicarboxylate exchange activity by incorporation into liposomes of a Triton-extract of mitochondrial rat liver membranes. Biochem Biophys Res Commun 113:205-211

Saint-Macary M, Foucher B (1985) Comparative partial purification of the active dicarboxylate transport system of rat liver, kidney and heart mitochondria. Biochem Biophys Res Commun 133:498-504

Sluse FE, Meijer AJ, Tager JM (1971) Anion translocators in rat-heart mitochondria. FEBS Lett 18:149-151

Szewczyk A, Nałęcz MJ, Broger C, Wojtczak L, Azzi A (1987) Purification by affinity chromatography of the dicarboxylate carrier from bovine heart mitochondria. Biochim Biophys Acta 894:252-260

Wojtczak L, Nałęcz MJ (1985) The surface potential of membranes: Its effect on membrane-bound enzymes and transport processes. In: Benga G (ed) Structure and Properties of Cell Membranes, vol 2. CRC Press, Boca Raton, pp 215-242

New Photoaffinity Derivatives of Malonate and Succinate to Study Mitochondrial Carrier Systems

ADAM SZEWCZYK AND MACIEJ J. NAŁĘCZ

Department of Cellular Biochemistry, Nencki Institute of Experimental Biology, Pasteur str. 3, 02-093 Warsaw, Poland
and
Institut für Biochemie und Molekularbiologie, Universität Bern, Bühlstrasse 28, CH-3012 Bern, Switzerland

It is now well established that the inner mitochondrial membrane contains at least twelve functionally different carriers (LaNoue & Schoolwerth, 1979). Some of the mitochondrial carriers have already been isolated in a pure form, some others have been extracted and partially purified and their activity has been measured after reconstitution into proteoliposomes (for review see Nałęcz, 1986).

Photoaffinity labeling has been used successfully to identify substrate or inhibitor binding sites in enzymes and membrane receptors (Chowdhry & Westheimer, 1979; Bayley, 1983; Schäfer, 1987). Aryl azides, as nitrene source, are now the most commonly used reagents. The ease of synthesis of aryl azides and their useful physico-chemical properties are responsible for a great popularity of this group of compounds.

Till now two kinds of azido labels have been used for characterization of mitochondrial substrate carriers: aryl azido analogues of ADP and ATP (Lauquin et al., 1978) and of phosphate (Tommasino et al., 1987). The scope of the present paper is to describe the synthesis of a new family of azido derivatives of other substrates of mitochondrial carrier proteins. Namely, we present here methods to produce aryl azido derivatives of malonate and succinate, i.e. substrates of the three major mitochondrial transporting systems: the oxoglutarate, the dicarboxylate and the tricarboxylate translocators. Synthesized probes are different in their hydrophobicity and distance between

A. Azzi et al. (Eds.)
Anion Carriers of Mitochondrial Membranes
© Springer-Verlag Berlin Heidelberg 1989

dicarboxylic and azido groups and thus seem promising in the studies on identification of carrier proteins and/or on characterization of their substrate binding sites.

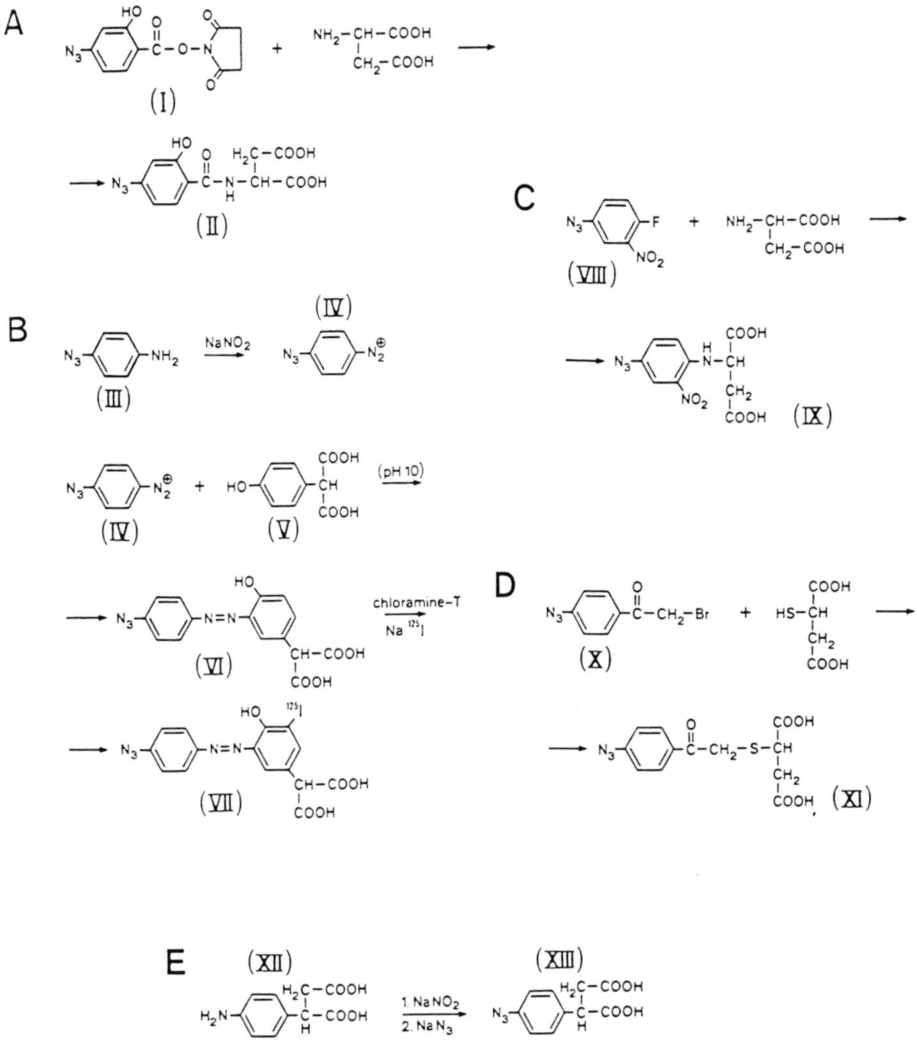

Fig. 1. Synthesis of: (A) N-(4-azidosalicylic)-aminosuccinic acid (II), (B) 3-(p-azi-dophenylazo)-4-hydroxyphenylmalonic acid (VI), (C) (p-azido-2-nitroanilino)-N-suc-cinic acid (IX), (D) p-azidophenacylthiosuccinic acid (XI), (E) p-azidophenylsuccinic acid (XIII).

Results of some preliminary experiments involving synthesized compounds are also presented.

MATERIAL AND METHODS

Chemicals

N-hydroxysuccinimidyl-4-azidosalycylic acid was from Pierce; 4-azidoaniline, 4-fluoro-3-nitrophenylazide and p-azidophenacyl bromide were from Fluka AG; mercaptosuccinic acid was from Merck and p-hydroxyphenylmalonic acid from Chemical Dynamics Corporation. All other chemicals were of analytical grade.

Biological materials

Extraction of the carrier proteins from the inner mitochondrial membrane, hydroxyapatite chromatography of this extract, reconstitution of the carriers into proteoliposomes, assay of the dicarboxylate carrier activity and SDS-polyacrylamide gel electrophoresis were performed as described elsewhere (see *e.g.* Nałęcz *et al.*, 1989).

Synthesis of photoaffinity derivatives of carrier substrates

All reactions were carried out under subdued light conditions. Merck silica gel F_{254} was used for confirmation of purity of final products. Two eluents were used: (A) chloroform/methanol/acetic acid (20:35:30, v/v) and (B) (96:4:2, v/v). Light sensitivity of the products was confirmed by ultraviolet spectroscopy. Spectra were recorded on Hewlett Packard 8451A diode array spectrophotometer.

1. N-(4-azidosalicylic)-aminosuccinic acid (Fig. 1A, II)

200 μl of aspartic acid solution (100 mg/ml, pH 8.0) was mixed with 300 μl of 0.1 M carbonate buffer (pH 8.2) and 500 μl of acetone. 18 mg of N-hydroxysuccinimidyl-4-azidosalicylic acid (I) dissolved in 100 μl acetone was added. After 1 h at room temperature the reaction mixture was supplemented with 1 ml of 1 M NaCl and the product of hydrolysis of (I), as well as other possible impurities, were taken up with ethyl ether. Subsequently, the water phase was acidified with 200 μl of 1 M HCl and ethyl ether was used to extract the reaction product (II). The extract was dried by rotary evaporation and the purity of the product (white powder) was confirmed by thin layer chromatography with (A) as eluent. The R_f value was found to be 0.72.

2. 3-(p-azidophenylazo)-4-hydroxyphenylmalonic acid (Fig. 1B, VI)

A solution of diazotized 4-azidoaniline (IV) was prepared by dropwise addition of 420 μl of 1 M sodium nitrite dissolved in methanol/water (1:1) to an ice-cold solution of 4-azido-aniline hydrochloride (III) prepared by dissolving 50 mg (0.29 mmol) of the compound in 6 ml of 1 M HCl/methanol (1:1). After 15 min at 0°C all of the above solution was added, with stirring, to the ice-cold solution of 49 mg (0.25 mmol) p-hydroxyphenylmalonic acid (V) dissolved in 3.7 ml of 1 M sodium carbonate (pH 11.7). The reaction mixture was incubated for 3 h at 0°C. The product (VI) was extracted with 200 ml of ethyl acetate and dryed by rotary evaporation. 28.4 mg of the dry yellow powder was obtained. The purity of the product was confirmed by thin layer chromatography with (B) as eluent. The R_f value of 0.85 was found. Iodinated product (VII) was prepared according to Ji *et al.* (1985).

3. (p-azido-2-nitroanilino)-N-succinic acid (Fig. 1C, IX)

1.35 ml of aspartic acid solution (100 mg/ml, pH 8.0) was mixed with 8.15 ml of 0.1 M sodium carbonate (pH 9.3) and 12.5 ml of ethanol. 50 mg of 4-fluoro-3-nitrophenyl azide (VIII) (in 1.5 ml ethanol) was added and the incubation was carried out with stirring for 70 h at 45°C (water bath). Subsequently, ethanol was removed with rotary evaporation and impurities were extracted three times with ethyl acetate. The mixture was then acidified with a few drops of 5 M HCl and the next portion of ethyl acetate was added in order to extract the reaction product (IX). After removing ethyl acetate with rotary evaporation 16 mg of the orange powder was obtained. Its purity was confirmed by thin layer chromatography with (B) as eluent. The R_f value of 0.76 was found.

4. p-azidophenacylthiosuccinic acid (Fig. 1D, XI)

15 mg (0.1 mmol) of mercaptosuccinic acid was dissolved in a mixture of 2 ml of 0.1 M sodium hydrogen carbonate (pH 8.2) and 2 ml of methanol. Subsequently, 10 mg (0.042 mmol) of p-azidophenacyl bromide (X) in 1 ml of methanol was added followed by incubation at 0°C for 1 h. Further steps of the procedure, *i.e.* removing of impurities and extraction of the product (XI) were performed as described above (method 3). The purity of the product (white powder) was confirmed by thin layer chromatography with (B) as eluent. The R_f value of 0.70 was found.

5. p-azidophenylsuccinic acid (Fig. 1E, XIII)

The starting compound for synthesis was p-amino-phenylsuccinic acid prepared from phenylsuccinic acid as described previously (Szewczyk *et al.*, 1987). 4 mg of p-aminophenylsuccinic acid (XII) was dissolved in 450 μl of dioxan/water (2:1, v/v) and mixed with 30 μl of 0.3 M H_2SO_4. After 15 min at 0°C the mixture was supplemented with 30 μl of ice-cold 0.1 M sodium nitrite and, after additional 5 min at 0°C, with 30 μl of 0.1 M sodium azide. The reaction was carried out for 10 min. The mixture was then additionally acidified with a few drops of 5 M HCl and extracted with ethyl ether. The solvent was removed by rotary evaporation. The purity of the final product (XIII, white powder) was confirmed by thin layer chromatography with (B) as eluent. The R_f value of 0.77 was found.

RESULTS AND DISCUSSION

Aryl azides are known to be useful photoreactive moieties in photoaffinity probes (Bayley & Knowles, 1977). When photolyzed, aryl azides are capable to form aryl nitrenes, extremely reactive species able to form stable products with practically all kinds of chemical bonds (Knowles, 1972). The five azido labels described here were synthesized according to well characterized reactions: 1. Product (II) was synthesized utilizing a spontaneous reaction occurring under mild conditions between N-hydroxysuccinimide ester of 4-azidosalycilic acid and primary amino groups of aspartic acid. 2. Product (VI) was obtained by coupling p-azidoaniline with p-hydroxyphenylmalonic acid *via* the diazo bond. 3. The azido label (IX) was the product of the reaction between 4-fluoro-3-nitrophenylazide and the primary amino group of aspartic acid. 4. Product (XI) was synthesized *via* the selective alkylation of sulfhydryl group of thiosuccinic acid by p-azidophenacyl bromide. In this case the high pH of the medium was used in order to prevent the reaction between p-azidophenacylbromide and carboxyl groups of thiosuccinate. 5. Compound (XII) was diazotized and converted to aryl azide (XIII).

All of these products are easy to synthesize and the reactions employed are well defined chemically. They also yield satisfactory amounts of the final compounds. The yield, however, may be strongly influenced by initial conditions of the synthesis under which all starting compounds should be in solution. We therefore strongly advise to watch this initial step carefully and, in case of any visible sediment, to increase the volume and/or to change the ratio of solvents accordingly.

The azido labels described in the present paper differ in the distance between dicarboxylic and azido groups: in (XIII) this distance is the shortest, in (VI) the longest. They are also different in their hydrophobicity and acidic moiety: malonate or succinate. Absorption spectra measured for all these products contain azido group absorption bands typical for aryl azido compounds. The absorption maxima of different products were found to be: 350 nm for (VI), 265 nm for (IX), 292 nm for (XI), 255 nm for (XIII). As predicted for the azido compounds, all of these bands disappeared upon ultraviolet illumination (Fig. 2, 3, 4, 5).

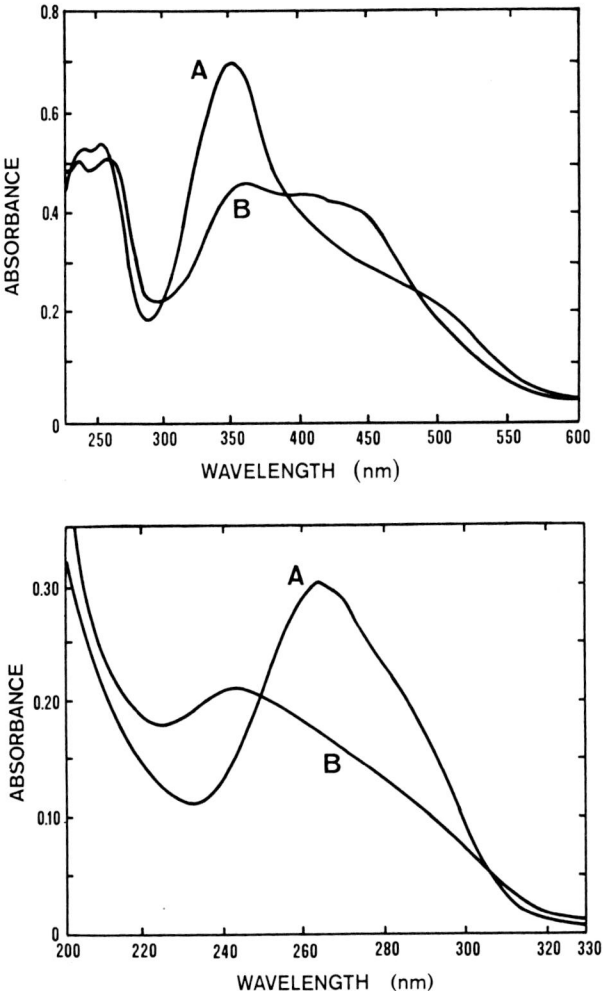

Fig. 2. Ultraviolet absorption spectra of (VI) before (A) and after (B) irradiation.
Fig. 3. Ultraviolet absorption spectra of (IX) before (A) and after (B) irradiation.

For a specific protein ligand (*e.g.* inhibitor or stimulator), the photoaffinity probe should reversibly affect protein function in the unphotolyzed state (competition for the binding site) but irreversibly inhibit the protein upon photolysis (covalent binding). Furthermore, natural substrates or effectors of the protein should protect against photoinactivation by the probe. For the group of newly synthesized compounds described above not all of these control experiments were performed yet.

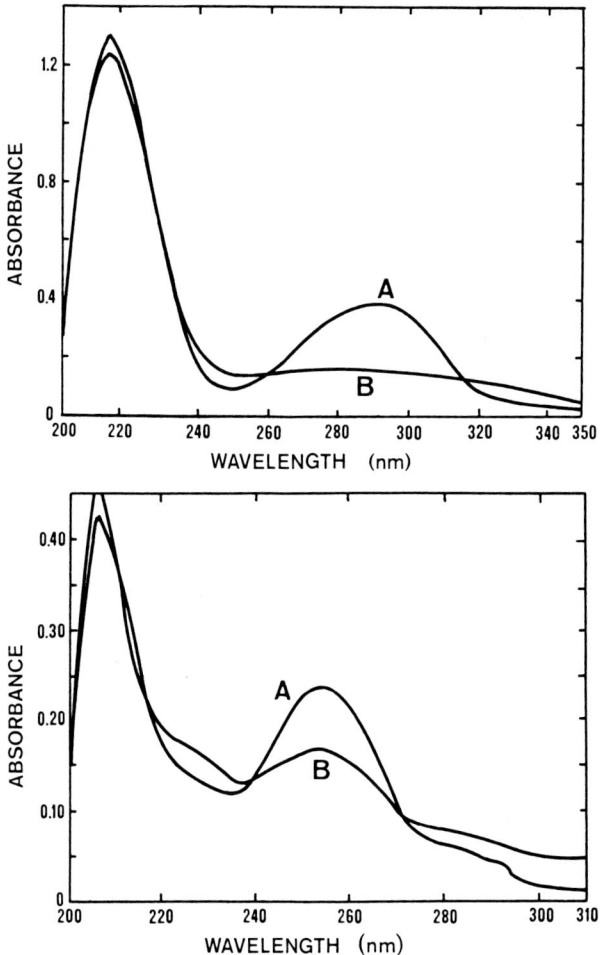

Fig. 4. Ultraviolet absorption spectra of (XI) before (A) and after (B) irradiation.
Fig. 5. Ultraviolet absorption spectra of (XIII) before (A) and after (B) irradiation.

However, we present here some preliminary results pointing to a possible usefulness of the products in studies on mitochondrial dicarboxylate translocators. For example, it was found that in the darkness products (VI) and (IX) inhibited the malonate-phosphate exchange catalyzed by the reconstituted dicarboxylate carrier, giving 50% inhibition at the concentration of approximately 0 μM (Fig. 6) and 250 μM (not shown), respectively. In order to analyze the inhibition of the reconstituted dicarboxylate carrier activity by product (IX) in dark, the influence of this inhibitor on the rate of malonate-phosphate exchange upon increasing concentration of externally added radioactive malonate was determined.

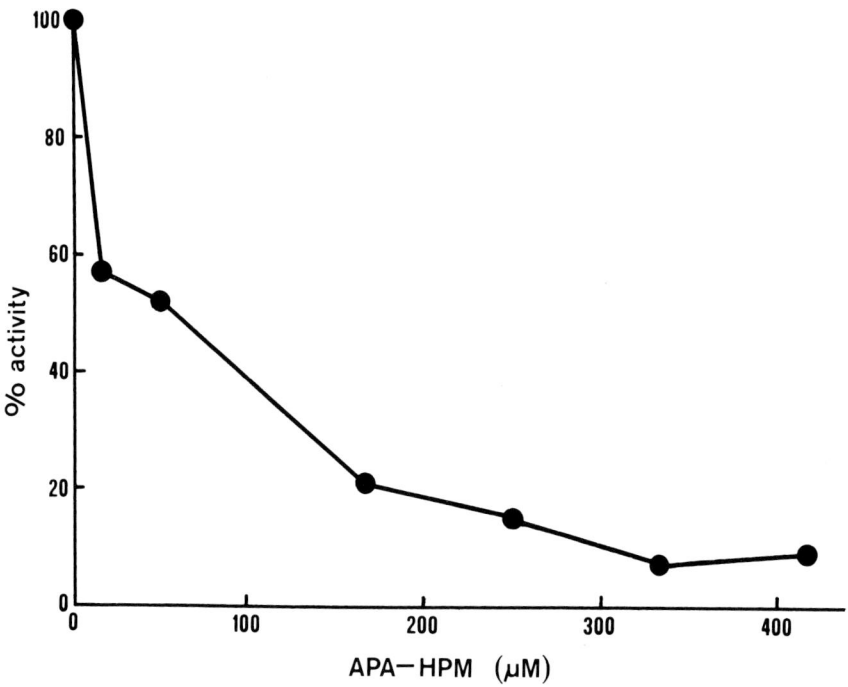

Fig. 6. Inhibition (in darkness) of reconstituted dicarboxylate carrier activity (malonate-phosphate exchange) by increasing concentrations of APA-HPM (VI). As 100% activity 62 nmol/min per mg protein was taken.

The reciprocal plot presented in Fig. 7 shows that product (IX) inhibited the transport of malonate in a competitive manner. Interestingly, when the same experiment was repeated under UV light (each sample was illuminated during 2 min incubation with the probe plus externally added substrate), a stronger inhibition was observed and the Lineweavera-Burk plot of the activity exhibited a mixed inhibition by (IX) (not shown). This suggests that, upon illumination, parts of the carrier became covalently modified and thus irreversibly inhibited by the probe.

In order to identify mitochondrial proteins labeled by product (VI) the latter was made radioactive (VII) by iodination (according to Ji et al., 1985). Rat liver mitochondria were solubilized with 4% Triton X-100 and the chromatography on dry hydroxyapatite column was performed in order to enrich the material in mitochondrial carriers (for details of this procedure see other articles of the present book).

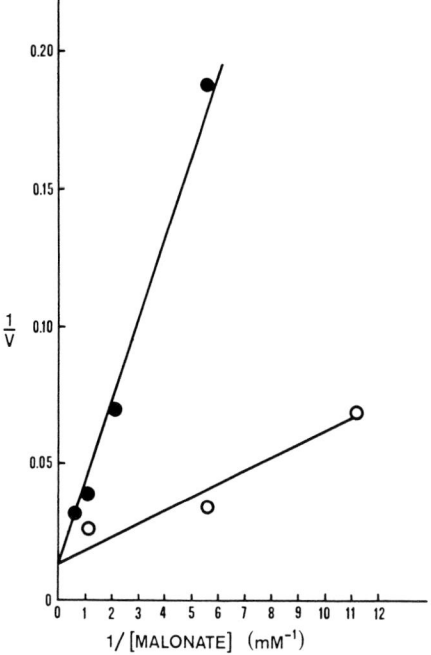

Fig. 7. Inhibition (in darkness) of dicarboxylate carrier activity (malonate-phosphate exchange) by (IX). The activity was measured in the absence ( ) and presence ( ) of 250 μM (XI). V is expressed in nmol/min per mg protein.

The hydroxyapatite eluate was mixed with radioactive product (VII) (0.2 mg of protein plus 10 μl of 1 μM solution of the probe), illuminated with ultraviolet lamp (Philips Tuv 4D 30 W lamp at a distance of 20 cm from the sample placed on ice and stirred continuously; the illumination time was 10 min) and then subjected to SDS-polyacrylamide gel electrophoresis. The radioactively labeled proteins were then identified by autoradiography (Fig. 8).

Lane A of Fig. 8 shows the electrophoretic pattern of the illuminated sample stained with silver nitrate. The strong band close to 90 kDa, not present in control samples of the usual hydroxyapatite eluate from rat liver mitochondria (see *e.g.* Nałęcz *et al.*, 1989), most likely represents the crosslinking product of some proteins from the eluate. Crosslinking phenomena are known often to accompany the photoreactions involving nitrene generation, most likely due to the ability of nitrenes to induce formation of carbon atom radicals in proteins (Knowles, 1972). However, the crosslinking induced by the UV light itself cannot be excluded either.

Lane B of Fig. 8 shows the result of autoradiography of the sample presented in lane A. It appears clear that the radioactive probe binds to more than one protein present in the hydroxyapatite eluate and that, under applied conditions, the majority of radioactivity is recovered at the level corresponding to the apparent molecular weight of 27-33 kDa. As known from the literature (Bisaccia *et al.*, 1988), the activity of the mitochondrial carriers able to react with the probe has been ascribed to polypeptides of the apparent molecular weight corresponding to the labeled region: the 2-oxoglutarate carrier (32.5 kDa) and the dicarboxylate carrier (28 kDa).

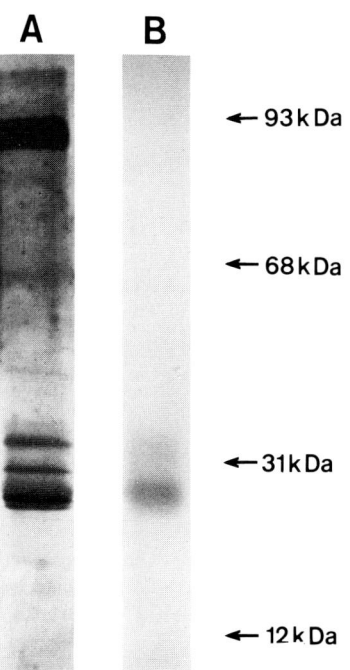

Fig. 8. Labeling of the hydroxyapatite chromatography eluate proteins with (VII). Lane A, electrophoretic pattern staining with silver nitrate. Lane B, autoradiography of lane A. Gel electrophoresis was performed as described by Szewczyk *et al.* (1987). The polypeptides were stained with silver nitrate according to the procedure described by Bio-Rad.

Further studies are in progress to clarify the usefulness of the sythesized aryl azido labels in identification and/or characterization of dicarboxylate, oxoglutarate and tricarboxylate carriers of the inner mitochondrial membrane.

ACKNOWLEDGMENTS

The first part of the present work, *i.e.* synthesis of aryl azido derivatives of malonate and succinate, was performed in the Laboratory of Professor Angelo Azzi in Bern, Switzerland. Warm thanks are addressed to Professor Azzi for his competent supervision and continuous support during realization of this project. A.S. was a recipient of the FEBS Fellowship and M.J.N. of the Hoffman-La Roche Fellowship. Both organizations are gratefully acknowledged for making this work possible. The second part of this study was performed in the Laboratory of Professor Lech Wojtczak in Warsaw to whom we are indebted for stimulating advice and discussion. This work was partially supported by the Polish National Research Program under grant No. CPBR 3.13. and partially by the Swiss National Science Foundation grant No. 3.525.086.

REFERENCES

Bayley H (1983) Photogenerated Reagents in Biochemistry and Molecular Biology. Elsevier, Amsterdam New York Oxford

Bayley H, Knowles JR (1977) Photoaffinity labeling. In: Jakoby WB, Wilchek M (eds) Methods in Enzymology, vol 46. Academic Press, New York, pp 69-114

Bisaccia F, Indiveri C, Palmieri F (1988) Purification and reconstitution of two anion carriers from rat liver mitochondria: the dicarboxylate and the 2-oxoglutarate carrier. Biochim Biophys Acta 933:229-240

Chowdhry V, Westheimer FH (1979) Photoaffinity labeling of biological systems. Ann Rev Biochem 48:293-325

Ji I, Shin J, Ji TH (1985) Radioiodination of a heterobifunctional reagent. Anal Biochem 151:348-349

Knowles JR (1972) Photogenerated reagents for biological receptor-site labeling. Acc Chem Res 5:155-160

LaNoue KF, Schoolwerth AC (1979) Metabolite transport in mitochondria. Ann Rev Biochem 48:871-922

Lauquin GJM, Brandolin G, Lunardi J, Vignais PV (1978) Photoaffinity labeling of the adenine nucleotide carrier in heart and yeast mitochondria by an arylazido ADP analog. Biochim Biophys Acta 501:10-19

Nałęcz MJ (1986) Metabolite transport systems of the inner mitochondrial membrane: Characterization and isolation attempt. In: Kuczera J, Przestalski S (eds) Biophysics of Membrane Transport, vol 2. Publ. Dept. of the Agricultural University of Wroclaw, Poland, pp 13-45

Nałęcz MJ, Szewczyk A, Broger C, Wojtczak L, Azzi A (1989) Isolation and functional reconstitution of the dicarboxylate carrier from bovine liver mitochondria. In: Azzi A, Nałęcz KA, Nałęcz MJ, Wojtczak L (eds) The Anion Carriers of the Mitochondrial Membranes. Springer-Verlag Heidelberg, pp 71-85

Schäfer HJ (1987) Photoaffinity labeling and photoaffinity crosslinking of enzymes. In: Eyzaguirre J (ed) Chemical Modification of Enzymes: Active Site Studies, John Wiley, New York Chichester Brisbane Toronto, p 45-62

Szewczyk A, Nałęcz MJ, Broger C, Azzi A, Wojtczak L (1987) Purification by affinity chromatography of dicarboxylate carrier from bovine heart mitochondria. Biochim Biophys Acta 894:252-260

Tommasino M, Prezioso G, Palmieri F (1987) Photoaffinity labeling of mitochondrial phosphate carrier by 4-azido-2-nitrophenyl phosphate. Biochim Biophys Acta 890:39-46

Reaction Mechanism of the Reconstituted Aspartate/Glutamate Antiporter from Mitochondria. Reversible Switching to Uniport Function

T. DIERKS AND R. KRÄMER

Institut für Biotechnologie 1 der KFA Jülich, Postfach 1913, 5170 Jülich, F.R.G.

Among the ensemble of carrier proteins found in the inner mitochondrial membrane the Asp/Glu carrier is one of the most interesting systems, because it is the only one to be regulated both by membrane potential and pH gradient as a consequence of proton co-transport together with glutamate but not with aspartate. Asp/Glu antiport is a central component of the malate/aspartate shuttle transferring reducing equivalents from the cytosol into the matrix. Thus $NADH/NAD^+$ ratios in the different compartments are modulated by this carrier. Furthermore, the Asp/Glu carrier connects intra- and extramitochondrial steps in gluconeogenesis from lactate and urea synthesis.

Mainly two groups have studied this antiporter in mitochondria trying to elucidate its complex mechanism in kinetic terms (LaNoue et al., 1979, Murphy et al., 1979). However, the derived models show a fundamental contradiction concerning the translocation of the countersubstrates (ping-pong or sequential mechanism) as well as the mode of proton co-transport (distinct proton binding site or transport of protonated glutamate). We addressed these unresolved issues by examining Asp/Glu antiport in a reconstituted system consisting of liposomes that carry partially purified Asp/Glu carrier (Krämer et al., 1986) of definite transmembrane orientation (inside-out as compared to mitochondria (Dierks & Krämer, 1988)).

Functional reconstitution of the carrier protein was achieved by hydrophobic chromatography of mixed micelles for removal of detergent (Dierks & Krämer, 1988,

A. Azzi et al. (Eds.)
Anion Carriers of Mitochondrial Membranes
© Springer-Verlag Berlin Heidelberg 1989

Krämer & Heberger, 1986). Antiport activity was determined by an inhibitor-stop method using high concentrations of pyridoxalphosphate as the stop inhibitor. In both the forward and the backward exchange the interior and the exterior of the liposomes are accessible to experimental variation, thus making possible studies of the mutual dependence of carrier activity on substrate concentrations in the two compartments (Fig.1). Since the backward exchange technique is more elaborate, this mode of flux measurement only was applied if internal substrate had to be varied under external substrate saturation (Fig. 2B). It could be demonstrated that forward and backward exchange rates show good agreement, if calculated according to Dierks and Krämer (1988).

RESULTS AND DISCUSSION

Bireactant mechanism of antiport

Counter-exchange of substrates across membranes catalyzed by antiport carriers can be described in terms of bisubstrate reactions. In order to gain insight into the sequence of binding and release of substrates during the catalytic cycle of exchange, bireactant initial velocity analyses were carried out varying substrate concentrations both in the internal and the external compartment of the liposomes. By plotting the transport velocities in double reciprocal plots in dependence of internal (Fig. 1A) or external substrate concentration (not shown), primary plots are obtained. The slopes and intercepts of these plots are then replotted in secondary plots, Fig. 1A thus leading to Fig. 1B. The following conclusions can be drawn from these results:

(a) The straight lines obtained in Lineweaver-Burk plots (Fig. 1A) form patterns that are not parallel, as would have been expected in the case of a ping-pong mechanism (one substrate binding site alternating between opposing membrane surfaces). This becomes even more obvious in the secondary plots (Fig. 1B), where the replot of the slope by no means is horizontal, which would be indicative for a ping-pong mechanism.

(b) Instead, the kinetic patterns show a common point of intersection. This result can only be explained by a sequential type of mechanism (Cleland, 1970). Characteristic for this mechanism is the formation of a ternary complex of the carrier with one aspartate and one glutamate molecule, which both have to bind before one of the 'products' is released from the protein.

(c) The fact that the same kinetic pattern is obtained whether the internal (Fig.1) or the external substrate (not shown) is chosen as the varied or the non-varied one, respectively, indicates the validity of the following (algebraicly symmetrical) bisubstrate

rate equation (Cleland, 1970). In this case, external aspartate is written as the varied substrate.

$$1/v = 1/V_{max} + K_a/(V_{max} \cdot A) + Kg/(V_{max} \cdot G) + K_{ia} \cdot Kg/(V_{max} \cdot A \cdot G)$$

A and G represent concentrations of external aspartate and internal glutamate, respectively. The Michaelis constants K are indexed accordingly. K_{ia} is the dissociation constant for the binary carrier-aspartate complex. The value of this particular constant in ping-pong mechanisms would be zero, since no ternary complex is involved. However, as clearly seen in Fig. 1, all parameters used in this equation have finite values, which can be calculated from the intercepts of the secondary plots.

(d) The observed points of intersection are located near the abscissa indicating that the binding affinity of both internal and external substrate to the carrier protein is similar, whether the carrier is available in the free or single-substrate occupied form ($K_m \approx K_{is}$). Thus it is conceivable that there is no obligate order in the binding steps of internal and external substrate (random mechanism).

Formation of the ternary complex and transport mechanism

Considering possible ways of ternary complex formation, one binding site at either membrane surface may be exposed at the same time or two binding sites at one side or even two sites at both sides. Models involving two classes of binding sites, one being specific for aspartate, the other for glutamate, can be tested in a study of transport inhibition by the second substrate when present in the same compartment. The second (unlabelled) substrate can be treated as an inhibitor, since not its own translocation is monitored but its effect on the transport of the labelled substrate. Fig. 2 shows the corresponding inhibition studies for the external (Fig. 2A) and for the internal membrane side (Fig. 2B) measured in the forward (aspartate uptake) or backward (glutamate export) exchange mode, respectively.

When the data are analyzed according to Eadie and Hofstee, straight lines are obtained that converge on the ordinate. This finding is consistent with a competitive inhibition mechanism indicating that both substrates bind to the same carrier form. The inhibition constants K_i determined in secondary plots (insets of Fig. 2) are 180 μM and 2.8 mM, hence showing good coincidence with the corresponding K_m values obtained for external glutamate (200 μM) and for internal aspartate (3 mM), respectively (Dierks & Krämer, 1988).

These results demonstrating substrate competition in both compartments can be explained by assuming the ternary complex to form as a transmembrane unit by substrate occupation of one binding site at either membrane surface. It would require a conformational change of this intermediate to channel both substrates through the membrane in a concerted manner.

The argument can be put forward that two carrier forms showing specificity for aspartate or glutamate, respectively, are in equilibrium with each other, thus explaining the observed competition of substrates (Fig. 2). In this alternative view binary carrier-substrate complexes represent the translocation intermediates, whereas the ternary

Table 1. Influence of the membrane potential on the exchange activity of the reconstituted Asp/Glu carrier at different substrate distributions. Transport activities were calculated from initial velocity measurements in the forward exchange mode. Data are compiled from two experiments carried out under different conditions: (a) external 50 μM aspartate/internal 16 mM aspartate or glutamate, respectively, pH 6.5; (b) external 100 μM glutamate/internal 24 mM aspartate or glutamate, respectively, pH 7.0. $\Delta \psi$ was generated as K^+-diffusion potential using valinomycin (500 ng/mg phospholipid) and potassium gradients as indicated. The asymmetry factor denotes the ratio of exchange activities at membrane potentials of opposite polarity (inside positive:inside negative).

substrate			exchange activity (μM/min) at K^+_{in}(mM)/K^+_{ex}(mM)				asymmetry factor
internal	external	1/100[a]		1/1	100/100	100/1[a]	
Asp	Asp (a)	4.4		4.0	4.6	3.6	1.2
Glu	Asp (a)	7.5		2.3	2.6	0.9	8.3
Asp	Glu (b)	0.8		n.d.[b]	2.1	3.1	0.26
Glu	Glu (b)	2.0		2.1	2.1	2.2	1.1

[a] theoretically corresponding to a value of 120 mV, inside positive or negative, respectively, if calculated according to the Nernst equation
[b] not determined

complex forms at one side of the membrane-inserted carrier when the second substrate binds before the first substrate (following translocation) is removed.

The experimental strategy for a decision, whether binary or ternary complexes are involved in the translocation step, is based on model calculations describing the electrophoretic control of the reconstituted nucleotide carrier. Krämer and Klingenberg, (1982) have shown that the membrane potential affects charged carrier-substrate intermediates attracting or repelling them during translocation. The established rate equations not only explain modulation of electrogenic heteroexchange by $\Delta \psi$; furthermore, an inhibition of electroneutral homoexchange is predicted if the translocation intermediates show asymmetric charge distribution, thus being shifted to the membrane surface of opposite polarity. The resulting imbalance of carrier binding sites causes a pronounced inhibition of the overall antiport reaction as was measured for ATP/ATP but not ADP/ADP homoexchange (Krämer & Klingenberg, 1982). This agrees with a functional model involving [carrier^{3+}-ATP^{4-}] and [carrier^{3+}-ADP^{3-}] complexes, respectively, during translocation. Transferring these conclusions to the Asp/Glu carrier, binary [carrier-Asp⁻] but not [carrier/H⁺-Glu⁻] complexes should be influenced by $\Delta\psi$. Therefore the analysis focussed on Asp/Asp homoexchange (Table 1).

Table 1 confirms the electrophoretic control of the heteroexchange reaction. Membrane potential when applied in opposite polarity has reciprocal effects, which is in accord with the conception outlined above. On the contrary, neither Glu/Glu nor Asp/Asp homoexchange is significantly influenced upon energization of the liposomal membrane. This means that glutamate must be transported together with a proton in either direction conferring electroneutrality to Glu/Glu antiport. Furthermore, the observation that the Asp/Asp homoexchange is not affected by $\Delta\psi$ argues against binary complexes as translocation intermediates. The only conceivable alternative is a model in which the ternary complex itself is the catalytic intermediate of the transport cycle responsible for the translocation step, since [Asp⁻-carrier-Asp⁻] can be assumed to show approximately symmetrical charge distribution providing no target for the membrane potential.

A minimal model for this synchronous antiport of two substrates is outlined in Fig. 3 showing the formation and translocation of the ternary complex.

The proton, a third substrate

Counter exchange of aspartate and glutamate/H^+ is also regulated by the pH component of the proton motive force. Experiments carried out varying the pH in both compartments ($\Delta pH=0$) led to the conclusion that the absolute H^+ concentration plays a regulatory role. In a bisubstrate analysis (Fig. 4), H^+ and external glutamate concentration were varied under internal glutamate saturation, hence treating the proton as a substrate of the Asp/Glu carrier.

As shown in Fig. 4, a dependence of Glu⁻/H^+ co-transport on H^+ concentration according to Michaelis-Menten can be found, if activity data are corrected for carrier inactivation (see legend of Fig. 4). In Lineweaver-Burk plots straight lines are obtained, which tend to become parallel to the abscissa with increasing glutamate concentration, indicating that transport activity is not influenced by pH under glutamate saturation. However, the affinity of the carrier for glutamate (but not for aspartate) increases considerably with higher H^+ concentrations (by 5-fold when lowering pH from 6.9 to 6.2). This kinetic pattern suggests a rapid-equilibrium-ordered mechanism of H^+ and glutamate binding (Cleland, 1970) implicating a protonation/deprotonation of the free carrier, that is much faster than the overall transport reaction.

$$\text{carrier} + H^+ \xrightleftharpoons{K_a} \text{carrier}/H^+ \xrightleftharpoons{Glu^-} [\text{carrier}/H^+\text{-Glu}^-] \longrightarrow (\text{translocation})$$

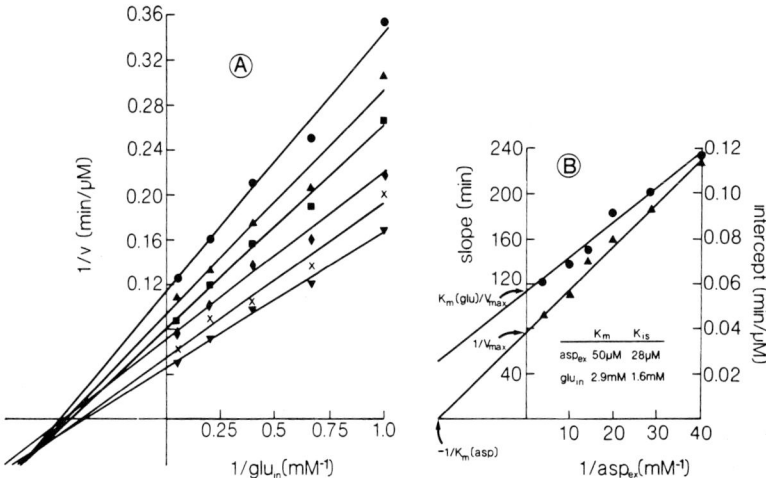

Fig. 1. Bisubstrate analysis of the Glu_{in}/Asp_{ex} exchange reaction mediated by the reconstituted Asp/Glu carrier (de-energized conditions). A: Dependence of uptake of labelled aspartate on internal glutamate (Lineweaver-Burk plot). The concentrations of the 'non-varied' external aspartate were 25, 35, 50, 70, 100 and 250 μM ($\bullet$, $\blacktriangle$, $\blacksquare$, $\blacklozenge$, $\times$ and $\blacktriangledown$, respectively). B: Slope ($\bullet$) and intercept ($\blacktriangle$) replots of primary plot A. The 'concentration independent' K_m values and the K_{is}, i.e. K_{ia} and K_{ig}, constants given in the inset were extracted graphically as indicated. V_{max} was 26 $\mu mol/min$ per l of proteoliposomes.

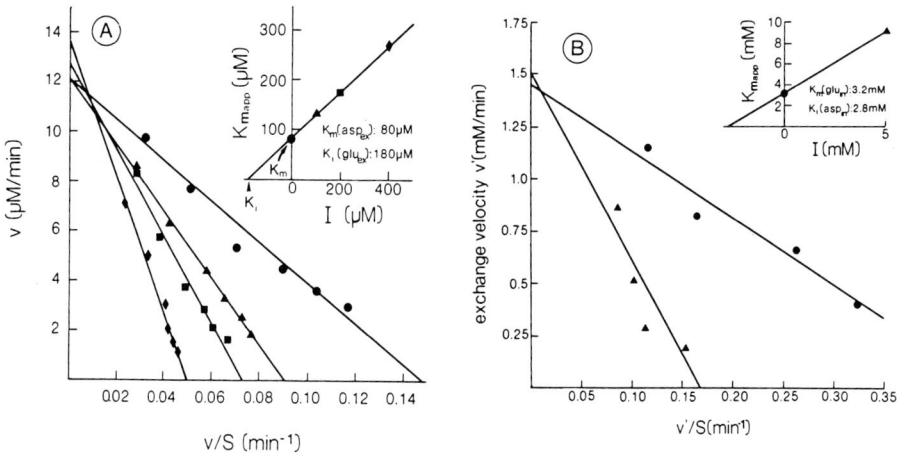

Fig. 2. Influence of the second substrate when present in the same compartment on reconstituted exchange activity (de-energized conditions). A: Influence of external glutamate on the uptake of aspartate (forward exchange). Inhibition kinetics (Eadie-Hofstee plot) are shown for different external glutamate concentrations: 0, 100, 200 and 400 μM ($\bullet$, $\blacktriangle$, $\blacksquare$, and $\blacklozenge$, respectively). The inset demonstrates the extrapolation of K_m and competitive inhibition constant (K_i) for substrate and inhibitor, respectively. B: Influence of internal aspartate on export of glutamate (backward exchange). Relative exchange velocities (v', see Dierks & Krämer, 1988) were determined in the absence ($\bullet$) or presence ($\blacktriangle$) of internal aspartate (5 mM).

The dissociation constant of the carrier/H^+ complex was estimated from secondary plots yielding a pK_a equivalent of 6.5.

These findings argue in favor of a distinct proton binding site possibly involving a histidine residue of the carrier. Since the pK_a of glutamate is very low (about 4.2), it seems improbable to assume that H^+ binds associated to glutamate. According to the rapid-equilibrium mechanism, an infinite glutamate concentration would transfer all carrier molecules to the [carrier/H^+-Glu$^-$] complex (regarding only one membrane side) independent of pH. Equilibration of protonated and unprotonated carrier molecules consequently explains the observed competitive nature of substrate inhibition (Fig. 2), though aspartate and glutamate are assumed to bind to two different carrier forms.

The physiological significance of this regulation may lie in the possibility of discriminating between aspartate and glutamate in order to avoid futile homoexchange cycles. Glutamate binding would be favored at the cytosolic, but impeded at the matrix side.

Reversible induction of uniport activity

Comprehensive studies of the impact of SH-reagents on transport function of the reconstituted Asp/Glu carrier led to a surprising discovery: Mercuric chloride and several organic mercurials such as mersalylic acid (mersalyl) and p-chloro-mercuribenzenesulfonic acid (pCMS) are able to convert the antiport carrier, normally catalyzing transmembrane counterexchange of substrate, into a uniport carrier. This activity can be measured as efflux of labelled substrate from the interior of proteoliposomes in absence of external substrate. Maximum induction leads to uniport rates, which are comparable to those measured for antiport activity. There are several lines of evidence that this efflux is not due to unspecific leakage of the liposomes, but in fact is mediated by the Asp/Glu carrier. In the following the basic characteristics of the induced uniport function are described.

a) Proteoliposomes are stable against the applied reagents. Liposomes carrying no protein remain totally unaffected by the applied reagents even when used in high concentrations (Fig. 5). Likewise proteoliposomes, if prepared with protein, which was previously denatured with 4% SDS at high pH, show only little "leakage". On the other hand uniport induction is possible in liposomes carrying intact Asp/Glu carrier. The results shown in Fig. 5 clearly prove that uniport activity is essentially dependent on active carrier protein.

b) The uniporter needs high activation energy. This is demonstrated by a corresponding Arrhenius plot (Fig. 6). The activation energy E_a was determined to be 86 kJ/mole (15-30°C), which is in the range of other values found for carrier-mediated

106

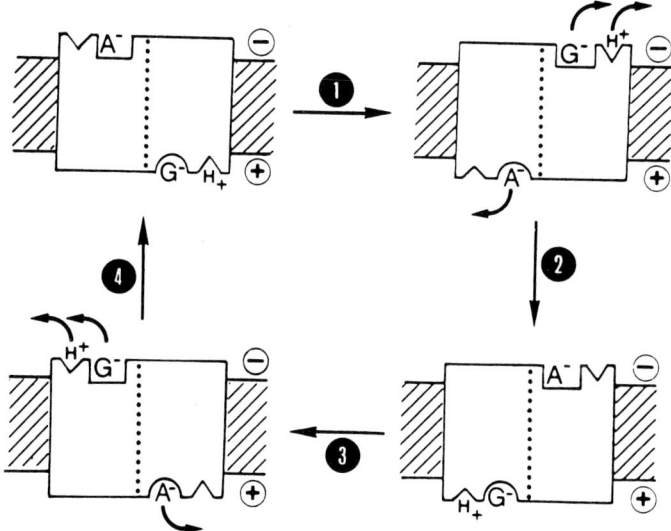

Fig. 3. Functional model of Asp/Glu antiport. The carrier molecule is schematically depicted as an integral unit in the mitochondrial membrane. Two functional 'subunits' are tentatively indicated. Either 'subunit' has one binding site for aspartate (A^-) or glutamate (G^-) and one for a proton. Four states of a complete heteroexchange cycle are shown connected either by translocation steps (1, 3) or by substrate release/binding steps (2, 4). The free carrier intermediates resulting after dissociation of transported substrates are omitted for simplicity. The translocation steps take place in a concerted manner between 'subunits' and are under electrophoretic control of the membrane potential.

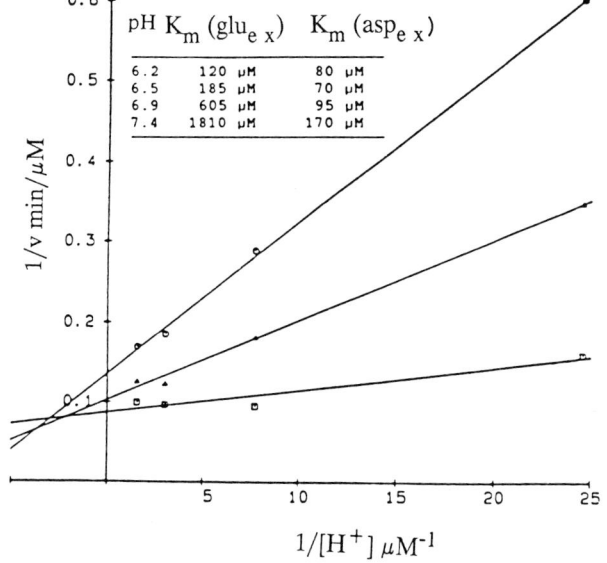

Fig. 4. Bisubstrate analysis of Glu/H^+ co-transport (de-energized conditions).
Glu/Glu and Asp/Asp (data not shown) homoexchange was analyzed varying external substrate (○ 100, ▲ 200, □ 400 μM) and proton concentration (pH 6.2-7.4, pH_{in} = pH_{ex}); internal substrate was saturating. The observed influence of pH on external K_m values for glutamate and aspartate is shown in the inset. When activity data were normalized on the basis of exchange equilibrium determinations in order to correct for irreversible pH effects, the proton could be treated like a Michaelis-Menten substrate in the case of Glu transport, as is demonstrated in form of a Lineweaver-Burk plot.

transport processes; for the antiporter a value of 77 kJ/mole was measured in the reconstituted system (Fig. 6). Such high activation energies clearly argue against simple holes in the liposomal membrane.

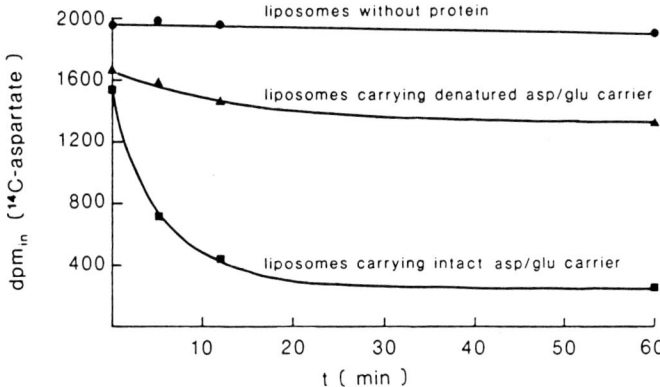

Fig. 5. Uniport induction is dependent on active carrier protein. Liposomes were loaded with labelled aspartate during reconstitution, subjected to Sephadex G-75 chromatography and treated with high concentration of mersalyl (1 mM). Liposomes carrying no protein (●) or protein, which was previously denatured by 4% SDS at pH 9 (▲) show no or only little efflux activity, respectively, as compared to intact proteoliposomes (■).

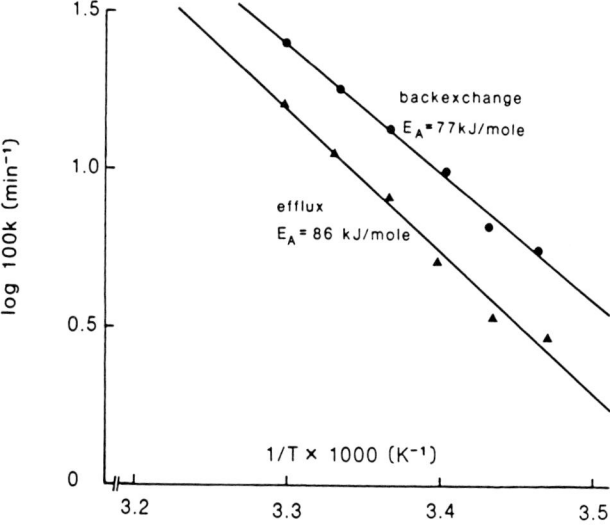

Fig. 6. Antiport and uniport show comparable temperature dependence. From the Arrhenius plots shown the activation energy E_A for either transport process was calculated. Similar values for back exchange (●) and efflux (▲), induced by 200 μM mersalyl, are obtained.

c) Antiport-uniport conversion is completely reversible. Fig. 7 shows the induction of uniport activity by 5 μM HgCl$_2$ (△△). In parallel, after 2.5 min of efflux, mercury was removed from the protein by an excess of the uninhibitory thiolic compound dithioery-thritol (DTE). As a result the export of internal substrate is reduced to very low rates (□) comparable to the "uninduced uniport activity" (○). If external substrate is added to these liposomes, antiport is reactivated, as is demonstrated by the observed export of label from the interior of the proteoliposomes (■) proceeding almost as rapidly as in the untreated backexchange control (●). On the contrary, antiport is significantly inhibited, if HgCl$_2$ is added simultaneously with (or previously to) the substrate and the addition of DTE is omitted (▲).

d) Efflux of internal substrate is correlated with the Asp/Glu carrier. The reconstituted protein preparation used in these experiments still contains an excess of the adenine nucleotide carrier. Therefore it was tested which of the two carrier proteins is involved in efflux activity. Uniport induction was not possible in liposomes carrying purified nucleotide carrier. Likewise porin was shown to be not involved, since no voltage gating could be found. Instead, uniport is correlated with the Asp/Glu carrier as becomes evident when further purifying this protein (results not shown).

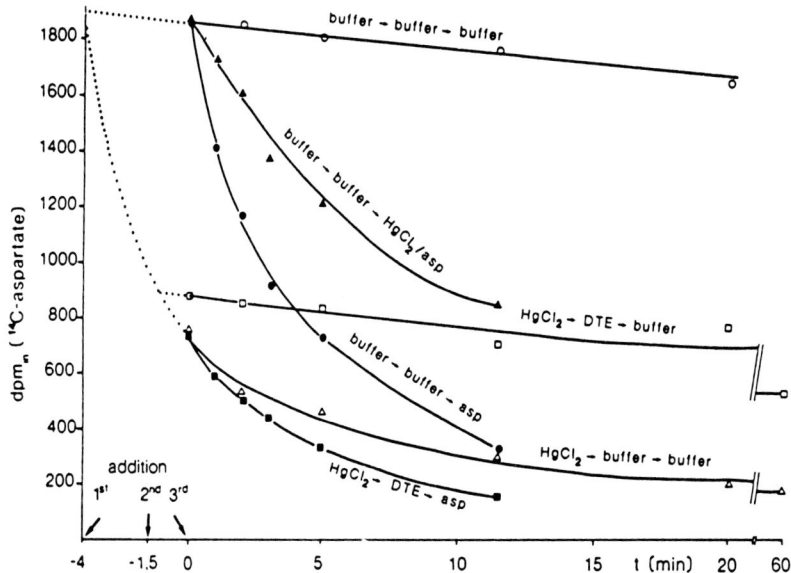

Fig. 7. Reversible antiport-uniport conversion. The effect of HgCl$_2$ on the export of labelled aspartate from proteoliposomes is observed. To the liposomes, which previously were loaded with labelled substrate, three consecutive additions were made at indicated time points. ○ uniport control; △ uniport induced by 5 μM HgCl$_2$; □ reversion of induced uniport by 4 mM DTE; ■ like □, but followed by reactivation of antiport by 2.5 mM aspartate (backexchange); ● back exchange control; ▲ inhibition of back exchange by 5 1μM HgCl$_2$.

e) Alteration of the substrate binding site by the applied mercury reagents. The uniport carrier clearly discriminates between aspartate and glutamate on the one hand and cationic amino acids like lysine and ornithine or non-substrate anions such as ATP, phosphate and sulphate on the other hand. However, the binding site of the carrier is obviously altered by the applied mercurials, since also 2-oxoglutarate, malate, malonate and even taurine, chloride and sodium ions (low conductance) are accepted as substrates by the uniporter (not shown).

An unmeasurable high K_m value is found, which suggests the formation of a transmembrane channel through the Asp/Glu carrier. The reduced substrate discrimination in the case of small ions supports this view. However, residual recognition of appropriate substrates is observed (glutamate/ornithine). Thus it has to be concluded that mercurials affect the substrate binding site leading to a somewhat unspecific flux through the carrier.

f) Antiport - uniport interconversion. Uniport induction always is coupled to inhibition of antiport. However, some SH-reagents (DTNB, pCMB, the benzoic acid analogue of uniport inducing pCMS) exist, which are not capable of inducing uniport, but do exert the inhibitory effect on carrier function. The action of all tested reagents could be reversed by an excess of uninhibitory thiolic compounds like DTE. In a working model it is proposed that two classes of reactive cysteines have to be modified by the reagents in order to achieve a weakened coupling of inward and outward directed transport. If the reagent has access to only one class, the result will be inhibition of transport function. Interestingly, uniport induction is also observed when proteoliposomes are passed through organomercurial agarose (Affi-Gel 501, BioRad). Since in these experiments free Hg^{++} ions were complexed by EDTA in excess, no covalent crosslinking between mercury and carrier cysteins is possible in the eluted liposomes. Hence secondary reactions have to be postulated which switch the carrier conformation from the antiport to the uniport state. This state shows some properties of an ion channel; however, the conductance is very low and the activation energy is too high. Possibly, due to alterations of the gating mechanisms by the Hg-treatment, the transport by the Asp/Glu carrier acquires more pore-like characteristics.

CONCLUSIONS

In order to elucidate the kinetic mechanism of the aspartate/glutamate antiport carrier from bovine heart mitochondria, bireactant initial velocity analyses were carried out. The kinetic patterns obtained demonstrate the formation of a ternary complex as a consequence of sequential binding of one internal and one external substrate molecule to the carrier. Results of transport inhibition studies when aspartate and glutamate were present in the same compartment, indicate that during exchange only one form of the carrier at either membrane surface exposes its binding sites. Investigation of the electrical properties of Asp/Asp homo-exchange (as compared to the electrogenic hetero-exchange) led to the conclusion that the translocating carrier-substrate intermediate exhibits a transmembrane symmetry with respect to the (negative) charge, which again only is conceivable assuming a ternary complex. Furthermore a distinct proton binding site has to be implicated, as can be derived from the different influence of pH observed on transport affinity and transport velocity, respectively, when glutamate is used as a substrate.

Interestingly, several mercury reagents are able to convert the antiporter into a uniport carrier. This change can completely be reverted by treatment with an excess of dithioerythritol. In the uniport state the protein exhibits reduced substrate specificity and can hardly be saturated with substrate. Nevertheless, the uniport function needs high activation energy, comparable to the antiport, which is inconsistent with the formation of an unspecific pore. Uniport induction in any case is connected with antiport inhibition. However, under certain conditions only inhibition of transport, but no uniport induction is observed. Therefore, the carrier has at least two different classes of SH-groups reacting with the mercurials.

ACKNOWLEDGEMENTS

The authors appreciate the technical assistance of Miss E. Riemer and Miss A. Salentin. We like to thank Prof. M. Klingenberg for helpful discussions. The work was supported by the Deutsche Forschungsgemeinschaft and by the Fonds der Chemischen Industrie.

REFERENCES

Cleland WW (1970) The steady state kinetics. In: Boyer PD (ed) The Enzymes, Vol.2, Academic Press, New York, pp 1-65
Dierks T, Krämer R (1988) Asymmetric orientation of the reconstituted aspartate/glutamate carrier from mitochondria. Biochim Biophys Acta 937:112-126
Krämer R, Klingenberg M (1982) Electrophoretic control of reconstituted adenine nucleotide translocation. Biochemistry 21:1082-1089
Krämer R, Heberger C (1986) Functional reconstitution of carrier proteins by removal of detergent with a hydrophobic ion exchange column. Biochim Biophys Acta 863:289-296
Krämer R, Kürzinger G, Heberger C (1986) Isolation and functional reconstitution of the aspartate glutamate carrier from mitochondria. Arch Biochem Biophys 251:166-174
LaNoue KF, Duszynski J, Watts JA, McKee E (1979) Kinetic properties of aspartate transport in rat heart mitochondrial inner membranes. Arch Biochem Biophys 195:578-590
Murphy E, Coll KE, Viale RO, Tischler ME, Williamson JR (1979) Kinetics and regulation of the glutamate-aspartate translocator in rat liver mitochondria. J Biol Chem 254:8369-8376

II. Functional Evidence and Characterization of Various Carriers

Recent Studies on the Mitochondrial Phosphate Transport Protein (PTP) and on its Relationship to the ADP/ATP Translocase (AAC) and the Uncoupling Protein (UCP)

HARTMUT WOHLRAB, CUNEYT BUKUSOGLU AND HEATHER DEFOE

Department of Cell Physiology, Boston Biomedical Research Institute
and
Department of Biological Chemistry and Molecular Pharmacology, Harvard Medical School, 20 Staniford Street, Boston, MA 02114, USA

The metabolite carriers of the inner mitochondrial membrane perform essential functions by making the transport of metabolites compatible with the chemiosmotic hypothesis and providing a link for metabolic pathways separated by the mitochondrial inner membrane. Active research programs are being carried out to characterize these transport proteins. Studies have been reported on ten of these (Table 1) and have recently been summarized (Krämer & Palmieri, 1988). Some time ago, we embarked on a research program to identify the protein responsible for the transport of inorganic phosphate and to determine the mechanism by which the protein is able to carry out this transport. This phosphate transport protein (PTP) has been identified, purified, and its activity reconstituted (Wohlrab, 1986) and the sequence of the protein has been determined (Kolbe & Wohlrab, 1985; Runswick *et al.*, 1987; Aquila *et al.*, 1987). Our partial protein sequence permitted us to identify homologies between the PTP and the ADP/ATP carrier. At the same time, similar homologies between the sequence of the uncoupling protein (UCP) and that of the ADP/ATP translocase were also identified (Aquila *et al.*, 1985). Thus while there are homologies among these three transport proteins, it will be of interest to determine whether non-phosphate (neither inorganic like PTP or organic like UCP or AAC) carriers such as the dicarboxylate, oxoglutarate, or monocarboxylate carriers will have similar homologies.

A. Azzi et al. (Eds.)
Anion Carriers of Mitochondrial Membranes
© Springer-Verlag Berlin Heidelberg 1989

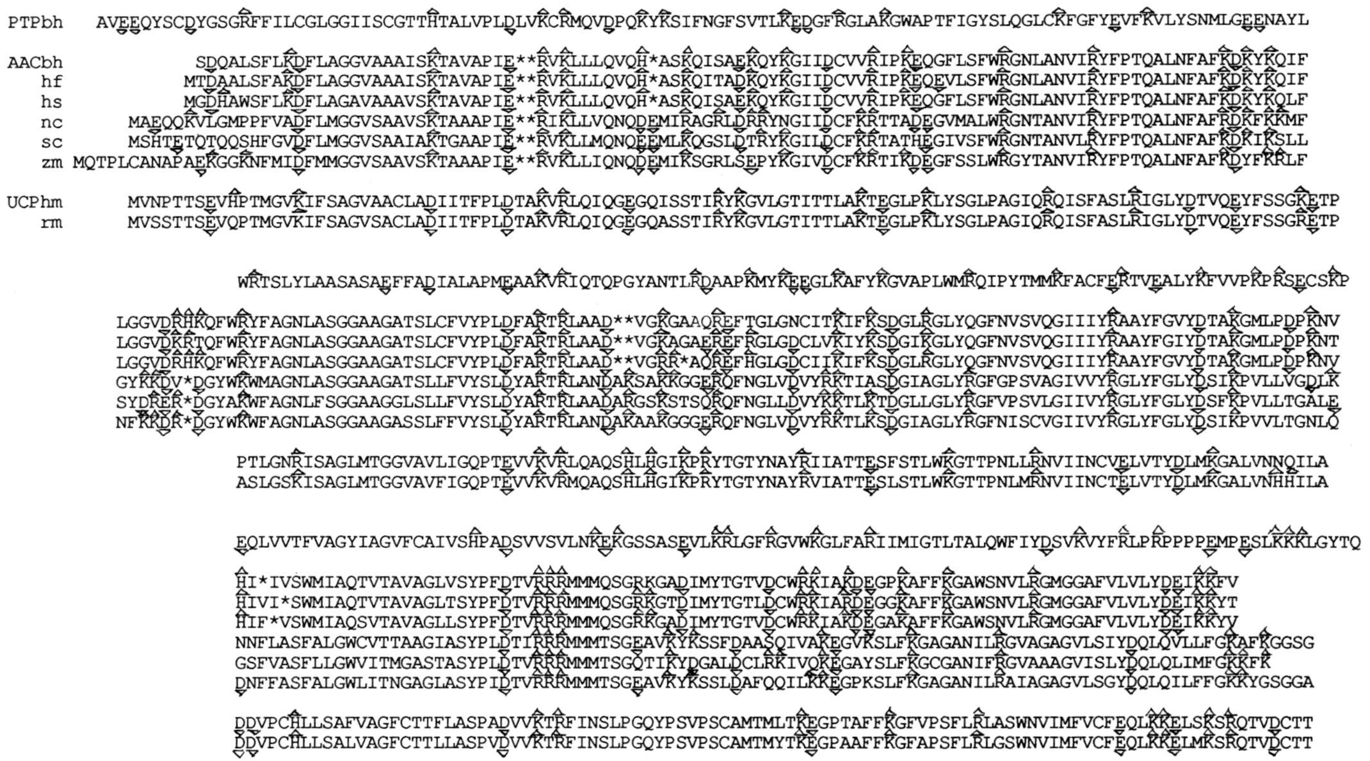

PTPbh AVEEQYSCDYGSGRFFILCGLGGIISCGTTHTALVPLLDLVKCRMQVDPQQKYKSIFNGFSVTLKEDGFGLAKGWAPTFIGYSLQGLCKFGFYEVFKVLYSNMLGEENAYL

AACbh SDQAALSFLKDFLAGGVAAAISKTAVAPIE**RVKLLLQVQH*ASKQISAEKQYKGIIDCVVRIPKEQGFLSFWRGNLANVIRYFPTQALNFAFKDKYKQIF
hf   MTDAALSFAKDFLAGGVAAAISKTAVAPIE**RVKLLLQVQH*ASKQITADKQYKGIIDCVVRIPKEQEVLSFWRGNLANVIRYFPTQALNFAFKDKYKQIF
hs   MGDHAWSFLKDFLAGGVAAVSKTAVAPIE**RVKLLLQVQH*ASKQISAEKQYKGIIDCVVRIPKEQGFLSFWRGNLANVIRYFPTQALNFAFKDKYKQLF
nc   MAEQQKTLGMPPFVADFLMGGVSAAVSKTAAAPIE**RIKLLVQNQDEMIKAGRLDRRYNGIIDCFKKTTADEGVMALWKGNTANVIRYFPTQALNFAFKDKFKKMF
sc   MSHTETQTQQSHFGVDFLMGGVSAAIAKTGAAPIE**RVKLLMQNQDEMLKQGSLDTRYKGIIDCFKRTATHEGIVSFWRGNTANVIRYFPTQALNFAFKDYIKSLL
zm   MQTPLCANAPAEKGGKNFMIDFMMGGVSAAVSKTAAAPIE**RVKLLIQNQDEMIKSGRLSEPYKGIVDCFKRTIKDEGFSSLWKGYTANVIRYFPTQALNFAFKDYFKRLF

UCPhm MVNPTTSEVHPTMGVKIFSAGVAACLADIITFPLDTAKVRLQIQEGQISSTIRYKGVLGTITTLAKTEGLPKLYSGLPAGIQRQISFASLRIGLYDTVQEYFSSGKETP
rm   MVSSTTSEVQPTMGVKIFSAGVSAACLADIITFPLDTAKVRLQIQEGQASSTIRYKGVLGTITTLAKTEGLPKLYSGLPAGIQRQISFASLRIGLYDTVQEYFSSGRETP

            WKTSLYLAASASAEFFADIALAPMEAAKVRIQTQPGYANTLRQAAPKMYKEEGLRAFVKGVAPLWMRQIPYTMMRFACFERTVEALYKFVVPRPRSECSKP

LGGVDKHHQFWRTFAGNLASGGAAGATSLCFVYPLDFARTRLAAD***VGKAAQEFTGLGNCITKIFASDGLGLYQGFNVSVQGIIIYRAAYFGVYDTAGMLPDPKNV
LGGVDKRTQFWRYFAGNLASGGAAGATSLCFVYPLDFARTRLAAD***VGKAGAEPEFRGLGDCLVKIYKSDGIKGLYQGFNVSVQGIIIYRAAYFGIYDTAKGMPDPKNT
LGGVDRHHQFWRTFAGNLASGGAAGATSLCFVYPLDFAKTRLAAD***VGKAQAQEFHGLGDCIIKIFKSDGLKGLYQGFNVSVQGIIIYRAAYFGVYDTAGMLPDPKNV
GYKQDY*DGYWKWMAGNLASGGAAGATSLLFVYSLDYARTRLANDAKSAKKGGEEQFNGLVVYRKTIASDGIAGLYRGFGPSVAGIVVYRGLYFGLYDSIKPVLLVGDLIK
SYDREEK*DGYAKWFAGNLFSGGAAGGLSSLFVYSLDYARTRLAADAFGKSTSQVEQFNGLLDVVRKTIKDGLLLGLWRGFVPSVLGIIVYRGLYFGLYDSFRPVLLTGALE
NFRKQRKDGYWKWFAGNLASGGAAGASSLFFVYSLDYARTRLANDAKAARGGEEQFNGLVDVVRKTIKSDGIAGLYRGFNISCVGIIVYRGLYFGLYDSIKPVVLTGNLQ

PTIGNRTSAGLMTGGVAVLLGQPTEVVKVRLQAQSHLEGIKPRYTGTYNAYRTIATTEGSFSTIWKGTTPNLLRNVIINCVELVTYDLIMKGALVNNQILA
ASLGSKTSAGLMTGGVAVFIGQPTEVVKVRLQMQAQSHLHGIKPRYTGTYNAYRVIATTESLSTLWKGTTPNLMRNVIINCTELVTYDLIMKGALVNHHILA

EQLVVTFVAGYIAGVFCAIVSHPADSVVSVLNEKGSSASEVLKRLGFRGVWKGLFARIIMIGTLIFALQWFIYDSVKVYFRLPRPPPEMPESLKKKLGYTQ

HI*IVSWMIAQTVTAVAGLVSYPFDTVRRRMMMOSGRKGADIMYTGTVDCWRKIAKDEGPKAFFKGAWSNVLRGMGGAFVLVLYDEIKKFV
HIVI*SWMIAQTVTAVAGLTSYPFDTVRRRMMMOSGRKGTDIMYTGTILDCWRKIARDEGGKAFFKGAWSNVLRGMGGAFVLVLYDEIKKYT
HIF*VSWMIAQSVTAVAGLLSYPFDTVRRRMMMOSGRKGADIMYTGTVDCWRKIAKDEGAKAFFKGAWSNVLRGMGGAFVLVLYDEIKKYY
NNFLASFALGWCVTTAAGIASYPITTTRRMMMTSGEAVKFKSSFDAASQIVAKEGVKSLFKGAGANILRGVAAGVLSIYDQLVLLFGKKAFFKGGSSG
GSFVASFLLGWVITMGASTASYPLDTVRRRMMMTSGQTIKYDGALDCLRKIVQKEGAYSLFKGCGANIFRGVAAAGVISLYDQLQIIMFGKKFK
DNFFASFALGWLITNGAGALASYPITVRRRMMMTSGEAVKYKSSLDAFQQIIKREGPKSLFKGAGANILRAIAGAGVLSGYDQLQILFFGKKYGSGGA

DDVPCHLLSAFVAGFCTTLASAPADVVKTRFINSLPGQYPSVPSCAMTMLTEGPTAFFKGFVPSFLRLASWNVIMFVCFEQLKKELSKSRQTVDCTT
DDVPCHLLSALVAGFCTTLIASPVDVVKTRFINSLPGQYPSVPSCAMTMYTEGPAAFFKGFAPSFLRLGSWNVIMFVCFEQLKKELMKSRQTVDCTT

Fig. 1. Sequences of PTP, AAC, and UCP from various sources. Alignment is primarily based on D(38)..K(41). R(43) of PTP (Wohlrab, 1986). Beef heart (bh), human fibroblasts (hf), human skeletal muscle (hs), Neurospora crassa (nc), Saccharomyces cerevisiae (sc), Zea mays L. (zm), brown fat hamster mitochondria (hm), and brown fat rat mitochondria (rm) ▽ indicates an acidic residue and △ a basic residue.

Table 1.

Mitochondrial Metabolite Transport Proteins Currently Being Characterized by Purification and reconstitution

ADP/ATP Translocase	electrogenic
Phosphate Transport Protein	electronutral/proton compensated
Monocarboxylate (Pyruvate) Carrier	electronutral/proton compensated
Uncoupling (GDP-Binding) Protein	electrogenic
Dicarboxylate Carrier	electronutral/exchange/divalent
Tricarboxylate (Citrate) carrier	electronutral/proton compensated
Aspartate-Glutamate Carrier	electrogenic
Oxoglutarate Carrier	electronutral/exchange/divalent
Ornithine Carrier	electronutral/proton compensated
Carnitine-Acylcarnitine Carrier	electronutral

In addition, with or without such homologies, how are the concentrations of these carriers in the membrane controlled? Liver mitochondria appear to possess the largest variety of these metabolite transporters, while flight muscle mitochondria are at the other extreme with only those transporters essential for steady state oxidative phosphorylation (Wohlrab, 1974; Wohlrab & Greaney, 1978). Neither possesses the UCP. The study of inorganic phosphate transport is thus related to those of the other metabolite carriers.

Fig. 1 shows the published sequences of the three transport proteins (PTP, AAC, UCP) from various sources. These are proteins with which one can be reasonably certain that they are solely responsible for catalyzing the transport of their respective substrates. We have aligned them just on the basis of PLD**R*R (Wohlrab, 1986; Kolbe & Wohlrab, 1985). Thus while they show this homology, the PTP has an extensive N-terminal (signal) sequence, while the ADP/ATP translocase has no N-terminal extension but a basic signal type sequence between amino acids 72 and 112 (yeast) (Smagula & Douglas, 1988a; Smagula & Douglas, 1988b). A similar basic region does not exist in the sequence of PTP or UCP. The UCP does not have an N-terminal extension or an internal basic sequence. Thus these three proteins are in fact also quite different. This difference extends also to the number of mRNAs that code for them. Table 2 shows the species that have been identified so far.

To briefly cite highlights of our past studies, purification of PTP from beef heart mitochondria is carried out with Triton X-100 in the presence of high salt to yield a fraction that passes through a hydroxylapatite column and consists only of PTP and the ADP/ATP translocase (Wohlrab, 1980). Net phosphate uptake is readily achieved with this preparation by incorporation into a mixture of highly purified phosphatidylcholine and phosphatidylethanolamine from plant with some calcium phosphatidate (Wohlrab et al., 1984a). The inorganic phosphate transport catalyzed by these proteoliposomes is ΔpH dependent with an internal alkaline pH at external phosphate concentrations of

1 mM (less than the transport K_m of about 1.6 mM) and sensitive to N-ethylmaleimide and mercurials (mersalyl). The calcium salt of phosphatidic acid is clearly required and much less uptake is detectable with the sodium salt. The concentration of transport-active PTP was estimated by centrifugation of the reconstituted proteoliposomes and assaying the supernatant (Wohlrab *et al.*, 1984a). This centrifugation yields approximately 50% less protein without a loss of transport activity. The turnover number of this preparation (22°C) is about 1.4×10^4 min^{-1} for the PTP dimer (Wohlrab, 1986).

Table 2

Identified mRNA Species for Mitochondrial Metabolite Transport Proteins

	mRNAs	Source	Reference
PTP	one	bovine heart	Runswick *et al.*, 1987
			Wohlrab *et al.*, 1988
AAC	two	bovine heart	Rasmussen & Wohlrab, 1986
	three	human	Houldsworth & Attardi, 1988
	one	*Zea mays L.*	Baker & Leaver, 1985
	one	*N. crassa*	Arends & Sebald, 1984
UCP	two	rat	Bouilleaud *et al.*, 1985

The turnover number of PTP in the mitochondrial membrane has been estimated in two different manners. PTP has been purified from rat liver as a reconstitutively active preparation with only one of several proteins showing the same peptide map as the beef heart PTP (Wohlrab, *et al.*, 1984b). The staining intensity of this band (assumed to stain the same as beef heart PTP) was used to quantitate PTP in rat liver mitochondria. Together with the (0°C) kinetic results (Coty & Pedersen, 1974), one obtains a turnover number of 1460 min^{-1}, using published (20-25°C) kinetic results (Ligeti *et al.*, 1985), a turnover number of 2.1×10^4 min^{-1} for the PTP dimer is obtained. Using the same kinetic numbers but a quantitation of PTP per mg mitochondrial protein using inhibitor binding results, a turnover number of 5-6×10^4 min^{-1} is obtained (Ligeti *et al.*, 1985). Most likely, the inhibitor binding results reflect a dimeric nature of the functional PTP. These turnover numbers are clearly higher than those determined for the ADP/ATP carrier at 18°C in rat liver mitochondria (600 min^{-1}) (Klingenberg, 1980). Another aspect of PTP that we have investigated is the unfolding, refolding, and renaturation of PTP. Our procedure for the very high degree of purification of PTP consists of exposing the protein to SDS/urea/phosphate/DTT and passing the preparation through a room temperature hydroxylapatite column (Kolbe & Wohlrab, 1985; Wohlrab & Kolbe, 1986).

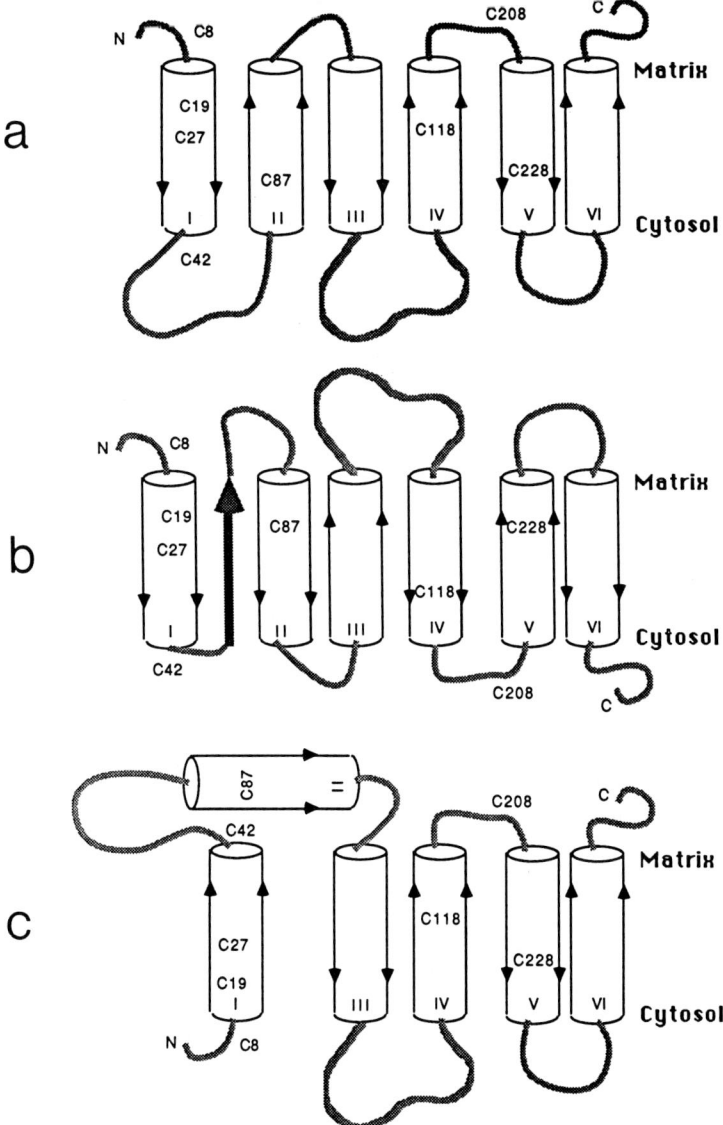

Fig. 2. Topological models of PTP based on those of AAC. (a) Based on original hydropathy plot (Runswick *et al.*, 1987); (b) model suggested by Aquila *et al.*, 1986; (c) model according to Vignais and coworkers (this meeting). The cylinders represent predicted α-helical structures.

The resulting highly purified preparation can be reconstituted with partial recovery of activity (Wohlrab & Kolbe, 1986). Why the recovery of activity is not complete has not yet been established.

The SH groups of PTP have always been assumed to play an important role in the activity of PTP. The reason is that PTP is extremely sensitive to N-ethylmaleimide (NEM), while other mitochondrial transport proteins, especially the ADP/ATP translocase, have a low sensitivity towards NEM. NEM reacts with cys-42 of beef heart PTP (Kolbe & Wohlrab, 1985). This cys is adjacent to the basic residues lys-41 and arg-43, which most likely are responsible for making cys-42 so highly reactive towards N-ethylmaleimide. It is of interest that this cys-42 is very sensitive to autoxidation since autoxidation inactivates the protein and makes it insensitive to N-ethylmaleimide (Wohlrab et al., 1984a). From the proposed transmembrane structure of PTP, it appears likely that cys-42 is very close to another cys, possibly cys-42 of another subunit (Fig. 2a), cys-208 in another topological arrangement (Fig. 2b), or cys-87 as well as cys-208 in a third model (Fig. 2c). Fig. 2 is based on topological results obtained by several laboratories for the AAC. The primary structure homologies between PTP and AAC suggest that topologies of the two membrane proteins will be very much the same.

Another type of cys interaction has been observed in SDS gels. All indications are that the dual protein band of beef (and pig) heart PTP (Kolbe et al., 1984) is due to an intramolecular disulfide bond. The size of the stacking gel shifts the relative concentration between the two bands (Wohlrab, unpublished observation) and the addition of N-ethylmaleimide to the purified, denatured, and reduced protein results in only a single PTP band (Kolbe et al., 1984). The PTP from rat liver and rat heart consists of a single protein band (Wohlrab et al., 1984b). It is most likely that the distribution of cys in the sequence of these PTP's is different from that in beef heart. It is expected that the availability of the primary sequence of the rat enzyme will help explain the double protein band of the beef heart preparation.

Most recently we have designed experiments to identify the active sites of the PTP. For this purpose we have prepared (Lauquin et al., 1980) and labeled with ^{32}P the inorganic phosphate analogue 4-azido-2-nitrophenyl phosphate (ANPP) (Bukusoglu & Wohlrab, 1987; Bukusoglu & Wohlrab, 1988). We have titrated beef heart mitochondria and sonic submitochondrial particles with ANPP, purified the PTP and reconstituted its activity as proteoliposomes.

Fig. 3 shows that PTP is much more sensitive to ANPP from the matrix side. To demonstrate the membrane impermeability of ANPP, a characteristic essential for differentiating between the matrix and cytosolic active sites of PTP, we reacted mitochondria and sonic submitochondrial particles with [32P]ANPP and separated the proteins on an SDS gel. Fig. 4 shows that the beta subunit of the mitochondrial F1-ATPase is only labeled with [^{32}P]ANPP in the sonic submitochondrial particles.

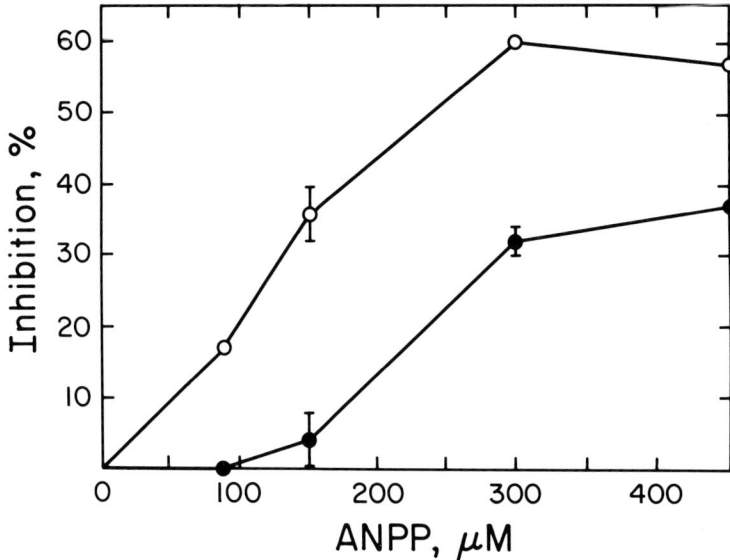

Fig. 3. Inhibition of PTP in beef heart mitochondria (o) and sonic submitochondrial particles (o) by ANPP at various concentrations.

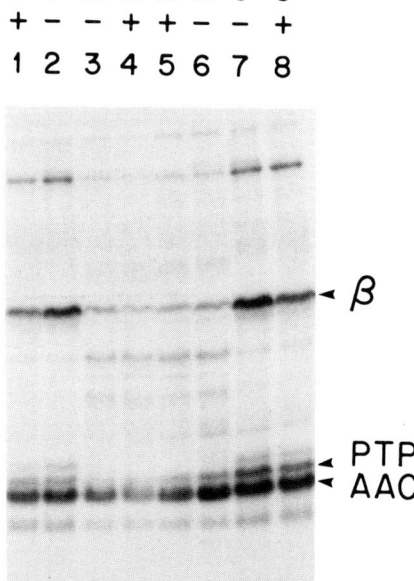

Fig. 4. Autoradiography of SDS gels of [^{32}P]ANPP-labeled beef heart mitochondria (M) and submitochondrial particles (S). Labeling was done in the presence (+) and absence (-) of P_i. β is the beta subunit of F_1-ATPase.

These results demonstrate that ANPP is impermeable to the mitochondrial inner membrane and thus suitable for studies of the PTP epitopes facing the matrix and the cytosol.

With active site studies in mind, we have now demonstrated, that PTP can be protected from inhibition by ANPP with inorganic phosphate from both sides of the membrane (Bukusoglu & Wohlrab, 1987).

Interestingly, succinate protects also very well, however only from the matrix side of the membrane (Bukusoglu & Wohlrab, 1988). We are currently digesting labeled PTP with various proteases in order to identify peptides that are labeled only from one side of the membrane and other peptides that are not labeled in the presence of phosphate or other anions and that protect PTP from inhibition by ANPP. It should be noted that in all our experiments, we use low ANPP concentrations to yield only partial inhibition of PTP in order to minimize nonspecific labeling of PTP epitopes.

REFERENCES

Aquila H, Link TA, Klingenberg M (1985) The uncoupling protein from brown fat mitochondria is related to the mitochondrial ADP/ATP carrier. Analysis of sequence homologies and of folding of the protein in the membrane. EMBO J 4:2369-2376

Aquila H, Link TA, Klingenberg M (1987) Solute carriers involved in the energy transfer of mitochondria form a homologous protein family. FEBS Let 212:1-9

Arends H, Sebald W (1984) Nucleotide sequence of the cloned mRNA and gene of the ADP/ATP carrier from *Neurospora crassa*. EMBO J 3:377-382

Baker A, Leaver CJ (1985) Isolation and sequence analysis of a cDNA encoding the ADP/ATP translocator of *Zea mays L*. Nucl Acids Res 13:5857-5867

Bouillaud F, Weissenbach J, Ricquier D (1986) Complete cDNA-derived amino acid sequence of rat brown fat uncoupling protein. J Biol Chem 261:1487-1490

Bukusoglu C, Wohlrab H (1987) Phosphate-sensitive and membrane side-specific inhibition of the mitochondrial phosphate transport protein by 4-azido-2-nitrophenyl phosphate. Biophys J 51:183a

Bukusoglu C, Wohlrab H (1988) Side-specific peptide maps from 4-azido-2-nitrophenyl phosphate-labeled mitochondrial phosphate transport protein. FASEB J 2:A1026

Coty WA, Pedersen PL (1974) Phosphate transport in rat liver mitochondria. Kinetic and energy requirements. J Biol Chem 249:2593-2598

Houldsworth J, Attardi G (1988) Two distinct genes for ADP/ATP translocase are expressed at the mRNA level in adult human liver. Proc Natl Acad Sci (USA) 85:377-381

Klingenberg M (1980) The ADP-ATP translocation in mitochondria, a membrane potential controlled transport. J Membr Biol 56:97-105

Kolbe HVJ, Costello D, Wong A, Lu RC, Wohlrab H (1984) Mitochondrial phosphate transport. Large scale isolation and characterization of the phosphate transport protein from beef heart mitochondria. J Biol Chem 259:9115-9120

Kolbe HVJ, Wohlrab H (1985) Sequence of the N-terminal formic acid fragment and location of the N-ethylmaleimide-binding site of the phosphate transport protein from beef heart mitochondria. J Biol Chem 260:15899-15906

Krämer R, Palmieri F (1988) Molecular aspects of isolated and reconstituted carrier proteins from animal mitochondria. Biochim Biophys Acta (in press)

Lauquin G, Pougeois R, Vignais PV (1980) 4-Azido-2-nitrophenyl phosphate, a new photoaffinity derivative of inorganic phosphate. Study of its interaction with the inorganic phosphate binding site of beef heart mitochondrial adenosine triphosphatase. Biochem 19:4620-4626

Ligeti E, Brandolin G, Dupont Y, Vignais PV (1985) Kinetics of P_i-P_i exchange in rat liver mitochondria. Rapid filtration experiments in the millisecond time range. Biochem 24:4423-4428

Rasmussen UB, Wohlrab H (1986) Bovine cardiac mitochondrial ADP/ATP-carrier: Two distinct mRNAs and an unusually short 3'-noncoding sequence. Biochem Biophys Res Comm 138:850-857

Runswick MJ, Powell SJ, Nyren P, Walker JE (1987) Sequence of the bovine mitochondrial phosphate carrier protein: Structural relationship to ADP/ATP translocase and the brown fat mitochondria uncoupling protein. EMBO J 6:1367-1373

Smagula CS, Douglas MG (1988a) ADP-ATP Carrier of *Saccharomyces cerevisiae* contains a mitochondrial import signal between amino acids 72 and 111. J Cell Biochem 36:323-327

Smagula C, Douglas MG (1988b) Mitochondrial import of the ADP/ATP carrier protein in *Saccharomyces cerevisiae*. Sequence required for receptor binding and membrane translocation. J Biol Chem 263:6783-6790

Wohlrab H (1974) Respiration-linked calcium ion uptake by flight muscle mitochondria from the blowfly *Sarcophaga bullata*. Biochem 13:4014-4018

Wohlrab H (1980) Purification of a reconstitutively active mitochondrial phosphate transport protein. J Biol Chem 255:8170-8173

Wohlrab H (1986) Molecular aspects of inorganic phosphate transport in mitochondria. Biochim Biophys Acta 853:115-134

Wohlrab H, Greaney J (1978) Mitochondrial phosphate transport and the N-ethylmaleimide binding proteins of the inner membrane. Biochim Biophys Acta 503:425-436

Wohlrab H, Kolbe, HVJ (1986) Isolation and reconstitution of the phosphate transport protein from mitochondria. Methods Enzymol 25:697-705

Wohlrab H, Collins A, Costello D (1984a) Purified mitochondrial phosphate transport protein. Improved proteoliposomes and some properties of the transport protein sulfhydryl group(s). Biochem 23:1057-1064

Wohlrab H, Kolbe HVJ, Rasmussen UB, Collins A (1984b) Mitochondrial phosphate transport protein: Purification from different tissues, immunological cross-reactivities, reconstitution, and turnover number in the rat liver mitochondrial membrane. In: Bronner F, Peterlik M (eds) Epithelial Calcium: Phosphate Transport. Alan R. Liss, New York, pp 211-216

Wohlrab H, Bukusoglu C, Abuerreish G, Rasmussen UB, Kolbe HVJ (1988) The mitochondrial phosphate transport protein and the kidney and intestine sodium/phosphate cotransporters - Recent progress and a comparison. In: Bronner F, Peterlik M (eds) Cellular Calcium and Phosphate Transport in Health and Disease. Alan R. Liss, New York, pp 177-182

Mitochondrial Phosphate Carrier: Relation of its SH Groups to Oligomeric Organization

Erzsébet Ligeti, Edgár Brázda and Attila Fonyó

Department of Physiology, Semmelweis University of Medicine, Budapest 8, POB 259, H-1444, Hungary

Free sulfhydryl groups are a prerequisite for the function of the phosphate transport protein (PTP) of the inner mitochondrial membrane as the translocation process can be inhibited by a large variety of SH-reagents. Sequential inhibition of phosphate transport by reversibly and irreversibly acting thiol reagents revealed a state of the PTP where the transport function was intact yet the protein was protected against irreversible inactivation. On the basis of these experiments it was suggested that the functional unit of PTP carried two sulfhydryl groups accessible from the external side and the two SH groups would by equivalent in the transport function but react with SH reagents sequentially (Fonyó, 1974). However, investigation of the structure of the purified PTP demonstrated that in the 35 kDa polypeptide chain only one single cysteine (out of a total of 6 or 8) reacted with radiolabelled N-ethylmaleimide (NEM) (Kolbe & Wohlrab, 1985). Data obtained in transport studies and by protein structure analysis could be reconciled by suggesting an oligomeric organization of the functional unit of PTP, consisting of two 35 kDa monomers with one essential SH group belonging to each subunit.

Cooperativity between the SH groups of the transport unit is supported by the experiment shown in Fig. 1. Mitochondria were prepared in which either both SH groups of the PTP or only the first-reacting or the second-reacting SH group was free (for details see Ligeti & Fonyó, 1987). Protection of both thiol groups against NEM inhibition

A. Azzi et al. (Eds.)
Anion Carriers of Mitochondrial Membranes
© Springer-Verlag Berlin Heidelberg 1989

124

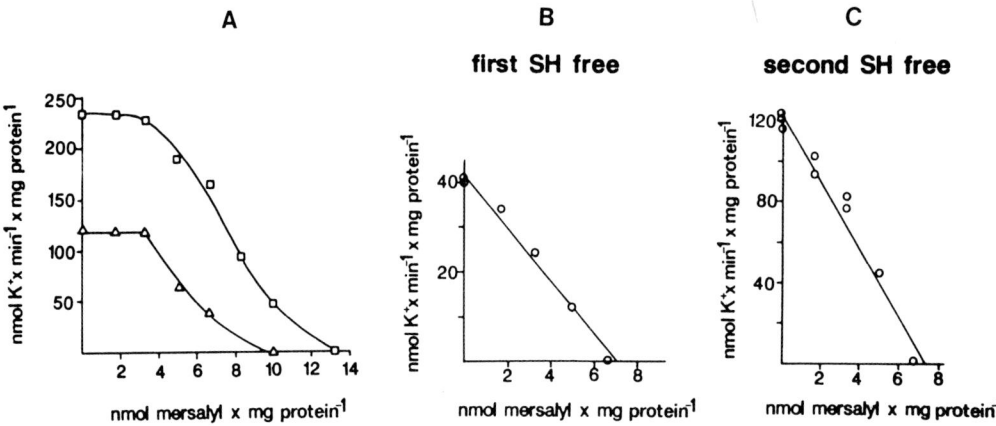

Fig. 1. Disappearance of the cooperativity of the SH groups in the single-SH-free preparations. Mersalyl titration was carried out in (A) non-treated mitochondria (□) or in mitochondria where both SH groups of PTP were protected (△) or (B) where only the first or (C) the second SH group of PTP was protected. From Ligeti and Fonyó (1987).

and the subsequent washing and reactivation procedure decreased the rate of phosphate transport but did not alter the sigmoidal titration with mersalyl (Fig. 1A). In contrast, when only one single SH group was protected and the same washing and reactivation procedure was carried out, the resulting preparations exhibited linear titration with mersalyl (Figs. 1B and C).

Oxidizing agents were shown to inhibit several bacterial transport systems. Similarly to the mitochondrial PTP, the function of these carriers depended on free sulfhydryl groups. Oxidizing agents also protected all these proteins (lactose, proline, glucose and mannitol specific carriers) against irreversible inhibition by NEM, and transport activity was fully reversible after the addition of dithiothreitol (Robillard & Konings, 1981; Konings & Robillard, 1982; Roossien & Robillard, 1984a). It was thus suggested that oxidizing compounds brought about a reversible dithiol-disulfide interconversion between two adjacent protein molecules, i.e. the two parts of a dimer. The hypothesis was put forward also in a generalized form, proposing dithiol-disulfide interchange as a basic mechanism in membrane transport processes (Robillard & Konings, 1982). Recently the mannitol specific transport protein of E.coli was demonstrated to

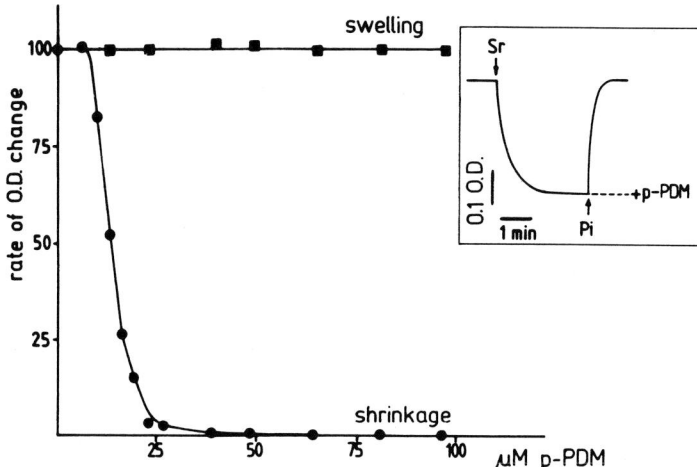

Fig. 2. Effect of p-PDM on mitochondrial energy coupling (■) and phosphate transport (●). The insert shows an original photometric recording.

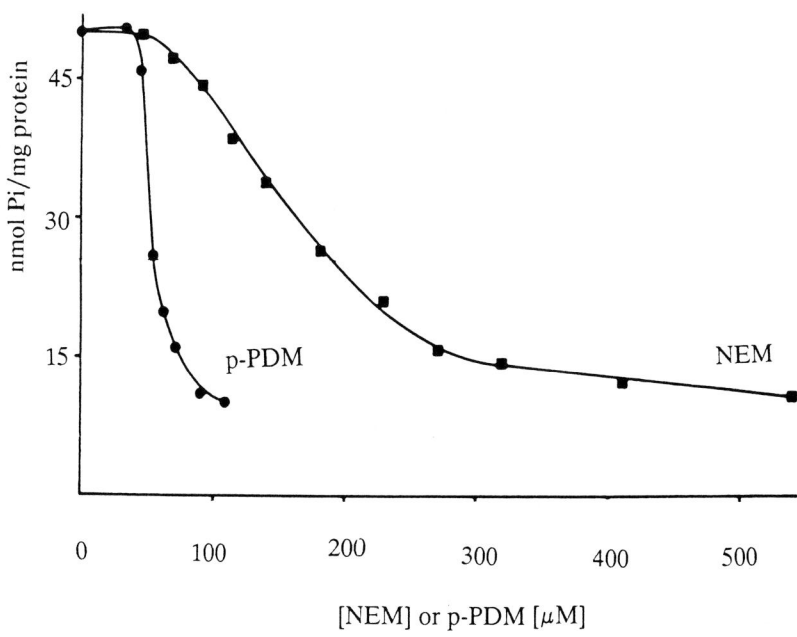

Fig. 3. Parallel titration of mitochondrial phosphate transport by mono- and dimaleimide.

occur in the membrane in dimeric form (Roossien & Robillard, 1984b) and the purified protein could be cross-linked both by bifunctional maleimide derivatives and some oxidizing agents (Roossien et al., 1986).

Cross-linking of the two monomers would be a way of providing experimental support also in favour of the oligomeric organization of PTP. Therefore we started to test the effect of various oxidizing agents and of bifunctional SH reagents on mitochondrial phosphate transport in rat liver mitochondria.

EFFECT OF OXIDIZING AGENTS ON PHOSPHATE TRANSPORT

The effect of the oxidizing agents ferricyanide, phenazine methosulfate and Cu-phenantroline and of the dithiol specific reagent phenylarsine oxide was investigated. No unequivocal indication of inhibition of phosphate transport in intact mitochondria could be observed with any of the tested substances applied in concentrations which did not yet damage mitochondria irreversibly.

These results are in accord with the data of Hüther and Kadenbach (1984) who reported the lack of inhibitory effect of Cu-phenanthroline on PTP reconstituted into proteoliposomes. However, it should be noted that proteoliposomes do not obligatorily reflect the state in the intact membrane as the weak interactions between monomers were shown to be interrupted by different isolation procedures (Roossien & Robillard, 1984b).

EFFECT OF DIMALEIMIDE DERIVATIVES

Bifunctional dimaleimide derivatives were effective not only in the inhibition but also in cross-linking of the isolated mannitol specific transport protein of the phosphotransferase system from E.coli (Roossien et al., 1986). Therefore we investigated the effects of o-phenylenedimaleimide (o-PDM), p-phenylenedimaleimide (p-PDM) and 1,6-hexanedimaleimide (HDM) on various mitochondrial processes.

The maleimide derivatives decreased mitochondrial phosphate content which precluded ^{32}P-phosphate exchange measurements. They also inhibited respiration and all coupled processes when succinate was used as respiratory substrate. However, they did not interfere with energy coupling if electrons were fed to complex IV via ascorbate plus TMPD. In the experiment shown in Fig. 2 volume changes of mitochondria energized with ascorbate plus TMPD were recorded (see the insert). Swelling (apparent as a decrease of optical density) started in the presence of 10 mM acetate when Sr^{2+} ions were added and accumulated together with acetate ions. Increased osmotic activity of the internal space induced water movement. Addition of phosphate was followed by its exchange for acetate, the phosphate ions being transported by the PTP (Fonyó, 1974).

The process is visible as a rapid shrinkage because the solubility of the Sr^{2+}-phosphate salt is lower than that of Sr^{2+}-acetate. As shown in the main part of Fig. 2 p-PDM

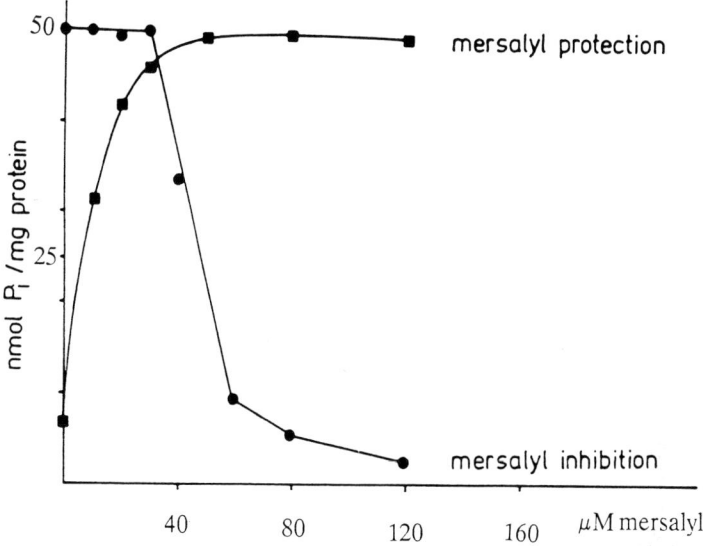

Fig. 4. Demonstration of the protective effect of mersalyl against irreversible inhibition by dimaleimide. Mitochondria were treated either with the indicated amount of mersalyl alone (●) or mersalyl was followed by 120 μM p-PDM and after 60 sec by 1 mM dithiothreitol (■).

did not interfere with the swelling phase, proving that neither substrate oxidation nor energy coupling were affected. In contrast, shrinkage after the addition of phosphate was completely inhibited by low concentration of p-PDM. The effect of o-PDM and HDM was similar: they did not interfere with energy coupling but inhibited phosphate transport completely.

In the following experiments phosphate transport was measured as phosphate uptake accompanying Sr^{2+} accumulation into energized mitochondria (respiring on ascorbate plus TMPD). Phosphate uptake was terminated by rapid separation of mitochondria from the medium by spinning down in a bench centrifuge; the pellet was extracted by chilled perchloric acid and phosphate content was determined. Phosphate uptake was calculated as the difference of phosphate content in the tested samples and mersalyl-inhibited samples.

Measurement of the inhibitory effect of mono- and dimaleimides on the same mitochondrial preparations showed that p-PDM was more effective than NEM (Fig. 3). When allowed to react for the same time, 50% inhibition of phosphate transport was achieved by 200 μM NEM and 65 μM p-PDM, respectively. The inhibitory potency of the two other derivatives was similar to p-PDM. Non-inhibitory concentrations of mersalyl did fully protect the carrier against the action of p-PDM (Fig. 4) just as it was reported previously for NEM.

Table 1. Alterations of the reaction rate of p-PDM as a function of the intramitochondrial pH

p-PDM concentration (μM)	Inhibition of P_i transport (%) when intramitochondrial pH	
	alkaline	acidic
40	4	0
60	43	0
80	57	0
100	65	8
120	72	17
160	76	61

The inhibitory effect of p-PDM depended on the intramitochondrial pH similarly as it was observed in the case of NEM (Ligeti & Fonyó, 1984). As shown in Table 1, inhibition of phosphate transport began at 40 μM p-PDM and 50% inhibition was attained around 80 μM when the internal pH was rendered alkaline (by previous accumulation of Sr^{2+} in the absence of phosphate). These values should be compared to 100 μM and approximately 150 μM in the case of acidic matrix pH (in the presence of nigericin). Both the protective effect of mersalyl and the variations of the effective concentration in the function of internal pH were similar when o-PDM or HDM were used.

All the three bifunctional dimaleimide derivatives inhibited phosphate transport in lower concentration but otherwise with the same characteristics as NEM did. It can thus be suggested that they react with the same SH group(s) as NEM. Pretreatment of mitochondria with low concentration of mersalyl (which did not yet reduce phosphate transport) allowed subsequent inhibition of PTP by dimaleimides, but decreased the concentration of dimaleimide required for complete inhibition (Fig. 5). Thus the effects of mersalyl and dimaleimides seemed to be additive. These observations indicate clearly that dimaleimides are able to react with one of the SH groups of PTP, when the other one is already occupied by mersalyl. Whether dimaleimides, when present alone, react simultaneously with both or only with one SH group of the PTP could not be unequivocally decided from transport studies.

ISOLATION OF PTP FROM DIMALEIMIDE-TREATED MITOCHONDRIA

Isolation of PTP proceeded as described earlier (Kolbe *et al.*, 1982; Kaplan *et al.*, 1986) and consisted of the following basic steps: rat liver mitochondria were solubilized by 3% Triton X-100 (final concentration); after 10 min at 0°C the suspension was centrifuged for 60 min at 25000 g; 400 μl of the supernatant was layered on the top of 0.5 g dry hydroxyapatite (HTP) powder; proteins not bound to the column were eluted by approximately 2.5 ml solubilizing solution containing 0.5% Triton X-100.

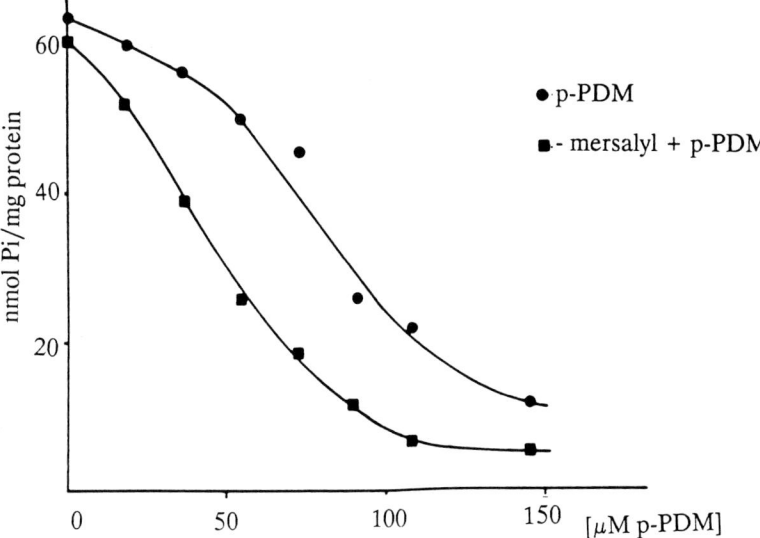

Fig. 5. Additive effect of mersalyl and p-PDM on phosphate transport. Mitochondria were treated by the indicated concentration of p-PDM in the absence (●) or in the presence of 30 μM mersalyl (■).

An aliquot (400 μl) of the HTP-eluate was precipitated by acetone and the pellet dissolved in sample buffer was subjected to SDS-PAGE.

The HTP eluate of control mitochondria (Fig. 6, lanes 1, 2) contained 4 to 5 major bands in the 35-30 kDa region and 3 to 4 weaker bands around 67 kDa. The presence or absence of dithiothreitol during the isolation procedure did not influence the number of bands appearing, however some bands of higher molecular mass became more intensive in its presence. The presence of PTP in the eluate was tested by determining radiolabelled NEM bound to the different proteins. Mitochondria were labelled as follows: both groups of PTP were protected by mersalyl, the rest of thiols allowed to react with cold NEM, PTP reactivated by mercaptoethanol, and after sufficient washing, mitochondria were treated with ^{14}C-NEM. Isolation of PTP was carried out as above, and distribution of radioactivity in the visible protein bands of SDS-PAGE is shown in Fig. 7. The peak of radioactivity corresponds to the second, rather weak band of the 35-30 kDa region (marked by an arrow in Fig. 6). No radioactivity above the ackground could be detected in the 65-70 kDa region, pointing to the absence of PTP-dimers under the given preparation and migration conditions.

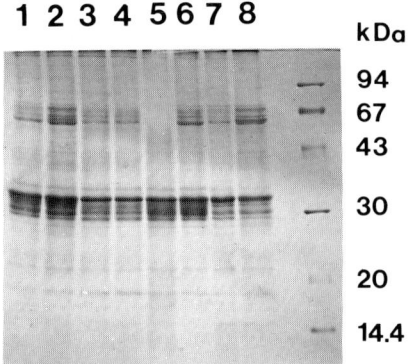

Fig. 6. SDS-PAGE (12% gel, Coomassie staining) of the hydroxyapatite eluates obtained from control (lanes 1,2,5,6) or dimaleimide-treated (lanes 3, 4) or NEM-treated (lanes 7, 8) rat liver mitochondria. Dithiothreitol (5 mM) was either absent (lanes 1, 3, 5, 7) or present (lanes 2, 4, 6, 8) during the entire procedure.

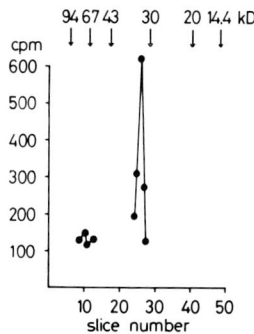

Fig. 7. Distribution of radioactivity in the visible protein bands after SDS-PAGE of the hydroxyapatite eluate obtained from ^{14}C-NEM-labelled rat liver mitochondria.

When HTP eluate was prepared from mitochondria treated with mono- or dimaleimides, only slight changes could be observed with SDS-PAGE. The major band which was shown to bind ^{14}C-NEM (Fig. 7) became fainter (Fig. 6 lanes 3,4) but sharper (lanes 7,8). The changes occurring in the 35-30 kDa range need better resolution in order to be clearly interpretable. One thing is however evident: the HTP eluate issuing from mitochondria in which phosphate transport was completely blocked by dimaleimides is not enriched in protein bands in the region of 65-70 kDa. Thus no evidence of dimerization of PTP due to crosslinking by the bifunctional maleimide derivatives could be demonstrated.

Considering the lack of effect of various oxidizing agents on phosphate transport in intact mitochondria and the fact that inhibition by the dimaleimide derivative does not result in the formation of detectable dimers lead us to the conclusion that the SH groups of the PTP are not in suitable steric position to form a disulfide or to be crosslinked by bifunctional maleimides. We want to stress that our present results do not exclude the possibility of an oligomeric organization of PTP in the native membrane. What they argue against, is the occurrence of spontaneous disulfide-dithiol interchange as part of the transport process.

ACKNOWLEDGEMENTS

Dimaleimide derivatives were a generous gift of Dr. G.T. Robillard. The authors are indebted to Ms. Edit Fedina and Erzsébet Seres-Horváth for their excellent technical assistance and Ms. Zsuzsa Oláh for typing the manuscript. Experimental work was supported by grants of Ministry of Health and OTKA and OKKFT projects, Hungary.

REFERENCES

Fonyó A (1974) Phosphate carrier of liver mitochondria: two equivalent SH-groups in the carrier unit. Biochem Biophys Res Comm 57:1069-1073
Hüther FJ, Kadenbach B (1984) Reactivity of the -SH groups of the mitochondrial phosphate carrier under native, solubilized and reconstituted conditions. Eur J Biochem 143:79-82
Kaplan RS, Pratt RD, Pedersen PL (1986) Purification and characterization of the reconstitutively active phosphate transporter from rat liver mitochondria. J Biol Chem 261:12767-12773
Kolbe HVJ, Mende P, Kadenbach B (1982) The protein components(s) of the isolated phosphate-transport system of mitochondria. Eur J Biochem 128:97-105
Kolbe HVJ, Wohlrab H (1985) Sequence of the N-terminal formic acid fragment and location of the N-ethylmaleimide-binding site of the phosphate transport protein from beef heart mitochondria. J Biol Chem 260:15899-15906
Konings WN, Robillard GT (1982) Physical mechanism for regulation of proton solute symport in *Escherichia coli*. Proc Natl Acad Sci (USA) 79:5480-5484
Ligeti E, Fonyó A (1984) Reactivity of the sulphydryl groups of the mitochondrial phosphate carrier. Eur J Biochem 139:279-285
Ligeti E, Fonyó A (1987) Mitochondrial phosphate carrier: functional role of its SH groups and interrelations within the carrier unit. Eur J Biochem 167:167-173
Robillard GT, Konings WN (1981) Physical mechanism for regulation of phosphoenolpyruvate-dependent glucose transport activity in *Escherichia coli*. Biochemistry 20:5025-5032
Robillard GT, Konings WN (1982) A hypothesis for the role of dithiol-disulfide interchange in solute transport and energy-transducing processes. Eur J Biochem 127:597-604
Roossien FF, Robillard GT (1984a) Vicinal dithiol-disulfide distribution in the *Escherichia coli* mannitol specific carrier enzyme II^{mtl}. Biochemistry 23:211-215
Roossien FF, Robillard GT (1984b) Mannitol-specific carrier protein from the *Escherichia coli* phosphoenolpyruvate-dependent phosphotransferase system can be extracted as a dimer from the membrane. Biochemistry 23:5682-5685
Roossien FF, van Es-Spiekman W, Robillard GT (1986) Dimeric enzyme II^{mtl} of the *E.coli* phosphoenolpyruvate-dependent phosphotransferase system. FEBS Lett 196:284-290

Recent Developments in the Study of the Conformational States and the Nucleotide Binding Sites of the ADP/ATP Carrier

P.V. VIGNAIS, G. BRANDOLIN, F. BOULAY, P. DALBON, M.R. BLOCK AND
I. GAUCHE

DRF/LBio/Biochimie, Centre d'Etudes Nucléaires, 85X, 38041 Grenoble cedex,
France

The function of the ADP/ATP carrier is to provide the cell with ATP synthesized in mitochondria by oxidative phosphorylation. It does so by carrying out the exchange between the extramitochondrial and intramitochondrial ADP and ATP through the inner mitochondrial membrane. A wealth of structural and physiological data has accumulated during the last twenty years. These data have been recently reviewed (Klingenberg, 1985; Vignais et al., 1985). This paper and the following one (Brandolin et al., 1988) will mainly deal with recent results pertaining to the topography of the ADP/ATP carrier. Results obtained by the photolabeling approach will be summarized and discussed in this paper, whereas immunological and enzymatic approaches of the topography of the carrier will be presented in the following paper. In this introduction, we shall essentially recall some of the salient structural features related to the ADP/ATP carrier. Although much interest is given now to the genetic control of this key protein, in particular the presence of two distinct genes (Houldsworth & Attardi 1988) explaining multiple isoforms of the carrier, we have deliberately omitted to review this rapidly expanding field.

The ADP/ATP carrier is a small protein. In the case of beef heart mitochondria, the minimum molecular weight calculated from the amino acid sequence is close to 32.800, corresponding to 297 amino acid residues (Aquila et al., 1982). The N-terminal amino acid, serine, is blocked by an acetyl group. The amino acid sequence contains

A. Azzi et al. (Eds.)
Anion Carriers of Mitochondrial Membranes
© Springer-Verlag Berlin Heidelberg 1989

three repeats, each repeat corresponding to about 100 amino acid residues (Saraste & Walker, 1982). Interestingly, two other bovine mitochondrial proteins, namely the phosphate carrier (Runswick *et al.*, 1987) and the uncoupling protein of the brown fat adipose tissue (Casteilla L, Bouillaud F, Forest C, & Ricquier D, private communication), also contain three repeats which are related to each other.

The beef heart ADP/ATP carrier complexed with carboxyatractyloside (CATR), a specific inhibitor of ADP/ATP transport, has been purified from the mitochondrial membrane as an homodimer (Hackenberg & Klingenberg, 1980; Block *et al.*, 1982). However, the degree of oligomerization of the carrier in the mitochondrial membrane is possibly higher (Block *et al.*, 1984).

Under the conditions of oxidative phosphorylation, the ADP/ATP carrier catalyzes the one to one exchange of external ADP against mitochondrial ATP. Only the free forms of ADP and ATP, and not their Mg^{++}-complexed forms, are transported. As transport of ADP^{3-} against ATP^{4-} is not charge-compensated by a concomitant movement of protons, the resulting charge imbalance is compensated by the membrane potential generated by the mitochondrial respiration. In other words, under the conditions of oxidative phosphorylation, part of the proton motive force, about 30%, arising from the functioning of the respiratory chain is used to drive the efflux of mitochondrial ATP^{4-} in exchange of imported ADP^{3-} (Duszynski *et al.*, 1981).

Specific natural inhibitors have been invaluable in exploring the structure and mechanism of the ADP/ATP carrier. CATR and its decarboxylated derivative, ATR, are extracted from the Mediterranean thistle, *Atractylis gummifera*. On the other hand, bongkrekic acid, BA, and its isomer iso-BA, are fermentation products from *Bacillus cocovenenans*. Using those inhibitors in their radiolabeled forms, it has been shown that CATR and ATR are non permanent ligands. They inhibit ADP/ATP transport by binding to the cytosolic face of the carrier. In contrast, BA and isoBA inhibit ADP/ATP transport only under conditions which favor their penetration into the matrix space of mitochondria, for example after protonation at slightly acid pH. In inside-out particles, the reverse situation holds, indicating that the binding asymmetry of the ADP/ATP carrier with respect to ATR or CATR, and BA or isoBA, is in fact inherent in the topography of the carrier (for review see Vignais, 1976). In reconstitution experiments, the functional carrier protein is incorporated into liposomes with the CATR site exposed to the outside, *i.e.* with the same orientation as in mitochondria (Brandolin *et al.*, 1980). This is probably due to a specific distribution of polar residues in the carrier protein, with more polar residues present on the face exposed to cytosol, *i.e.* on the face which contains the CATR site.

The binding of CATR (ATR) and that of BA (isoBA) to the carrier are mutually exclusive. As elaborated in more detail hereafter, this is explained by two different conformations of the carrier, the CATR and BA conformations. The results of covalent modifications provided evidence that ATR and BA bind to amino acid sequences which may overlap, but are not identical (Block *et al.*, 1981a,b). As a consequence, a common binding center with possibly overlapping sites for CATR and BA (for review see Vignais *et al.*, 1985) appears to be more likely than the single site for ATR and BA advocated by Klingenberg (1985). Furthermore, the postulate of the single reorientatable binding site which was applied to all ligands of the carrier, *i.e.* the inhibitors CATR, ATR, BA, isoBA, and the substrates ADP and ATP (for review, see Klingenberg *et al.*, 1985) is not compatible with recent data, as discussed in the following sections. An interesting result from chemical modifications of the carrier (Block *et al.*, 1981a) was that modification of one mol of arginine per carrier dimer following reaction with phenylglyoxal inhibits ATR binding fully. This half site reactivity has also been demonstrated for CATR binding (Klingenberg *et al.*, 1978; Block *et al.*, 1986). This means that the binding of a ligand to one subunit of the carrier is able to influence the juxtaposed subunit(s) by a propagated change of conformation. In the case of phenylglyoxal and CATR, the result is an inhibition of transport.

Besides ATR (CATR) and BA, long chain acyl CoAs have been found to inhibit ADP/ATP transport in mitochondria (Morel *et al.*, 1974), and in inside-out submito-chondrial particles (Lauquin *et al.*, 1977). Inhibition was competitive and the K_i values were 0.15 μM and 1.6 μM respectively. Although in some pathological situations, *e.g.* myocardial ischemia, the large accumulation of long chain acyl CoAs in the mitochondrial matrix (Idell-Venger *et al.*, 1978) may be responsible for inhibition of ADP/ATP transport, in physiological conditions long chain acyl CoAs are not likely to act as inhibitors. These aspects have been discussed in a recent review (Vignais *et al.*, 1985).

CATR AND BA CONFORMATIONS OF THE ADP/ATP CARRIER ATP/ADP INDUCED TRANSITION BETWEEN THE CATR AND BA CONFORMATIONS

The demonstration of two interconvertible forms of the ADP/ATP carrier, characterized their ability to bind CATR or ATR (CATR conformation), or BA or isoBA (BA conformation), and the evidence that ADP or ATP triggers this transconformation, were based mainly on the use of fluorescent probes. Here we shall summarize some of the most salient results.

a) Evidence for the existence of CATR and BA conforms of the ADP/ATP carrier, based on changes of fluorescence of tryptophanyl residues

The solubilized ADP/ATP carrier in detergent is able to respond specifically to the addition of transportable nucleotides by modification of the environment of trypto-phanyl residues as reflected by changes of the intrinsic fluorescence (Brandolin *et al.*, 1985). By kinetic analysis of fluorescence changes upon addition of ADP or ATP at different temperatures, it has been possible to differentiate, at temperatures lower than

10^0 C, a rapid binding step followed by a relatively slow change of conformation of the carrier protein. In brief, the fluorescence increase elicited by ADP or ATP was further enhanced by BA and powerfully counteracted by CATR. These changes were interpreted on the basis of a simple scheme (Fig. 1). In this scheme, the two conforms of the carrier denoted by C (CATR conformer) and C^* (BA conformer) differ by the extent of the intrinsic fluorescence. For the transition between C and C^* to occur, the carrier must bind ADP or ATP. Micromolar concentrations of ADP or ATP (or any transportable nucleotides) are sufficient to trigger the transition between the two conformations. Non-transportable nucleotides are inefficient.

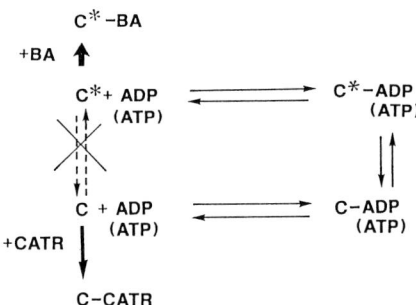

Fig. 1. The transition from the CATR conformation (C) of the carrier to the BA conformation (C^*) requires ADP or ATP. The direct transition from C to C^* is forbidden.

Interestingly, the nature of the detergent determines the basal conformational state of the solubilized carrier in the absence of added nucleotide. This was typically shown with cholamidopropyl-dimethylammoniopropane sulfonic acid (CHAPS) and lauryl amido dimethyl propylamineoxide (LAPAO) (Block & Vignais, 1986). LAPAO appeared to stabilize the CATR conformation, and CHAPS, the BA conformation. This might be related to the fact that CHAPS is a zwitterion and LAPAO a non ionic molecule. With CHAPS, as with LAPAO, ADP and ATP were required for the transition to occur between the two conformations. The activation energy of the ADP (ATP) induced transition of the BA conformation to the CATR conformation has been measured for the carrier in CHAPS. It was 10 Kcal/mol between 0^0 and 10^0 C (Block & Vignais, 1986), a value relatively modest, compared to the activation energy of ADP/ATP transport in mitochondria which is 50 Kcal/mol below 10^0 C and 13 Kcal/mol above 10^0 C (*cf* Vignais, 1976). This indicates that transport of ADP or ATP depends not only on conformational changes of the carrier protein itself, but also on the fluidity of the lipid bulk of the mitochondrial membrane. Indeed, in early experiments carried out with preparation of *Candida utilis* mitochondria differing in the degree of saturation of their membrane lipids, the rate of transport per carrier unit was found to be markedly increased when the lipids were less saturated and the membrane more fluid (Lauquin & Vignais. 1973).

b) Use of naphthoyl-ADP (ATP) to probe the transition between the CATR and BA confirmations of the membrane-bound ADP/ATP carrier

Naphthoyl-ADP (N-ADP) and naphthoyl-ATP (N-ATP), two fluorescent derivatives of ADP or ATP bind with high affinity to the ADP/ATP carrier, but are not transported (Block *et al.,* 1982). These fluorescent analogs have been used to explore the properties of the transition between the CATR and BA conformations (Block *et al.,* 1983). The binding of N-ADP (or N-ATP) to the membrane-bound carrier, and its specific release upon addition of CATR or BA is associated with opposing changes in the fluorescence intensity of the fluorophore: fluorescence of N-ADP is decreased upon binding and increased upon release. The carrier binds N-ADP either in the CATR or BA conformation, but only in the CATR conformation is N-ADP released by CATR, and in the BA conformation is N-ADP released by BA.

By lowering the temperature, it has been possible to resolve the kinetics of the fluorescence increase in two phases, a rapid one corresponding to the binding of the probe and a slow one corresponding to the transition between the two conformations of the carrier. In a typical experiment, performed at 10^0 C, addition of CATR to N-ADP-loaded mitochondria resulted in a rapid release of about half of the bound N-ADP which was complete in less than 0.5 s. This was followed by a very slow release of the remaining bound N-ADP, which could be considerably accelerated either by addition of BA (Fig. 2A) or that of ADP (or ATP) (Fig. 2B).

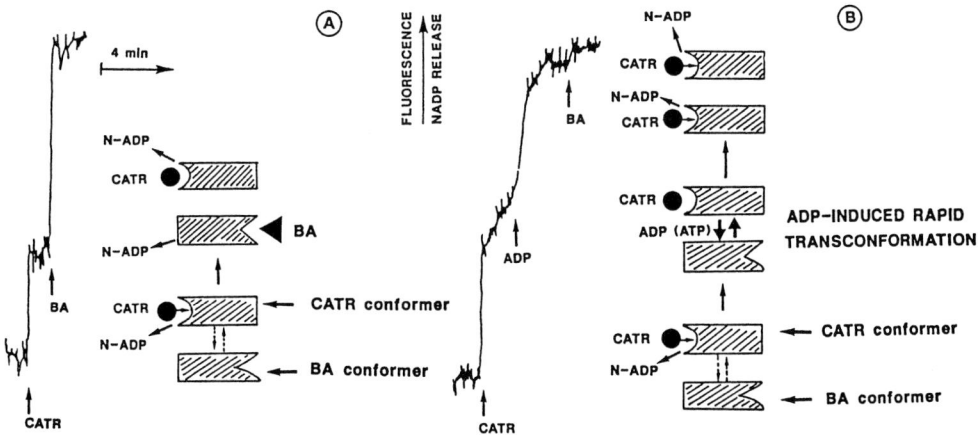

Fig. 2. Fluorescence modification corresponding to the release of naphthoyl-ADP (N-ADP) bound to the ADP/ATP carrier in mitochondria at 10^0 C. (A) Release induced by CATR and BA; (B) Release induced by CATR and ADP. For detail see Text (adapted from Block MJ, Lauquin GJM, Vignais PV (1982) Biochemistry 22:2202-2208).

The simplest interpretation of these data in terms of carrier conformation is that in intact mitochondria roughly half of the ADP/ATP carriers are in the CATR conformation and the other half in the BA conformation. Only the carriers in the

CATR conformation can bind CATR, and in response to CATR binding they rapidly release their bound N-ADP. The remaining loaded carriers in the BA conformation keep their bound N-ADP. They release it rapidly upon addition of BA (Fig. 2A) or upon addition of a minute concentration of ADP or ATP (Fig. 2B). It must be kept in mind that the concentration of ADP or ATP required for N-ADP release under the above conditions is much lower (μM) than that needed (mM) for the displacement of N-ADP by direct competition (Block et al., 1982). As explained in the scheme of Fig. 2B, ADP (or ATP) facilitates the transition from the BA conformation to the CATR conformation. As CATR is present in the medium, the BA conformers which undergo the ADP (or ATP) induced transition to the CATR conformation are trapped in this conformation, and the binding of CATR results in the release of bound N-ADP. Whatever the sequence of additions, i.e. CATR added prior to BA or BA added prior to CATR, and the type of particles used, i.e. mitochondria or inside out submitochondrial particles, the results obtained were the same, provided ADP or ATP were used. Non transportable nucleotides were ineffective. It is therefore clear that the transition from the CATR conformation to the BA conformation is reversible and is triggered by micromolar concentrations of transportable nucleotides. These experiments call for several comments:

1) They provide a strong argument against the single site mechanism advocated by Klingenberg (1985). In fact, not only are the carriers in the CATR conformation or the BA conformation capable of binding external ADP or ATP, but the binding of ADP or ATP is absolutely required for rapid transition between the two conformations. This contradicts the single reorientating site model for ADP/ATP transport, which postulates that when the carrier is in the BA conformation, its single site is opened to the inside and compelled to bind internal ADP or ATP.

2) It is unlikely that each carrier subunit contains more than one nucleotide binding site. To explain how externally added ADP reacts with N-ADP-loaded carriers in the BA conformation, one has to postulate that, in a given carrier, the subunit liganded by ADP is not the same as the subunit occupied by N-ADP. This reinforces the idea that the ADP/ATP carrier is a multisite oligomeric protein.

3) The denomination of cytosolic state or c state, and matrix state or m state (Klingenberg, 1985) is based on the postulate that a single site per carrier unit moves alternately between the cytosol and the matrix space. As the c state and the m state are believed to correspond to two conformations of the carrier during transport, and since CATR binds to the cytosolic face of the carrier and BA to the matrix face, it is clear that the c and m states are equivalent to the CATR and BA conformations respectively. However, because of the presence of several nucleotide binding sites per carrier unit (Dupont et al., 1982; Brandolin et al., 1982; Block & Vignais, 1984), we prefer to keep the operational terms of CATR and BA conformations.

The BA conformer, but not the CATR conformer has a cysteinyl residue readily accessible to permeant alkylating reagents

It has been known for a long time that membrane permeant alkylating reagents, like N-ethylmaleimide (NEM), inhibit ADP/ATP transport and also ATR binding in mitochondria, provided the mitochondria are preincubated with a minute amount of ADP or ATP (for review Vignais 1976). No inhibition was observed when external ADP or ATP was omitted. No inhibition was found either with non-permeant SH-group reagents, like mersalyl. A further step in the elucidation of the mechanism of action of permeant alkylating reagents was the discovery that NEM does not inhibit BA binding, and that the reactivity of the carrier to NEM is abolished by preincubation with CATR, but enhanced by preincubation with BA (Block *et al.,* 1981a).

Of the four cysteinyl residues present in the beef heart carrier, only one, Cys 56, is alkylable by the membrane permeant NEM, in the presence of ADP, or when the carrier is in the BA conformation (Boulay & Vignais, 1984). Cys 56 in the beef heart carrier can therefore be considered as a topographical probe of the transition from the CATR conformation to the BA conformation.

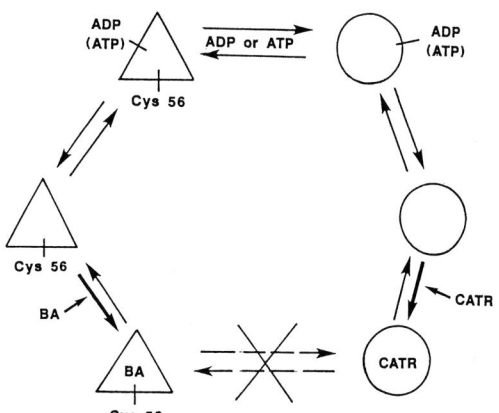

Fig. 3. Unmasking of Cys 56 in the ADP/ATP carrier during the transition from the CATR conformation to the BA conformation. The BA conformation is characterized by a higher accessibility or reactivity of Cys 56 to membrane permeant SH-reagents.

Additional topographical arising during the transition between the CATR and BA conformations

As will be reported in detail in the following paper (Brandolin *et al.,* 1988), a sequence of about ten amino acid residues belonging to the N terminal sequence of the carrier is accessible to specific antibodies when the carrier is in the CATR conformation. This finding together with the demonstration of the reactivity of Cys 56 to NEM when the carrier is in the BA conformation (see above) suggests that the transition from the CATR conformation to the BA conformation includes a limited pulling motion of the N-terminal region of the carrier towards the mitochondrial matrix, through the membrane-embedded α-helix which spans residues 9-28.

THE NUCLEOTIDE BINDING SITES OF THE ADP/ATP CARRIER.
TITRATION AND PHOTOLABELING DATA

Titration data. The half site reactivity of the ADP/ATP carrier

The nucleotide binding sites of the ADP/ATP carrier have been titrated with fluorescent and radiolabeled derivatives of ADP and ATP, that were either transportable or non transportable. Some experiments were performed with the isolated carrier, in the detergent LAPAO, and others with the membrane-bound carrier using mitochondria or inverted submitochondrial particles. A brief summary of the data makes it clear that more than one site is present per carrier unit.

Using formycin triphosphate (FTP) as a transportable nucleotide (Brandolin *et al.*, 1982) and N-ATP as a non-transportable nucleotide (Dupont *et al.*, 1982) to titrate the nucleotide binding sites of the isolated carrier in LAPAO, the following results were obtained. The same numbers of N-ATP and FTP binding sites per carrier were found; yet the distribution of high and low affinity sites was strikingly different. Half of the N-ATP binding sites were of a high-affinity type (Kd ≤ 10 nM) and the other half of a low affinity type (Kd ≈ 0.5 μM), whereas in the case of FTP, the number of high affinity sites (Kd ≤ 10 nM) was one third the number of the low affinity ones (Kd ≈ 0.1 to 2 μM). It is surprising that not only the "transportable" nature of a nucleotide is recognized by the solubilized carrier in detergent, but also a change of conformation is elicited upon binding of the transportable nucleotides, which results in negative co-operation. When mitochondria and inverted submitochondrial particles were used for titration of carrier nucleotide binding sites (Block & Vignais, 1984), negative co-operation was also observed for transportable nucleotides like ADP or ATP, and absence of co-operation for non transportable nucleotides like Br-ATP and N-ATP. The scheme of Fig. 4 illustrates the possible interactions between two nucleotide binding sites located on the same face of the carrier, either the cytosolic face or the matrix face.

Each nucleotide binding site is supposed to belong to a different subunit of the carrier. When the carrier in mitochondria is titrated with a non-transportable nucleotide, like N-ADP, the sites on juxtaposed subunits behave as if they were independent: they bind N-ADP with the same affinity. When the carrier is titrated with a transportable nucleotide like ADP, the curvilinear Scatchard plots can be interpreted to indicate that the binding of one ADP induces a conformational change in the carrier, which excludes binding of the other.

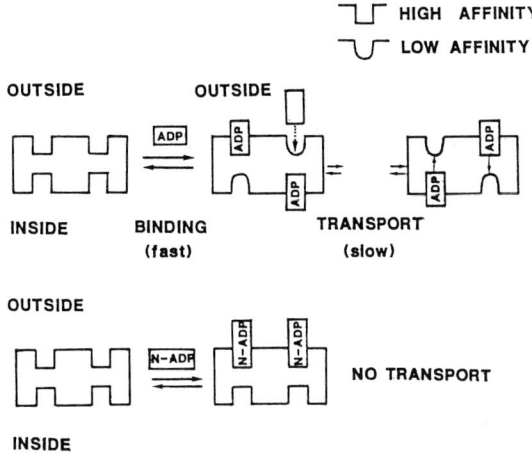

Fig. 4. Scheme illustrating the interactions between nucleotide binding sites in the ADP/ATP carrier, depending on the transportable or non transportable nature of the nucleotides. Transportable nucleotides: upper part of the Fig.; non transportable nucleotides:lower part of the Fig. Transport involves the formation of a ternary complex, corresponding to the binding of an external nucleotide to the outer face of the carrier and the binding of an internal nucleotide to the inner face.

In a dimer as illustrated in Fig. 4, the binding of ADP (or ATP) on the outer and inner faces of the carrier results in the formation of a ternary complex, which would be the prerequisite for transport, and is in agreement with the sequential mechanism of ADP/ATP transport based on kinetic data (Duyckaerts *et al.*, 1980; Barbour & Chan, 1981).

In the course of titration of the ADP/ATP carrier nucleotide binding sites, we were faced with the following paradoxical situation. The total number of CATR binding sites was found to be twice the number of the high affinity sites for ADP or ATP (Block & Vignais, 1984). In other experiments dealing with the isolated CATR-carrier complex, 1 mol of CATR was found to bind to 1 mol of carrier dimer (Klingenberg *et al.*, 1978; Block *et al.*, 1986). Taken together, these results led us to formulate the hypothesis that the membrane-embedded carrier might have a tetrameric structure (Block & Vignais 1984) as illustrated in Fig. 5.

Photolabeling data

Mapping of the nucleotide binding sites of the ADP/ATP carrier was recently undertaken with 2-azido[α-^{32}p]ADP, a photoactive derivative of ADP which is predominantly in the anti-conformation, like natural ADP (Czarnecki *et al.*, 1982). In the absence of photoirradiation, 2-azido ADP was found to inhibit reversibly ADP transport (Dalbon *et al.*, 1985) and to mimic ADP in binding assays with heart mitochondria (Dalbon *et al.*, 1985), *i.e.* two binding sites were displayed, one of high affinity and the other of low affinity. Photoirradiation of beef heart mitochondria in the presence of 2-azido[α-^{32}p]ADP led to the photolabeling of two segments of the peptide chain of the carrier.

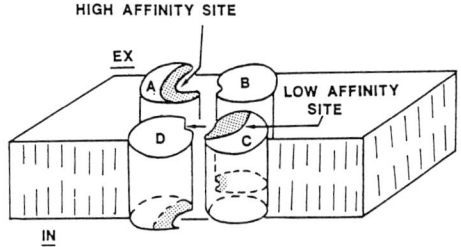

Fig. 5. Possible tetrameric organization of the ADP/ATP carrier subunits in the mitochondrial membrane. At a given time, of the four subunits of the tetramer, two of them exhibit functional nucleotide binding sites exposed to the outside: one site of high affinity on one subunit, and one site of low affinity on the other subunit. Likewise, the other two subunits exhibit nucleotide binding sites (high and low affinity) exposed to the inside. Functional binding sites are indicated by dots.

These segments spanned the sequence Phe 153-Met 200 with the probe predominantly attached to Lys 162, Lys 165 and Ile 183, and the sequence Tyr 250-Met 281, with Val 254 and Lys 259 being predominantly labeled. As the 2-azido[α-^{32}p]ADP is non permeant, it was concluded that these two segments are exposed to the cytosolic face of the carrier or are part of the hydrophilic translocation path. The scheme of Fig. 6A illustrates a model of arrangement of the polypeptide chain of the ADP/ATP carrier in the mitochondrial membrane, based on hydropathy plots reported in the literature (Runswick et al., 1987). This model includes five α helices of 20-30 amino acid residues, which cross the membrane, and three extra loops which extend from amino acids 28 to 105, 137 to 170 and 234 to 266. The nucleotide binding sites reside in the second and third loops. The prediction that the N terminus of the ADP/ATP carrier protein is exposed to the cytoplasmic face of the mitochondrial membrane is derived from the results of immunological experiments (see Brandolin et al., accompanying paper 1988). The arrangement of the second and third loops facing the cytoplasm is consistent with cross-linking data, showing that reaction of copper-O-phenanthroline with the carrier forms an intramolecular disulfide bridge between Cys 159 and Cys 256 (Torok & Joshi 1985).

It is noteworthy that the segment spanning Cys 159-Met 200, photolabeled by 2-azido ADP, is also photolabeled by two derivatives of ATR, [^{3}H]arylazido nitrophenyl ATR and [^{3}H]azidobenzoyl ATR (Boulay et al., 1983). Although the structures of ADP and ATR (or CATR) are different, their binding sites may overlap substantially. The competitive inhibition of ADP transport by ATR (Vignais et al., 1973) can be explained by these structural data.

Several possible explanations can be offered for the photolabeling of two distinct segments Phe 153-Met 200 and Tyr 250-Met 281 on the carrier chain by 2-azido ADP. 1. These segments might line a crevice which corresponds to the nucleotide binding site. This explanation would suppose that the purine ring is loosely adjusted in this crevice and moves freely, exposing its 2-azido group to both faces of the crevice. This is not consistent with the high affinity binding of 2-azido ADP to the carrier in the absence of photoirradiation. 2. Another explanation would be that the two sites are the high and low affinity sites demonstrated in reversible titration in the absence of photoirradiation, and that they reside in the same subunit. Although the presence of two binding sites on the same face of a given monomer is not likely for steric reasons, it cannot be definitely dismissed. 3. One is finally led to postulate that the two photolabeled fragments are located in two juxtaposed, interacting subunits of the carrier. These subunits are likely to assume different conformations, which are reflected by different spatial arrangements of the peptide chain. Assuming a tetrameric organization of the carrier, this may result into two functional distinct binding sites with different affinities on the cytoplasmic face or the matrix face of the carrier tetramer (*cf.* Fig. 5).

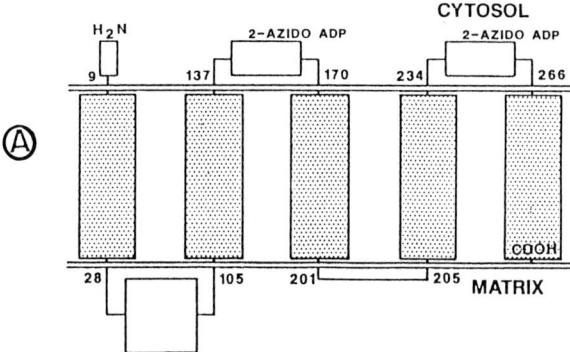

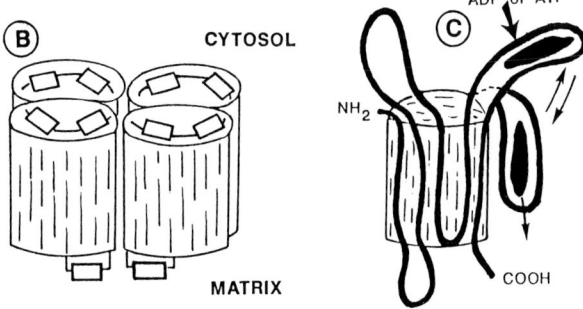

Fig. 6. (A) Possible arrangement of the peptide chain of the beef heart ADP/ATP carrier in the inner mitochondrial membrane. The membrane spanning α-helices are represented by dotted boxes, and the intervening extramembrane segments by open boxes. Schemes (B) and (C) illustrate the possible tetrameric organization of the ADP/ATP carrier, and the back-and-forth movement of an extramembrane loop carrying ADP or ATP.

In this model the two large loops, which protrude on the cytosolic face of each subunit of the tetramer (Fig. 6B) and are potential nucleotide binding sites might play a role not only in binding, but also in transport by moving the bound nucleotide towards the matrix face (Fig. 6C). A back-and-forth movement of the two loops, or one of them, in the space limited by the subunits of an oligomeric carrier might be the basis for the transmembrane exchange of ADP and ATP.

Conclusions and Perspectives

The structural basis for the CATR-BA conformational changes of the ADP/ATP carrier is being actively explored. In fact the transition between the two conformations is probably similar to the transition undergone by the carrier during the vectorial transport of ADP and ATP across the mitochondrial membrane. In the present survey, evidence was presented that the reversible transition between the CATR and BA conformations was undergone only by the ADP- or ATP-carrier complex, and not by the free form of the carrier. The environment of cysteinyl and tryptophanyl residue(s) was markedly modified during the transition between the CATR and BA conformations. The cysteinyl residue unmasked in the BA conformation was identified as Cys 56; however, the identity of the tryptophanyl residue(s) whose environment is modified during the transconformation, is as yet unknown.

Detailed exploration of the topography of the membrane-embedded carrier requires an immunological approach based on the use of antipeptide antibodies directed against the segments of the carrier chain presumably exposed to the inside or the outside of the mitochondrial membrane, and also an enzymatic approach based on the use of specific proteases able to attack peptide bonds at protruding segments of the carrier in right-side-out and inside-out particles. A typical example of the application of these two approaches is described in the accompanying paper (Brandolin *et al.*, 1988).

REFERENCES

Aquila H, Misra D, Eulitz M, Klingenberg M (1982) Complete amino acid sequence of the ADP/ATP carrier protein from beef mitochondria. Hoppe-Seyler's Z Physiol Chem 363:345-349
Barbour RL, Chan SHP (1981) Characterization of the kinetics and mechanism of the mitochondrial ADP/ATP carrier. J Biol Chem 256:1940-1948
Block MR, Lauquin GJM, Vignais PV (1981a) Chemical modifications of atractyloside and bongkrekic acid binding sites of the mitochondrial adenine nucleotide carrier. Are there distinct binding sites. Biochemistry 20:2692-2699
Block MR, Lauquin GJM, Vignais PV (1981b) Atractyloside and bongkrekic acid sites in the mitochondrial ADP/ATP carrier protein. An appraisal of their unicity by chemical modifications. FEBS Letters 131:213-218
Block MR, Lauquin GJM, Vignais PV (1982) Interaction of 3'-0(1-naphthoyl) adenosine 5'-diphosphate, a fluorescent adenosine 5'-diphosphate analogue, with the adenosine 5'-diphosphate/adenosine 5'-triphosphate carrier protein in the mitochondrial membrane. Biochemistry 21:5451-5457
Block MR, Zaccaï G, Lauquin GJM, Vignais PV (1982) Small angle neutron scattering of the mitochondrial ADP/ATP carrier protein in detergent. Biochem Biophys Res Commun 109:471-477
Block MR, Lauquin GJM, Vignais PV (1983) Use of 3'-0-naphthoyl adenosine 5'-diphosphate to probe distinct conformational states of membrane-bound adenosine 5'-diphosphate/adenosine 5'-triphosphate carrier. Biochemistry 22:2202-2208

Block MR, Vignais PV (1984) Substrate-site interactions in the membrane-bound adenine nucleotide carrier as disclosed by ADP and ATP analogs. Biochim Biophys Acta 767:369-376

Block MR, Vignais PV (1986) Dependence of the conformational state of the isolate adenine nucleotide carrier protein on the detergent used for solubilization. Biochemistry 25:374-379

Block MR, Boulay F, Brandolin G Lauquin GJM, Vignais PV (1986) Chemical modifications and active site labeling of the mitochondrial ADP/ATP carrier. Methods Enzymol 125:658-670

Boulay F, Lauquin GJM, Tsugita A, Vignais PV (1983) Photolabeling approach to the study of the topography of the atractyloside binding site in mitochondrial adenosine 5'-diphosphate/adenosine 5'-triphosphate carrier protein. Biochemistry 22:477-484

Boulay F, Vignais PV (1984) Localization of the N-ethylmaleimide reactive cysteine in beef heart mitochondrial ADP/ATP carrier protein. Biochemistry 23:4807-4812

Brandolin G, Dupont Y, Vignais PV (1982) Exploration of the nucleotide binding sites of the isolated ADP/ATP carrier protein from beef heart mitochondria. II. Probing of the nucleotide sites by formycin triphosphate, a fluorescent transportable analogue of ATP. Biochemistry 21:6348-6353

Brandolin G, Doussière J, Gulik A, Gulik-Kzywicki T, Lauquin GJM, Vignais PV (1980) Kinetic, binding and ultrastructural properties of the beef heart adenine nucleotide carrier protein after incorporation into phospholipid vesicles. Biochim Biophys Acta 592:592-614

Brandolin G, Dalbon P, Block MR, Gauche I, Vignais PV (1989) Immunological and enzymatic approaches of the orientation of the membrane-bound ADP/ATP carrier. In: Azzi A, Nałęcz KA, Nałęcz MJ, Wojtczak L (eds) (1989) The Anion Carriers of the Mitochondrial Membranes. Springer-Verlag Heidelberg, pp 147-157

Czarnecki JJ, Abbott MS, Selman B (1982) Photoaffinity labeling with 2-azido adenosine diphosphate of a tight nucleotide binding site on chloroplast coupling factor 1. Proc Natl Acad Sci (USA) 79:7744-7748

Dalbon P, Boulay F, Vignais PV (1985) Exploration of nucleotide binding sites in the mitochondrial membrane by 2-azido[α-^{32}p]ADP. FEBS Lett 180:212-218

Dalbon P, Brandolin G, Boulay F, Hoppe J, Vignais PV (1988) Mapping of the nucleotide binding sites in the ADP/ATP carrier of beef heart mitochondria by photolabeling with 2-azido[α-^{32}p] adenosine diphosphate. Biochemistry 27:5141-5149

Dupont Y, Brandolin G, Vignais PV (1982) Exploration of the nucleotide binding sites of the isolated ADP/ATP carrier protein from beef heart mitochondria. I. Probing of the nucleotide sites by naphthoyl-ATP, a fluorescent non transportable analogue of ATP. Biochemistry 21:6343-6347

Duyckaerts C, Sluse-Goffart CM, Fux JP, Sluse FE, Liebecq C (1980) Kinetic mechanism of the exchange catalyzed by the adenine nucleotide carrier. Eur J Biochem 106:1-6

Hackenberg H, Klingenberg M (1980) Molecular weight and hydrodynamic parameters of the adenosine 5'-diphosphate-adenosine 5'-triphosphate carrier in Triton X-100. Biochemistry 19:548-555

Houldsworth J, Attardi G (1988) Two distinct genes for ADP/ATP translocase are expressed at the mRNA level in adult human liver. Proc Natl Acad Sci (USA) 85:377-381

Idell-Wenger JA, Grotyohann LW, Neely JR (1978) Coenzyme A and carnitine distribution in normal and ischemic hearts. J Biol Chem 253:4310-4318

Klingenberg M, Riccio P, Aquila H (1978) Isolation of the ADP, ATP carrier as the carboxyatractylate protein complex from mitochondria. Biochim Biophys Acta 503:193-210

Klingenberg M (1985) The ADP/ATP carrier in mitochondrial membranes. In:Martonosi A (ed) The Enzymes of Biological Membranes: Membrane Transport, vol 4, John Wiley, New York, pp 511-553

Lauquin GJM, Vignais PV (1973) Adenine nucleotide translocation in yeast mitochondria. Effect of inhibitors of mitochondrial biogenesis on the ADP translocase. Biochim Biophys Acta 305:534-546

Lauquin GJM, Villiers C, Michejda JW, Hryniewiecka LV, Vignais PV (1977) Adenine nucleotide transport in sonic submitochondrial particles. Kinetic properties and binding of specific inhibitors. Biochim Biophys Acta 460:331-345

Morel F, Lauquin GJM, Lunardi J, Duszynski J, Vignais PV (1974) An appraisal of the functional significance of the inhibitory effect of long chain acyl CoAs on mitochondrial transports. FEBS Lett 39:133-138

Runswick MJ Powell SJ, Nyren P, Walker JE (1987) Sequence of the bovine mitochondrial phosphate carrier protein:structural relationship to ADP/ATP translocase and the brown fat mitochondrial uncoupling protein. Embo J 6:1367-1373

Saraste M, Walker JE (1982) Internal sequence repeats and the path of polypeptide in mitochondrial ADP/ATP translocase. FEBS Lett 144:250-254

Torok K, Joshi S (1985) Formation of an intramolecular disulfide bond in the mitochondrial adenine nucleotide translocase. FEBS Lett 182:340-344

Vignais PV (1976) Molecular and physiological aspects of adenine nucleotide transport in mitochondria. Biochim Biophys Acta 456:1-38

Vignais PV, Vignais PM, Defaye G (1973) Adenosine diphosphate translocation in mitochondria. Nature of the receptor site for carboxyatractyloside (gummiferin). Biochemistry 12:1508-1519

Vignais PV, Block MR, Boulay F, Brandolin G, Lauquin GJM (1985) Molecular aspects of structure-function relationships in mitochondrial adenine nucleotide carrier. In:Benga G (ed) Structure and Properties of Cell Membranes, Vol II, CRC Press, pp 139-179

Immunological and Enzymatic Approaches of the Orientation of the Membrane Bound ADP/ATP Carrier

G. BRANDOLIN, F. BOULAY, P. DALBON, M. BLOCK, I. GAUCHE AND P.V. VIGNAIS,

DRF/LBio/Biochimie, Centre d'Etudes Nucléaires, 85X, 38041 Grenoble cedex, France

The difficulties encountered in the crystallization of membrane proteins in a form suitable for X-ray analysis have stimulated the development of algorithms to predict, from primary amino acid sequences, the locations of transmembrane spans based upon assessment of local hydrophobicity. The proposed models for the folding of polypeptide chains of a variety of integral proteins predict membrane-spanning α-helices connected by hydrophilic, looplike segments. To test the validity of such models, much interest has been devoted to various probes that are important tools in studying the spatial orientation of membrane-spanning polypeptide chains (for review see Ovchinnikov, 1987).

Permeant and non-permeant chemical reagents have been widely used to probe either membrane-embedded or exposed parts of integral proteins. There are, however, severe drawbacks inherent in the use of chemical reagents, including uncontrolled permeability and absence of strict chemical specificity. Two other approaches, based on the identification of specific proteolytic cleavage sites or of epitopes recognized by antibodies, were found to be very informative in delineating the boundaries of transmembrane spans. High molecular weight probes, such as proteolytic enzymes and antibodies, are particularly interesting because they are nonpenetrant. Monoclonal antibodies are widely applied in structure-function studies of membrane proteins, but

A. Azzi et al. (Eds.)
Anion Carriers of Mitochondrial Membranes
© Springer-Verlag Berlin Heidelberg 1989

attempts to identify the epitopes recognized by these antibodies are often time-consuming and not always successful. Alternatively, models of the secondary structure can be used to predict hydrophilic regions in the protein which are likely exposed to the aqueous phase. Such peptides can be synthesized and used as antigens to prepare sequence-directed anibodies.

Combined with the use of proteolytic enzymes, we found the latter approach particularly useful in studying the topology of the polypeptide chain of the ADP/ATP carrier, a protein of the inner mitochondrial membrane whose function is to catalyze the import of cytosolic ADP into the matrix space of mitochondria and the concomitant export of mitochondrial ATP towards the cytosol. Two families of specific inhibitors, namely atractyloside (ATR) and carboxyatractyloside (CATR) on the one hand and bongkrekic acid (BA) on the other have been of great benefit in functional, conformational and structural studies of the ADP/ATP carrier protein. These inhibitors differ in their ability to recognize two different conformations of the carrier protein, and it is now well established that important structural changes occur during the transition from one to the other (for review see Vignais *et al.*, 1985).

The existence of two conformations of the carrier was first suggested by the results of immunological studies, in which antibodies raised against the CATR-protein complex were found to be less reactive against the BA-protein complex than against the CATR-protein complex (Buchanan *et al.*, 1976). The high sensitivity to protease of the protein in BA conformation, on the one hand, and its resistance to proteolysis in the CATR-associated form on the other hand, provided additional evidence for a large conformational difference between the two states. A less direct indication concerned the differential accessibility of tyrosine residue(s) of the membrane-bound carrier to iodine, the labeling being higher in the presence of CATR than in the presence of BA (Brdiczka & Schumacher, 1976). Particularly significant was the unmasking of Cys 56, which can be readily alkylated by N-ethylmaleimide (NEM) upon addition of transportable adenine nucleotides (Boulay & Vignais, 1984). This ADP or ATP induced -SH reactivity to NEM is abolished by CATR and strongly enhanced by BA.

The ADP/ATP carrier of beef heart mitochondria is ideally suited for a topographical study of the arrangement of its polypeptide chain, since a wealth of structural information is now available. The protein has been sequenced and is 297 amino acids long (Aquila *et al.*, 1982). A computer analysis of the primary structure revealed a striking pattern of three repeated related sequences, suggesting that the protein has evolved by a process involving two gene duplication (Saraste & Walker, 1982). On the basis of hydropathy plot analysis of the primary structure, these authors have predicted which segments are likely to traverse the membrane or form hydrophilic loops. According to the model they proposed, two hydrophilic loops that connect three membrane spanning α-helices are located on the same side of the lipid bilayer. Such arrangement was corroborated by Torok and Yoshi (1985) who showed that Cys 159 and Cys 256, though located on different loops, can form an intramolecular disulfide bond upon reaction with copper-O-phenantroline. Therefore, Cys 159 and Cys 256 must be located within a few angström of each other on the same side of the membrane.

The secondary structure of the ADP/ATP carrier has been probed by Klingenberg's group with the reagent pyridoxal 5-phosphate, a highly hydrophilic lysine reagent that is, therefore, considered to be membrane impermeable. However, the labeling by pyridoxal phosphate cannot readily discriminate the hydrophilic segments, exposed on either side of the lipid bilayer, from those involved in the hydrophilic translocation path. On the basis of the surface labeling by pyridoxal phosphate of the carrier embedded in the membrane of mitochondria or inverted submitochondrial particles, they have proposed three different models that differ from each other by the number of membrane spanning α-helices and the position of the N-terminus within the membrane (Bogner et al., 1982; Klingenberg, 1985; Bogner et al., 1986).

Two considerations led us to re-examine the arrangement of the N-terminal region by new approaches: (i) the ADP-induced alkylation of Cys 56 by N-ethyl-maleimide suggests that the N-terminal part of the ADP/ATP carrier is a highly mobile region during the transport processes; (ii) the N-terminal extremity is rather hydrophilic and should, therefore, be accessible to antibodies on one side or other of the membrane, rather than being embedded in the lipid bilayer.

In the present work, using sequence-directed antibodies and an arginine specific endoprotease, it was shown that the first nine amino acids of the ADP/ATP carrier are exposed on the cytoplasmic side of the mitochondrial membrane and that Arg 30 or Arg 59 is exposed on the matrix face. Our results also indicate that the accessibility of the N-terminal portion of the carrier protein depends on its conformational state.

DEMONSTRATION OF THE CYTOPLASMIC ORIENTATION OF THE NH$_2$-TERMINUS OF THE MEMBRANE-BOUND ADP/ATP CARRIER

In a number of proteins, the NH$_2$ and COOH terminal ends are flexible and exposed to the aqueous medium (Chavez & Sheraga, 1979; Thornton & Sibanda, 1983), as shown especially by region-directed antibodies (Kimura et al., 1982; Seckler et al., 1983; Young et al., 1985; Carrasco et al., 1984; Davies et al., 1987). In the present paper we describe how the arrangement of the N-terminal region of the ADP/ATP carrier polypeptide chain with respect to the plane of the membrane has been explored through the combined use of antipeptide antibodies and an arginine-specific endoprotease.

To test whether the N-terminal segment of the ADP/ATP carrier was exposed to the aqueous medium, polyclonal antibodies were raised in rabbits against an eleven amino acid peptide that corresponded to the N-terminal sequence of the protein (Fig. 1). Phe 11 was replaced by a tyrosine to allow coupling to ovalbumin with the bifunctional reagent bis-diazobenzidine.

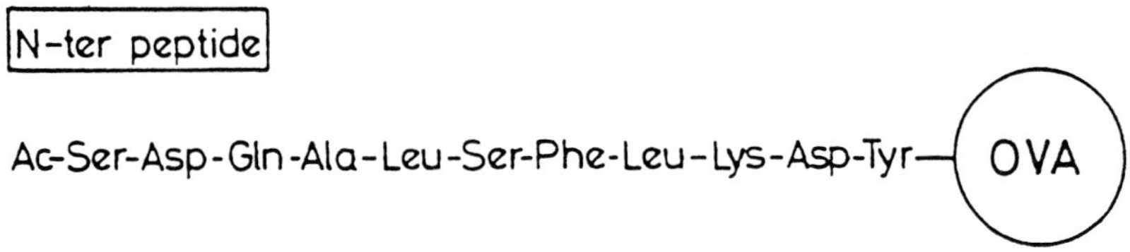

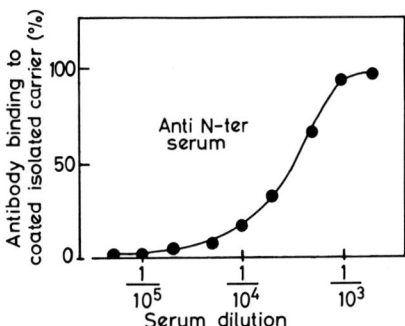

Fig. 1. Structure of the synthetic peptide used to generate antibodies for the immunochemical study of the NH_2-terminal region of the ADP/ATP carrier protein.

The anti N-ter antibodies were characterized (1) by an enzyme-linked immunosorbent assay (ELISA), using the peptide or the carrier protein as coated material (Fig. 2), and (2) by immunoblot analysis. The immunoblot revealed that only the whole carrier and the large fragment Ser 1 - Asp 203 obtained by acidolysis were able to interact with the N-ter antiserum (not shown).

Fig. 2. Immunoreactivity assessed by ELISA of the anti N-ter peptide antibodies against the isolated ADP/ATP carrier protein coated to microtiter plates.

Anti N-ter peptide antibodies reacted specifically with the ADP/ATP carrier in freeze-thawed beef heart mitochondria or in freshly prepared mitoplasts. This was first assessed by immunotitration experiments based on ELISA, performed with freeze-thawed mitochondria directly coated onto microtiter plates. The binding of anti N-ter peptide antibodies to the membrane-bound carrier was related to the amount of coated particles and it was inhibited by competing N-ter peptide (Fig. 3A). No binding to the coated freeze-thawed mitochondria was observed with the preimmune serum. These results demonstrated the specificity of the immunoreaction and the exposure of the N-terminal region of the carrier to the cytosolic face of the inner mitochondrial membrane.

In the face of these results, one could argue that, upon binding of mitoplasts or mitochondria to the microtiter plates, the membrane structure could be disorganized, resulting in access of the antibodies to initially non-exposed epitopes. We therefore performed another series of experiments using back-titration ELISA. We found that treatment of the anti N-ter peptide antiserum with increasing amounts of freshly prepared mitoplasts or freeze-thawed mitochondria reduced the amount of antibodies titrable by ELISA on N-ter peptide-coated microtiter plates (Fig. 3B).

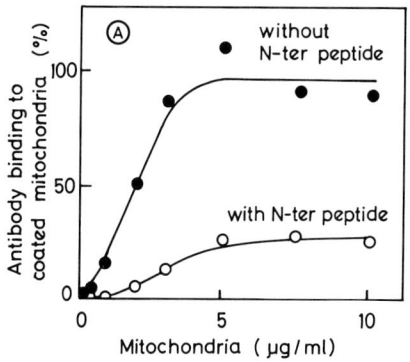 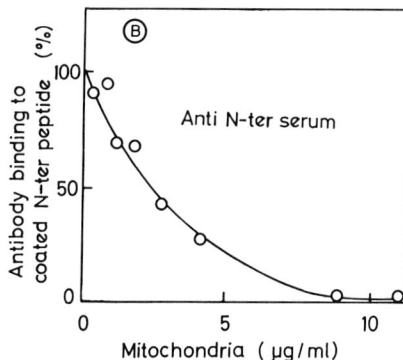

Fig. 3. (**A**) Reactivity of the anti-N-ter peptide antiserum towards the membrane-bound ADP/ATP carrier in freeze-thawed mitochondria, assessed by ELISA. Inhibitory effect of competing N-ter peptide. (**B**) Back titration by ELISA of anti N-ter antibodies after reaction to the membrane-bound ADP/ATP carrier in mitochondria. Antibodies were first allowed to react with a suspension of freeze-thawed mitochondria. After centrifugation the antibodies remaining in the supernatant were titrated against the N-ter peptide coated to microtiter plates.

Thus, in mitoplasts or freeze-thawed mitochondria in suspension the N-ter segment of the ADP/ATP carrier is also exposed to the antibodies added to the medium. Although the outer membrane was not removed from freeze-thawed mitochondria, our results showed that upon freezing this membrane is permeable enough to allow the access of antibodies to the intermembrane space. The possibility that the inner mitochondrial membrane of freeze-thawed mitochondria was permeable to antibodies was ruled out by the finding that the inner membrane is not permeable to small molecules such as 2-azido ADP (Dalbon et al., 1985). It can be therefore safely concluded that the membrane-bound carrier has its N-terminus exposed on the cytosolic face of the bilayer.

The finding that the NH_2-terminal part of the carrier protein is exposed to the cytoplasmic side of the inner mitochondrial membrane conflicts with the different models proposed by Klingenberg and co-workers for the transmembrane arrangement of the carrier (Bogner et al., 1982, 1986; Klingenberg, 1985). In fact, in these models the N-terminal end of the peptide chain of the carrier was located in the hydrophobic core of the membrane, based on the fact that Lys 9 was not modified by pyridoxal phosphate in mitochondria or in inside-out submitochondrial particles.

CLEAVAGE OF THE MEMBRANE-BOUND CARRIER BY AN ARGININE-SPECIFIC ENDOPEPTIDASE

For a further insight into the transmembrane arrangement of the carrier polypeptide chain, an enzymatic digestion of the ADP/ATP carrier in freshly prepared mitoplasts or inverted submitochondrial particles was combined with an immunological approach. The proteolytic fragments were detected by immunoblot analysis, using polyclonal antibodies that we had previously raised against the NaDodSO$_4$-solubilized ADP/ATP carrier (Boulay et al., 1986). The western blot analysis of mitoplasts treated for one hour with the arg-specific endoprotease demonstrated that the carrier protein had no arginyl residue accessible to the Arg-endoprotease from the cytosolic face. This was clearly evidenced by the absence of cleavage products as shown by immunoblot using anti NaDodSO$_4$ carrier antiserum (Fig. 4). In contrast, the carrier protein was cleaved when inverted submitochondrial particles were treated by the Arg-endoprotease. The proteolysis resulted in the accumulation of a large fragment of about 25 kDa, reactive towards the anti SDS-treated carrier antiserum but not towards the anti N-terminal peptide antibodies (Fig. 4). From these results the following conclusions could be drawn: (i) the Arg-endoprotease cleavage site is located close to an extremity of the polypeptide chain since the generated peptide is a large one; (ii) it is located in a region of the carrier which is exposed to the matrix since it is accessible only in inverted particles; (iii) as the large generated proteolytic fragment does not carry the N-terminal region of the carrier, the cleavage site is located close to the N-terminus of the carrier peptide chain. There are two arginine residues in the N-terminal region of the carrier, Arg 30 and Arg 59. Cleavage at either of these residues is compatible with the size of the generated proteolytic fragment. Taken together our results are consistent with a model in which the N-terminal end of the carrier faces the cytosol whereas residues Arg 30 or Arg 59 are exposed to the matrix face of the membrane. From the exclusive cleavage of the carrier in the inverted particles by the Arg-endoprotease we can exclude a model with an antiparallel organization of subunits.

THE ACCESS OF THE N-TERMINAL REGION OF THE MEMBRANE-BOUND ADP/ATP CARRIER TO SPECIFIC ANTIBODIES IS MODULATED BY THE CONFORMATIONAL STATES OF THE CARRIER

Taking advantage of the immunoreactivity of the N-terminal region of the membrane-bound carrier antibodies, we have investigated the binding of anti N-ter peptide antibodies to the carrier either in the CATR conformation or in the BA conformation. As mentioned above, the transition between the CATR and the BA conformation is thought to induce topographical changes of the carrier protein during transport. In the present experiments, the carrier in freeze-thawed mitochondria was trapped in the CATR conformation or in the BA conformation in the presence of

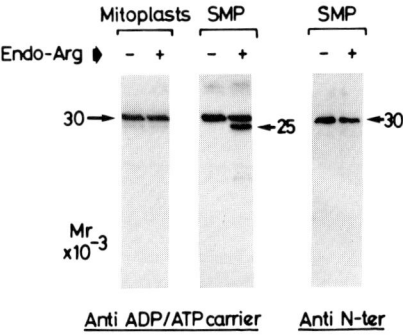

Fig. 4. Reactivity of purified anti N-ter antibodies towards the ADP/ATP carrier and the fragments generated by an Arg-endoprotease digestion, explored by Western blot analysis. When performed on inside-out submitochondrial particles, the proteolytic cleavage generated a large 25,000 M_r fragment which was not recognized by anti N-ter peptide antibodies. The membrane-bound carrier in mitoplasts was insensitive to the Arg-endoprotease digestion.

CATR *plus* ADP or BA *plus* ADP, respectively. Immunotitration, based on back titration by ELISA, indicated a much higher reactivity of the N-terminal region to anti N-ter peptide antibodies when the carrier was in the CATR conformation than when it was in the BA conformation (Fig. 5).

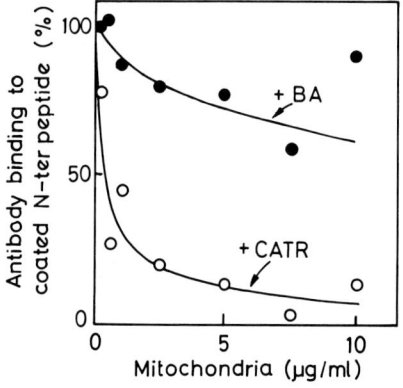

Fig. 5. Reactivities of the CATR and BA conformations of the membrane-bound ADP/ATP carrier in mitochondria towards the anti N-ter peptide antibodies, assessed by back titration of unreacted antibodies by ELISA. The carrier in the CATR conformation binds anti N-ter peptide antibodies to a much higher extent than the carrier in the BA conformation.

The weak reactivity of the N-terminal segment of the carrier in the BA conformation might be due to a restricted conformation or a partial embedment in the lipid bilayer.

154

One cannot, however, exclude interaction of the N-terminal segment with other cytosolic regions of the carrier peptide chain that would restrict or even block the access of antibodies. Thus, although the segment spanning Cys 159 - Met 200, which contains the CATR binding site (Boulay *et al.*, 1983), is quite remote from the N-terminus of the carrier, the conformational change which follows binding of CATR is sensed by the first ten amino acids. This result is not surprising in view of the much higher reactivity of Cys 56 to alkylating reagents when the carrier is in the BA conformation (Boulay *et al.*, 1984). These observations indicate that the N-terminal region of the carrier is mobile and flexible, and furthermore is able to sense the conformational changes undergone by the carrier during the course of the transition between the CATR conformation and the BA conformation. Surprisingly, the conformational change, which most probably involves the Lys 9 residue, was not probed by pyridoxal phosphate (Bogner *et al.*, 1982, 1986).

TRANSMEMBRANE ARRANGEMENT OF THE N-TER REGION OF THE
ADP/ATP CARRIER

The higher reactivity of the Cys 56 in the BA conformation is concomitant with a lower accessibility of the N-terminal region to antibodies from the cytosol. The reverse situation holds for the CATR conformation. One can therefore imagine that the transitions of the carrier between the CATR and the BA conformations involve a slight push-pull motion of the N-terminus segment propagated by a membrane spanning α-helix (Fig. 6).

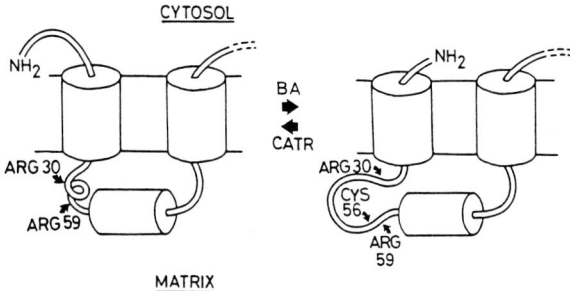

Fig. 6. Possible arrangement of the N-terminal region of the ADP/ATP carrier polypeptide chain. When the carrier adopts the BA conformation, the N-terminal segment is less accessible to antibodies from the cytosol; on the other hand the Cys 56 residue becomes unmasked and can be alkylated by permeant SH reagents. The matrix exposure of Arg 30 and Arg 59 is based on Arg-endoprotease cleavage experiments.

This interpretation would be consistent with previous results from freeze-fracture studies which indicate that the carrier molecules were anchored preferentially at the external surface of the vesicles in the presence of CATR and at the internal surface of liposomes in the presence of BA.

The model proposed in Fig. 6 is based on hydropathy plot reported by Runswick *et al.* (1987) and on our present data. It includes the helices 9-39 (or possibly residues 9-29 if the Arg-endoprotease cleavage site is located at Arg 30-Val 31 bond) and 105-137 which were postulated to cross the membrane. As already mentioned by Saraste and Walker (1982), the segment spanning residues 64 to 90 is rather weakly hydrophobic. The enzymatic and immunological data presented in this paper and the selective reactivity of Cys 56 to permeant alkylating reagents strongly suggest that the amino acid sequence 64-90 protrudes in the matrix space and that the nine N-ter residues are exposed to cytosol.

In the models proposed by Klingenberg and co-workers (Bogner *et al.*, 1982, 1986), the N-terminal region of the carrier is positioned within the membrane, based upon the lack of reactivity of Lys 9 to pyridoxal phosphate. Lys 22 is also located within the membrane, though strongly labeled by pyridoxal phosphate. To accomodate this result with the non permeability to pyridoxal phosphate, Klingenberg and co-workers attributed Lys 22 to the hydrophilic translocation path. They hypothesized that Lys 22 belonged to the binding center of the carrier because it was labeled by pyridoxal phosphate in the ATR-carrier complex but not in the CATR-carrier complex. In line with the "reorienting-gated-pore" model advocated by these authors, pyridoxal phosphate should be able to diffuse to the binding site of the carrier. However, two major results were not taken into account: (i) pyridoxal phosphate is not incorporated into Lys 22 in inverted submitochondrial particles, which is not in agreement with the "reorienting gated pore" model in which a single site is accessible from both sides of the membrane; (ii) the photoactivable derivatives of atractyloside or ADP have never been found to label the N-terminal region of the polypeptide chain (Boulay *et al.*, 1983). Indeed, the same labeling pattern was observed with different photoactivable derivatives of atractyloside whatever of the spacer between the photoactivable group and the atractyloside moiety. Furthermore, the 2-azido ADP, a "no-spacer" derivative of ADP, was found to label two discrete regions extending from Cys 159 to Met 200 and from Tyr 250 to Met 281 (Dalbon *et al.*, 1988).

To reconcile our photolabeling and immunological results with the cytoplasmic orientation of Lys 22, drawn from the pyridoxal phosphate labeling reported by Klingenberg and co-workers, one should consider that the postulated α-helix, spanning residues 9-39 (Runswick *et al.*, 1987) might not cross the membrane as shown in Fig. 6. Possibly, this α-helix, together with the segments labeled by the different azido derivatives, is delineating a large hydrophilic pore involved in the transport processes. Furthermore, since the carrier is at least in a dimeric state, one should also consider that an adjacent subunit might take part in the formation of the hydrophilic path.

156

REFERENCES

Aquila H, Misra D, Eulitz M, Klingenberg M (1982) Complete amino acid sequence of the ADP/ATP carrier from beef heart mitochondria. Hoppe-Seyler's Z Physil Chem 363; 345-349

Bogner W, Aquila H, Klingenberg M (1982) Surface labeling of membrane-bound ADP/ATP carrier by pyridoxal phosphate. FEBS Lett 146:259-261

Bogner W, Aquila H, Klingenberg M (1986) The transmembrane arrangement of ADP/ATP carrier as elucidated by the lysine reagent pyridoxal 5-phosphate. Eur J Biochem 165:611-620

Boulay F, Lauquin GJM, Tsugita A, Vignais PV (1983) Photolabeling approach to the study of the topography of the atractyloside binding site in mitochondrial adenosine 5'-diphosphate/adenosine 5'-triphosphate carrier protein. Biochemistry 22:477-484

Boulay F, Vignais PV (1984) Localization of the N-ethylmaleimide reactive cysteine in the beef heart mitochondria ADP/ATP carrier protein. Biochemistry 23:4807-4812

Boulay F, Lauquin GJM, Vignais PV (1986) Localization of immunoreactive regions in the beef heart adenine nucleotide carrier using rabbit antisera against the carboxy-atractyloside-liganded and the sodium dodecyl sulfate denaturated carrier forms. Biochemistry 25:7567-7571

Brdiczka D, Schumacher D (1976) Iodination of peripheral mitochondrial membrane proteins in correlation to the functional state of the ADP/ATP carrier. Biochem Biophys Res Commun 73:823-832

Buchanan BB, Eiermann W, Riccio P, Aquila H, Klingenberg M (1976) Antibody evidence for different conformational states of ADP/ATP translocator protein isolated from mitochondria. Proc Natl Acad Sci USA 73:2280-2284

Carrasco N, Herzlinger D, Mitchell R, Dechiara S, Danho W, Gabriel TF, Kaback HR (1984) Intramolecular dislocation of the COOH terminus of the lac carrier protein in reconstituted proteoliposomes. Proc Natl Acad Sci USA 81:4672-4676

Chavez LG, Sheraga HA (1979) Location of the antigenic determinants of bovine pancreatic ribonuclease. Biochemistry 18:4386-4395

Dalbon P, Brandolin G, Boulay F, Hoppe J, Vignais PV (1988) Mapping of the nucleotide-binding sites in the ADP/ATP carrier of beef heart mitochondria by photolabeling with 2-azido[α-^{32}P]adenosine diphosphate. Biochemistry 27:5141-5149

Davies A, Meeran K, Cairns MT, Baldwin SA (1987) Peptide-specific antibodies as probes of the orientation of the glucose transporter in the human erythrocyte membrane. J Biol Chem 262:9347-9352

Kimura K, Masson J, Khotzana HG (1982) Immunological probes for bacteriorhodopsin. Identification of three distinct antigenic sites on the cytoplasmic surface. J Biol Chem 257:2859-2867

Klingenberg M (1985) The ADP/ATP carrier in mitochondrial membranes. In: Martonosi AN (ed) The Enzymes of Biological Membranes: Membrane Transport, vol 4. John Wiley, New York, pp 511-553

Ovchinnikov YA (1987) Probing of the folding of membrane proteins. Trends in Biol Sci 12:434-438

Runswick MJ, Powell SJ, Nyren P, Walker JE (1987) Sequence of the bovine mitochondrial phosphate carrier protein: structural relationship to ADP/ATP translocase and the brown fat mitochondria uncoupling protein. EMBO J 6:1367-1373

Saraste M, Walker JE (1982) Internal sequence repeats and the path of polypeptide in mitochondrial ADP/ATP translocase. FEBS Lett 144:250-254

Sekler R, Wright JK, Overath P (1983) Peptide-specific antibody locates the COOH terminus of the lactose carrier of *Escherichia coli* on the cytoplasmic side of the plasma membrane. J Biol Chem 258:10817-10820

Thornton JM, Sibanda BL (1983) Amino and carboxy-terminal regions in globular proteins. J Mol Biol 167:443-460

Torok K, Joshi S (1985) Formation of an intramolecular disulfide bond in the mitochondrial adenine nucleotide translocase. FEBS Lett 182:340-344

Vignais PV, Block MR, Boulay F, Brandolin G, Lauquin GJM (1985) Molecular aspects of structure-function relationships in mitochondrial adenine nucleotide carrier. In: Bengha G (ed) Structure and Properties of Cell Membranes, vol 2. CRC Press, Boca Raton, pp 139-179

Young EF, Ralston E, Blake J, Ramachandran J, Hall ZW, Strend RM (1985) Topological mapping of acetylcholine segments and a cytoplasmic COOII-terminal peptide. Proc Natl Acad Sci USA 82:626-630

The ATP/ADP Antiporter is Involved in the Uncoupling Effect of Fatty Acids

A.YU. ANDREYEV, T.O. BONDAREVA, V.I. DEDUKHOVA, E.N. MOKHOVA,
V.P. SKULACHEV, L.M. TSOFINA, N.I. VOLKOV AND T.V. VYGODINA

A.N. Belozersky Laboratory of Molecular Biology and Bioorganic Chemistry, Moscow
State University, Moscow 119899, U.S.S.R.

The uncoupling of oxidation and phosphorylation by free fatty acids has been
studied since 1956 (Pressman & Lardy, 1956). The first indication that such an effect
may be physiologically significant was obtained later by our group when it was shown
that the fatty acid-mediated uncoupling in skeletal muscle mitochondria is involved in
the burst of heat production by cold-exposed pigeons (Levachev et al., 1965). Later on
it was reported that thermogenin, the protein responsible for the thermoregulatory uncoupling in brown fat mitochondria, is activated by fatty acids (for review see Nicholls
& Locke, 1984). Since thermogenin is absent in muscle mitochondria, the following two
possibilities should be considered: (i) uncoupling by fatty acids in muscle occurs with no
protein involved, or (ii) it is catalyzed by a protein other than thermogenin. The latter
option seems more likely, since fatty acids have been shown to increase the conductance of model BLM only slightly or have no effect at all (Antonov et al., 1973;
Gutknecht, 1987).

Abbreviations: BLM, bilayer phospholipid membrane; BSA, bovine serum albumin;
CAtr, carboxyatractylate; DNP, 2,4-dinitrophenol; EGTA, ethylene glycol bis(2-
aminoethyl ether)-N,N,N',N'-tetraacetate; FCCP, p-trifluoromethoxycarbonyl cyanide
phenylhydrazone; MOPS, morpholinopropane sulphonate; Ph_4P^+, tetraphenyl phos-
phonium; Palm, palmitic acid; Hepes, 4-(2-hydroxyethyl)-1-piperazineethane sulphonic
acid.

A. Azzi et al. (Eds.)
Anion Carriers of Mitochondrial Membranes
© Springer-Verlag Berlin Heidelberg 1989

As it was reported by Azzi (1988), palmitate did enhance H^+ conductance of cytochrome oxidase proteoliposomes. This suggests that the fatty acid-mediated uncoupling is not a common feature of any protein-containing membrane; rather, a specific protein should be involved. Since BLM is permeable for the protonated (neutral) form of a fatty acid (RCOOH) (Walter & Gutknecht, 1984), one may expect that this protein facilitates the transfer of the anionic species of fatty acid (RCOO$^-$).

Considering a possible involvement of various anion carriers in the RCOO$^-$ transfer, we focused our attention on the mitochondrial adenine nucleotide antiporter exchanging ATP^{4-} for ADP^{3-}. Wojtczak and Zaluska (1967) observed the oleate-induced inhibition of this exchange. Fatty acyl CoA was also shown to inhibit the antiporter (for review see Vignais *et al.*, 1985).

Recently Klingenberg and co-workers (Aquila *et al.*, 1985), analyzing the amino acid sequence and domain composition of thermogenin, concluded that this protein appears to be the nearest relative to the ATP/ADP antiporter. In the present work we have summarized the results of experiments indicating that ATP/ADP antiporter is involved in the fatty acid uncoupling in muscle and liver mitochondria (for preliminary communication see Andreyev *et al.*, 1988).

EXPERIMENTAL EVIDENCE THAT ATP/ADP ANTIPORTER IS INVOLVED
IN THE FATTY ACID UNCOUPLING IN MUSCLE AND LIVER
MITOCHONDRIA

In experiments on rat skeletal muscle mitochondria, stimulation of respiration in the presence of oligomycin by a low concentration of palmitate is partially abolished by ADP and more strongly by CAtr. Subsequent addition of 4×10^{-5} M DNP stimulated the respiration rate to a level higher than before the addition of ADP. The recoupling effects of ADP and CAtr were accompanied by a reversal of the palmitate-induced decrease of the membrane potential. Again, 4×10^{-5} M DNP abolished the ADP and CAtr action (Fig. 1).

Similar relationships were also revealed in rat liver mitochondria (Fig. 2). Here CAtr also proved to be inhibitory for the palmitate-uncoupled respiration (traces 1 and 5). Low concentrations of DNP, gramicidin D or FCCP were added to obtain an effect similar to that of palmitate, *i.e.* 2-2.5-fold stimulation of controlled respiration. It is seen that CAtr caused partial inhibition of the DNP-stimulated respiration (trace 2) but had no effect on that stimulated by gramicidin D and FCCP (traces 3 and 4).

The ADP effect proved to be specific for this nucleotide. GDP failed to substitute for ADP. The ADP concentration, needed for exerting a 50% inhibitory effect, was found to be as low as 2×10^{-6} M, *i.e.* close to K_m of the ATP/ADP antiporter. Mg^{2+} decreased the inhibitory effect of ADP.

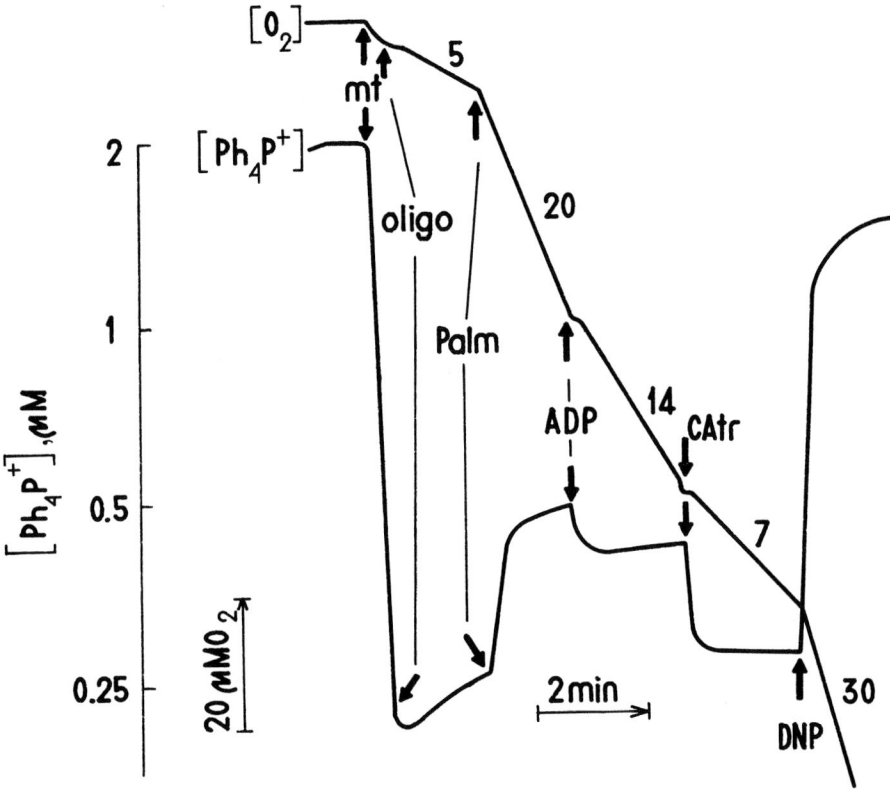

Fig. 1. Effect of palmitate, ADP and CAtr on membrane potential ($[Ph_4P^+]$) and respiration ($[O_2]$) of skeletal muscle mitochondria. Incubation mixture: 250 mM sucrose, 5 mM MOPS, 2 mM KH_2PO_4, 0.5 mM EGTA, 4 mM glutamate, 1 mM malate, BSA 0.2 mg/ml, 2×10^{-6} M Ph_4P^+ bromide, pH 7.2. Additions: mitochondria (1.1 mg protein/ml) oligomycin (1×10^{-6} g/ml), 1.5×10^{-5} M palmitate, 0.2 mM ADP, 1×10^{-6} M CAtr, 4×10^{-5} M DNP. Temperature, 22°C. Mitochondria were isolated from 150-250 g white rats. The isolation medium contained 250 mM sucrose, 50 mM Tris-HCl and 5 mM EDTA, pH 7.5. The muscle tissue was homogenized in a glass homogenizer for 4 min. After the first centrifugation (800 g, 10 min), the supernatant was decanted and filtered. After sedimentation of mitochondria (12,000 g, 10 min), the pellet was washed by the isolation medium supplemented with BSA (3 mg/ml). The final mitochondrial precipitate (12,000 g, 10 min) was suspended in the isolation medium with BSA.

CAtr-induced inhibition is stronger at lower concentration of the uncoupler (Fig. 3A). CAtr, in fact, abolishes the stimulation of respiration caused by 1×10^{-5} M palmitate and partially suppressed that induced by 2×10^{-5} M palmitate or by 3×10^{-6} DNP. In all cases, maximal effect was observed at 5×10^{-7} M CAtr. The same concentration of CAtr was observed to be required for complete inhibition of State 3 respiration (Fig. 3B). A further increase of the CAtr level failed to cause any additional inhibitory effect in samples with 2×10^{-5} M palmitate of 3×10^{-6} M DNP. FCCP-stimulated respiration was only slightly affected by any of the CAtr concentration used (see the respiration in the absence of uncouplers).

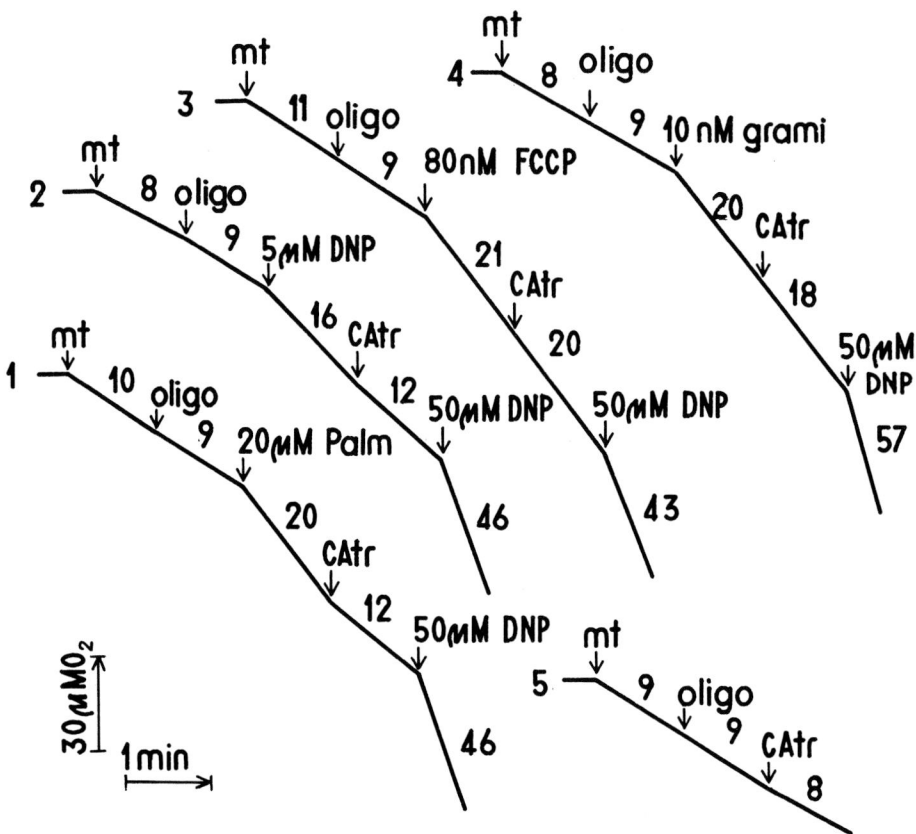

Fig. 2. Effect of ADP and CAtr on the respiration of liver mitochondria stimulated by low concentrations of various uncouplers. Incubation mixture: 250 mM sucrose, 2 mM KH$_2$PO$_4$, 5 mM MOPS, pH 7.4, BSA 0.2 mg/ml, 0.5 mM EGTA, 1 mM MgCl$_2$, 4 mM glutamate and 1 mM malate. Additions: rat liver mitochondria (2 mg protein/ml), oligomycin (2x10^{-6} g/ml), 1x10^{-6} M CAtr. Temperature, 37°C. Liver mitochondria were isolated essentially as those of muscle, but the tissue was homogenized for 30 s. The isolation medium contained 250 mM sucrose, 5 mM MOPS and 2 mM EDTA, pH 7.4.

Fig. 4 shows the effects of other ATP/ADP antiporter inhibitors, atractylate and bongkrekic acid. These compounds completely abolished State 3 respiration and partially the respiration stimulated by 2x10^{-5} M palmitate. The degrees of inhibition of the palmitate-stimulated respiration by atractylate and bongkrekic acid were lower than that by CAtr. Again, maximal inhibition was obtained at the concentration required for switching off State 3 respiration.

Like atractylate, palmitoyl-CoA induces small but reproducible inhibition of the palmitate-stimulated respiration at a concentration completely arresting State 3 respiration.

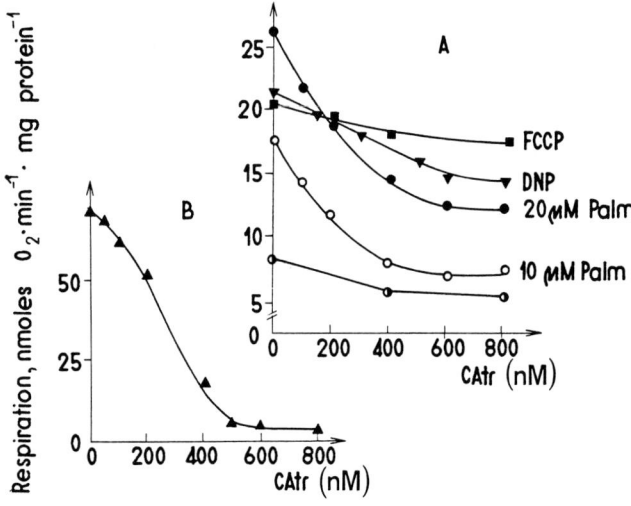

Fig. 3. Concentration dependence of the CAtr-induced inhibition of the uncoupled (**A**) and State 3 (**B**) respiration in skeletal muscle mitochondria. The incubation mixture (Fig. 1) was supplemented with 5 mM MgCl$_2$ (**A,B**), oligomycin (2×10^{-6} g/ml) (**A**) or 10 mM glucose, hexokinase (2 units/ml) and 0.1 mM ATP (**B**). Glutamate and malate concentrations were 5 mM each. Uncouplers: 4×10^{-8} M FCCP, 3×10^{-6} M DNP and 10 or 20 μM palmitate. Temperature: 30°C.

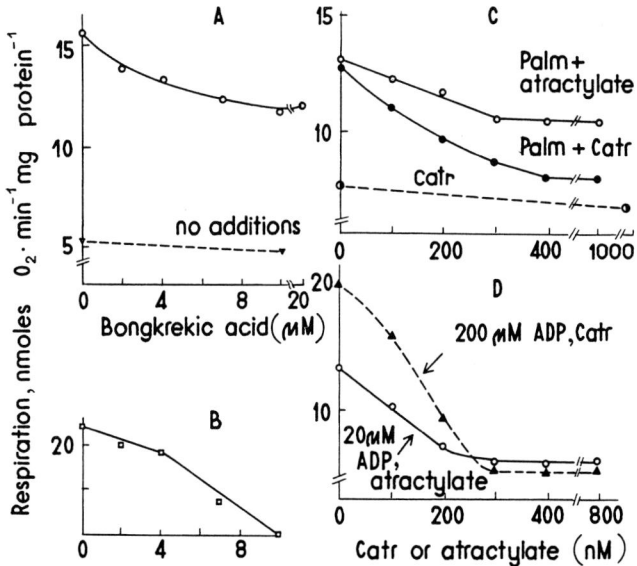

Fig. 4. Inhibition of the palmitate-stimulated (**A,C**) and State 3 (**B,D**) respiration of skeletal muscle (**A,B**) and liver (**C,D**) mitochondria by bongkrekic acid, atractylate and

CAtr. The incubation mixtures in **A** and **B** were as in Fig. 1, but pH was 7.0; the oxygen consumption rates are given for the 3rd minute after addition of bongkrekic acid. Incubation mixtures in **C** and **D** were as in Fig. 2 (2.9 mg protein/ml). Addition, 20 μM palmitate; temperature: 22°C (**A,B**) or 37°C (**C,D**).

The uncoupling effect of palmitate, as described above, was shown to appear immediately after palmitate addition. It should be therefore distinguished from another effect of palmitate which (i) had a lag, (ii) was stimulated by micromolar Ca^{2+}, and (iii) was prevented by EDTA, EGTA, the phospholipase A_2 inhibitor nupercain or the lipid peroxidation inhibitor ionol. K^+ ions proved to be favourable for this uncoupling. It was shown that ADP slowed down, whereas CAtr and atractylate accelerated, the uncoupling process and abolished the ADP effect. A similar pattern was also observed in the previous works (Panov *et al.*, 1980; Dragunova *et al.*, 1981).

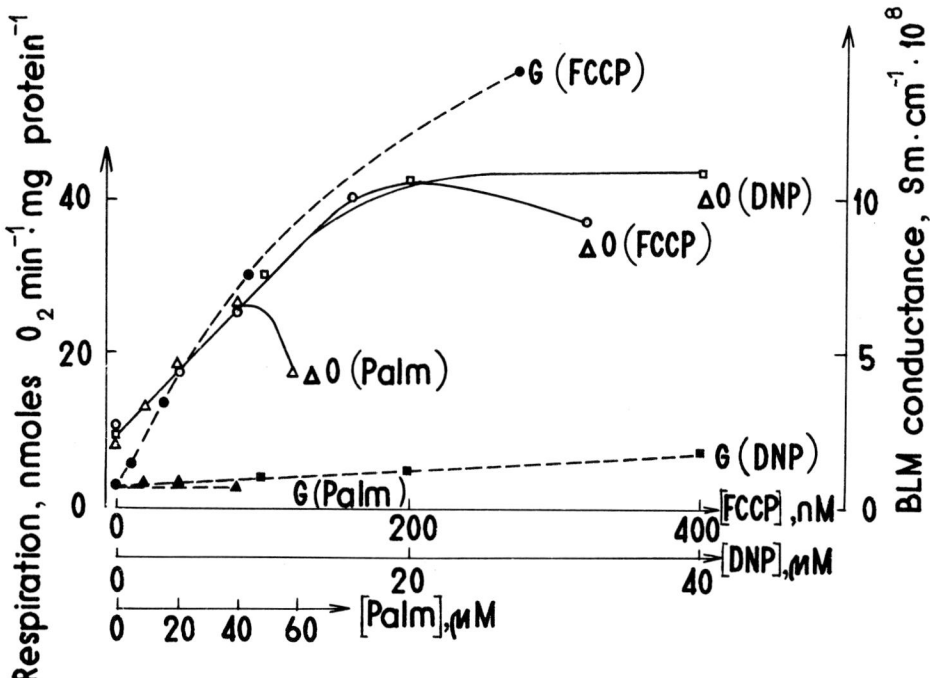

Fig. 5. Comparison of the effects of uncouplers on BLM and mitochondria. BLM was formed in a 0.8 mm aperture in the Teflon septum partitioning the chamber into two compartments. Two Ag/AgCl electrodes were put (*via* agar bridges) into solutions (0.1 M KCl, 30 mM Tris-HCl, pH 7.4) on both sides of the septum. BLM conductance was measured with a VA-J-51 electrometer. In experiments with mitochondria, the incubation mixture was as in Fig. 2; liver mitochondria, 1.7 mg protein/ml; temperature 37°C. O, respiration rate; G, BLM conductance.

Analysing the possible reasons for the above difference in ours and Panov's and Yaguzhinsky's (Dragunova *et al.*, 1981) results, we found that contaminations with Ca^{2+} ions were critical; addition of 5×10^{-6} M Ca^{2+} was sufficient to reproduce them. It is not excluded that this indirect effect of palmitate may be involved in the CAtr-resistant uncoupling observed with the increase of palmitate concentration. Involvement of anion carriers other than the ATP/ADP antiporter seems to be yet another possibility.

EFFECT OF PALMITATE ON MODEL MEMBRANES AND CYTOCHROME OXIDASE PROTEOLIPOSOMES

In Fig. 5 the effects of palmitate and of two artificial uncouplers on mitochondria and BLM are compared. One can see that the stimulation of controlled respiration by FCCP parallels the increase in the BLM conductance by this protonophore. Conversely, palmitate concentrations that stimulate respiration failed to induce a measurable increase in BLM conductance. An intermediate situation was revealed with DNP. Here we observed some increase in the conductance, but this effect was much smaller than that with FCCP.

Fig. 6 shows the effect of low concentrations of uncouplers on the membrane potential in cytochrome oxidase proteoliposomes oxidizing ascorbate *via* hexaminoruthenium.

A dramatic difference was revealed between the effects of palmitate, on one hand, and gramicidin, FCCP and DNP, on the other. Palmitate increased, whereas the three other uncouplers decreased, the potential (Ph_4P^+ uptake).

CONCLUSIONS

To uncouple, a protonophore must cross the hydrophobic barrier of the membrane twice: in its protonated and deprotonated forms. The inefficiency of palmitate as a protonophore in BLM may be due to the impermeability of BLM to the palmitate anion ($RCOO^-$). The membrane potential increase in cytochrome oxidase proteoliposomes, observed in this work (Fig. 6), can easily be explained by the ΔpH $\longrightarrow \Delta\psi$ conversion due to the ΔpH-dependent distribution of the protonated form of palmitate as a penetrating weak acid. In agreement with such reasoning, it was found that the Na^+/H^+ antiporter monensin, also converting ΔpH to $\Delta\psi$, can effectively substitute for palmitate in the same system.

Thus, the RCOOH form of fatty acid does not require any protein to traverse membrane. This means that it is the movement of $RCOO^-$ rather than that of RCOOH that is the rate-limiting step in the circulation of fatty acids across the biological membranes. Within the framework of such a concept, one may understand how the ATP/ADP antiporter facilitates the uncoupling effect of fatty acids. Most probably, this protein, exchanging anions of ATP^{4-}(in) for anions of ADP^{3-}(out), also increases in some way the membrane permeability for the anion $RCOO^-$ (Fig. 7).

One may assume that the ATP/ADP antiporter is composed of (i) a very specific gate discriminating *e.g.* ADP and GDP, and (ii) a rather unspecific anion translocator ("rotor", channel or relay of fixed cationic groups crossing the hydrophobic barrier of the membrane) (Fig. 7B). Fatty acid anions RCOO⁻ might reach, say, cationic groups of the translocator with no gate involved, or, alternatively, there is another gate specific for RCOO⁻. The former possibility seems more likely since not only palmitate but also small (μmolar) concentrations of DNP were shown to stimulate respiration in a CAtr-sensitive manner.

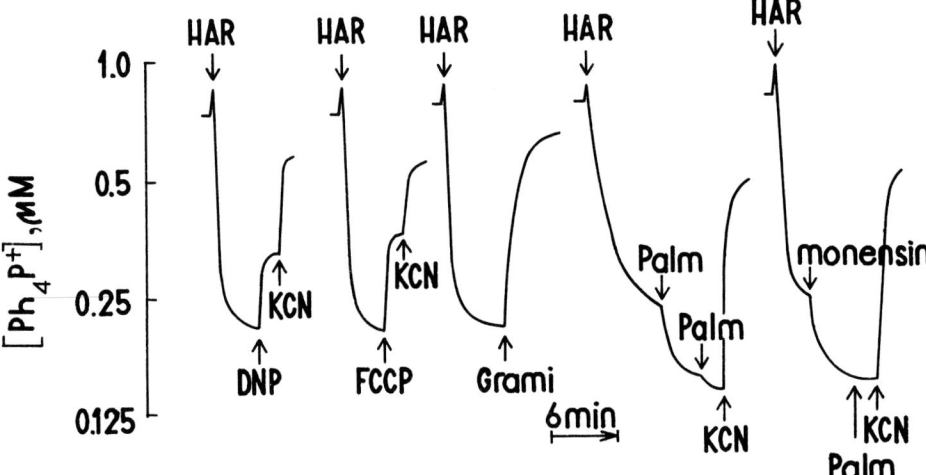

Fig. 6. Effect of uncouplers and palmitate on the membrane potential generation by cytochrome oxidase proteoliposomes. The incubation mixture (5 mM Hepes-KOH, pH 7.5, 5 mM ascorbate) was supplemented with proteoliposomes (6 mg protein/ml). Additions: 3×10^{-5} M Ru(NH$_3$)$_6$Cl$_3$ (HAR), 2 mM KCN, 5×10^{-5} M DNP, 5×10^{-8} FCCP, 3×10^{-8} gramicidin, 3×10^{-5} M palmitate, 2.5×10^{-8} M monensin; temperature 20°C. Phospholipids were isolated from beef heart mitochondria using extraction with chloroform-methanol (2:1) and subsequent washing with water and 0.1 M KCl. When isolated, phospholipids were stored under argon. Proteoliposomes were reconstituted by the dialysis method (Racker, 1972).

One may speculate that it is DNP⁻ that is transported by the ATP/ADP antiporter whereas DNPH, like RCOOH, crosses the phospholipid bilayer without the assistance of any protein. Since FCCP effectively decreases the resistance of BLM, we may conclude that FCCP can cross the phospholipid bilayer not only in the protonated but also in the anionic form. Therefore, no assistance of the ATP/ADP antiporter seems to be necessary. Similarly, the antiporter is not involved in the action of gramicidin operating as a channel permeable for H⁺, Na⁺ and K⁺.

The above suggestion that the ATP/ADP antiporter serves as a RCOO⁻ carrier may be extended to thermogenin, the brown fat uncoupling protein which proved to be the nearest relative of the ATP/ADP antiporter as to the amino acid sequence and domain composition. In fact, uncoupling of thermogenin, stimulated by low concentrations of fatty acids and inhibited by ADP, resembles in this respect the ATP/ADP antiporter. One may speculate that thermogenin is in fact an ATP/ADP antiporter derivative specializing in the translocation of fatty acid anions and not capable to translocate adenine nucleotides.

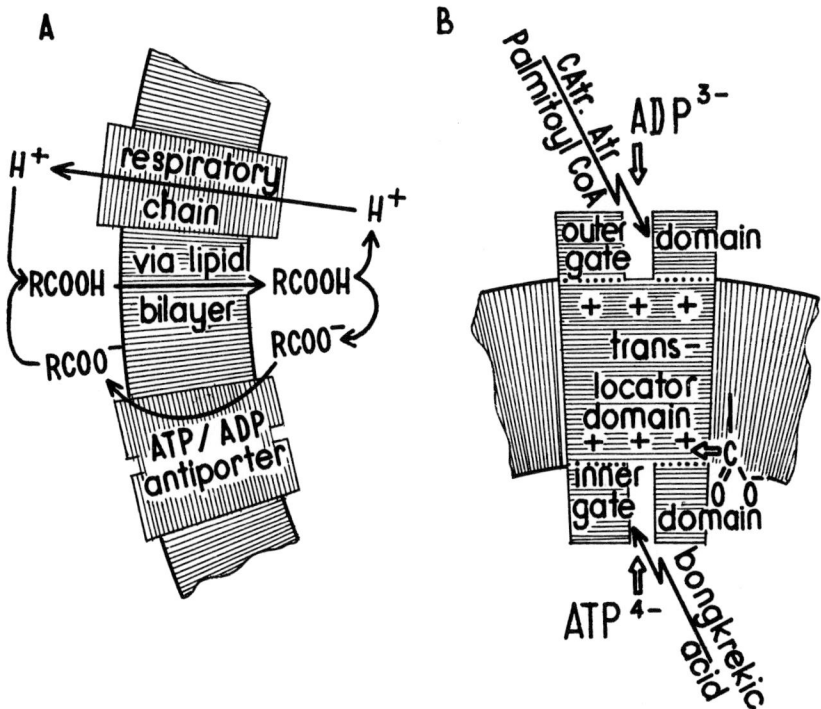

Fig. 7. A scheme illustrating how the ATP/ADP antiporter may facilitate the transmembrane flux of the ionized form of fatty acid (RCOO$^-$). **A**, Flow diagram; **B**, a tentative scheme of the ATP/ADP antiporter. It is suggested that the antiporter is composed of three functionally different parts: the outer and inner gates and the anion-translocating domain. Hydrophilic ADP^{3-} and ATP^{4-} preferentially reach the cationic groups of the translocator *via* outer and inner gates, respectively. As to the hydrophobic fatty acid anion, it attacks the translocator cations with no gates involved. The antiporter inhibitors (CAtr, Atr, palmitoyl-CoA or bongkrekic acid) combine with the outer or inner gate domains. Attachment of the inhibitors completely blocks the gate and suppresses, indirectly and therefore at different degrees, the operation of the translocator. At a low palmitate concentration, the latter effect results in almost complete (CAtr) or partial (bongkrekic acid, palmitoyl-CoA and Atr) reversal of the palmitate-induced uncoupling.

If it were the case, one should consider thermogenin not as a H$^+$ but rather as a RCOO$^-$ carrier (for review see Nicholls & Locke, 1984)

Brown fat possesses a special uncoupling protein. As to other tissues, we suggest that here the role of an uncoupling protein is performed by the ATP/ADP carrier the main function of which is the transport of adenine nucleotide anions.

REFERENCES

Andreyev AYu, Bondareva TO, Dedukhova VI, Mokhova EN, Skulachev VP, Volkov NI (1988) Carboxyatractylate inhibits the uncoupling effect of free fatty acids. FEBS Lett 226:265-269

Antonov VF, Vladimirov YuA, Rossels AN, Korkina LG, Korepanova EA, Trukhmanova KI (1973) Influence of unsaturated fatty acids peroxidation products on ion transport through bimolecular lipid membranes (in Russian). Biofizika 18:668-673

Aquila H, Lin TA, Klingenberg M (1985) The uncoupling protein from brown fat mitochondria is related to the mitochondrial ADP, ATP carrier. Analysis of sequence homologies and of folding the protein in the membrane. EMBO J 4:2369-2376

Dragunova SF, Novgorodov SA, Sharyshev AA, Yaguzhinsky LS (1981) Regulation of ion transport processes and ATP synthesis in mitochondria by nucleotides (in Russian, with English summary). Biokhimiya 46:1242-1247

Gutknecht J (1987) Proton/hydroxide conductance through phospholipid bilayer membranes: effects of phytanic acid. Biochim Biophys Acta 898:97-108

Levachev MM, Mishukova EA, Sivkova VG, Skulachev VP (1965) Energetics of pigeon at self-warming after hypothermia (in Russian, with English summary). Biokhimiya 30:864-874

Müller M, Labonia N, Schläpfer B and Azzi A (1987) Cytochrome c Oxidase: Past, Present and Future. In: Papa S , Chance B and Ernster L (eds) Cytochrome System: Molecular Biology and Bioenergetics. Plenum Press, Bari, pp 239-246

Nicholls DG, Locke RM (1984) Thermogenic mechanisms in brown fat. Physiol Rev 64:1-64

Panov A, Filippova S, Lyakhovich V (1980) Adenine nucleotide translocase as a site of regulation by ADP of the rat liver mitochondria permeability to H^+ and K^+ ions. Arch Biochem Biophys 199:420-426

Pressman BC, Lardy HA (1956) Effect of surface active agents on the latent ATPase of mitochondria. Biochim Biophys Acta 21:458-466

Racker E (1972) Reconstitution of cytochrome oxidase vesicles and conferral of sensitivity to energy inhibitors. J Embr Biol 10:221-235

Vignais PV, Block MR, Boulay F, Brandolin G, Lauquin GJ-M (1985) Molecular aspects of structure-function relationships in mitochondrial adenine nucleotide carrier. In: Benga G (ed) Structure and Properties of Cell Membranes, vol 2. CRC Press, Boca Raton, pp 139-179

Walter A, Gutknecht J (1984) Monocarboxylic acid permeation through lipid bilayer membranes. J Membrane Biol 77:255-264

Wojtczak L, Zaluska H (1967) The inhibition of translocation of adenine nucleotides through mitochondrial membranes by oleate. Biochem Biophys Res Commun 28:76-81

Molecular Aspects of the Adenine Nucleotide Carrier from Mitochondria

M. KLINGENBERG

Institut für Physikalische Biochemie, Universität München, Goethestrasse 33, D-8000 München 2, FRG

The primary structure of the ADP/ATP carrier (AAC), first established for the bovine heart carrier, showed a relatively wide distribution of hydrophilic residues which permits assignment of only two hydrophobic transmembrane stretches. However, a striking tripartition of the primary structure into three 100-residue long domains allows a more significant assignment of transmembrane elements. With proper alignment of these three domains for maximum conservation of structurally critical residues, each domain can be assigned to have two transmembrane α-elements which are in each case 18 and 22 residues long, respectively. The interdomain homology between these α-regions is low. The central regions flanked by these helices contain most of the polar residues and are significantly interdomain conserved. With lysine probes the central regions are assigned to the matrix side (m-side) and the two connecting regions as well as C- and N-terminal to the cytosolic side (c-side). Out of the central regions a loop is assumed to protrude through the membrane, probably for lining the translocation channel. This localization of a major protein mass within the membrane is in accordance with hydrodynamic evidence, the carrier being an oblate ellipsoid with only about 50 Å along the short axis. In accordance, the loops of domain 2 and 3 are affinity-labelled by azido-ADP or aryl-atractylate. Primary structures of AAC from other sources (fungi, plants) also exhibit the tripartition. The interdomain conserved residues are also inter-

---

Abbreviation: AAC, ADP/ATP carrier

A. Azzi et al. (Eds.)
Anion Carriers of Mitochondrial Membranes
© Springer-Verlag Berlin Heidelberg 1989

species conserved, thus showing that they are essential. These repeat domains have probably evolved from a common original gene of 100 residues length. Isoforms of the AAC exist, as shown by primary structure analysis of human cDNA libraries from different organs. Three different isoforms are identified in human organs. These isoforms may have developed in accordance with specific requirements of the various tissue metabolism on the ADP/ATP exchange.

This brief review will treat primarily the present knowledge of the structure of the ADP/ATP carrier (AAC), with the structure-function relationship in mind. Information on the structure of membrane proteins is mostly to a large degree tentative, since atomic resolution data are not available, except in one or two cases. The AAC seemed to be early an "ideal" target for this approach because of its good availability and its molecular simplicity, but efforts in several laboratories have failed so far to obtain any crystals.

PRIMARY STRUCTURE

The primary structure of the AAC from beef heart was the first one to become known for any membrane carrier (Aquila *et al.*, 1982). It was obtained by amino acid sequencing and the structure has recently been confirmed by cDNA analysis (Walker *et al.*, 1987).

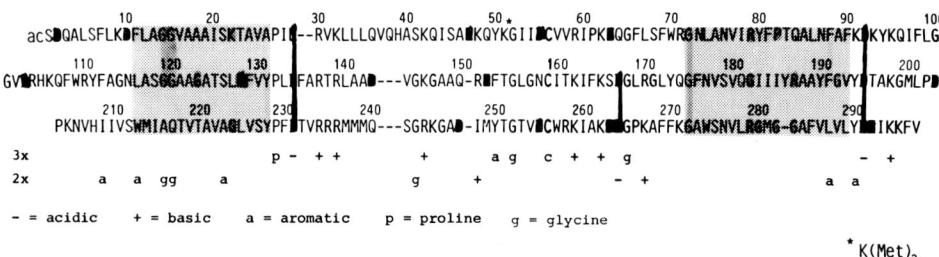

Fig. 1. The primary structure of the ADP/ATP carrier from bovine heart. The amino acid sequence is arranged according to a triplication of about 100 residue long domains and, by allowing for deletions and insertions, to obtain maximum interdomain conservation of critical residues.

In Fig. 1 the primary structure is presented in three sections, for reasons shown below. The first striking impression is the widespread distribution of polar residues. The chain contains 40 basic and about 21 acidic amino acids, resulting in an excess of 19 basic charges which render this protein highly basic. The many polar residues were highly surprising in view of the prevailing concept that the integral membrane protein must contain largely hydrophobic sections. This was actually later shown to be the case in the *E. coli* lactose permease (Büchel *et al.*, 1980). The widespread polarity in the AAC collided with the generally held view, primarily based on bacteriorhodopsin, that a membrane protein should be highly hydrophobic.

The folding of the primary structure of the AAC, based on the usual assignment of transmembrane α-helices, also met with difficulties. Only the stretches from residue 111 to 133 and from 209 to 230 contain at least 20 residues free of polar residues. At the time when the AAC sequence from beef heart was established, hydrophobicity averaging was introduced which could tolerate single polar residues (Kyte & Doolittle, 1982). With this admittedly crude method, several weakly hydrophobic stretches of 20 or more residues could be discerned (Aquila *et al.*, 1985; Saraste & Walker, 1982). However, the hydropathy significance level exceeds only in two locations the level known for bacteriorhodopsin and some other membrane proteins. On the basis of this plot one might conclude that there exist about five membrane-spanning α-helices (Saraste & Walker, 1982). Another procedure searching for amphipathic helices allowed the definition of 4 additional, about 20 residue long transmembrane α-helices (Aquila *et al.*, 1985). As we shall see below, there is another, more compelling procedure of assigning α-helical stretches to the AAC.

The problem of protein folding in the membrane is intimately linked to the secondary structure since transmembrane segments are generally considered to consist of α-helices. The various secondary structure predictions for soluble proteins fail to agree on the position of α- and β-structures in the AAC (unpublished data). Obviously the cooperative hydrophobic forces important for the intermembrane localizations are more dominant. A hybrid procedure according to the empirical occurrence of residues within membrane-localized α-structures gives a similar but less distinct pattern of potential α-structures than the mere hydropathy averages (Rao & Argos, 1986). The actual location, *i.e.* start and end of the transmembrane section cannot be assigned with any precision from the too wide hydrophobic segments.

These predictions have to be in accordance with the amount of α-structures which actually was determined for the AAC in appropriate detergents by circular dichroism. A content of 41.5% α-structure was measured for the intact solubilized carboxyatractylate-AAC complex (unpublished results). This content is considerably lower than that of bacteriorhodopsin of about 75% (Urry, 1972). It is similar to the α-content, for example, of some soluble nucleotide binding enzymes as determined by their 3-D structure. Much more powerful evidence for the folding arrangement of the AAC is based on the finding that the protein appears to be internally triplicated, as will be elucidated in the following.

TRIPLICATE STRUCTURE

A striking feature of the AAC structure is the apparent triplication within the primary structure. As first seen in internal homology plots, there are significant stretches of homologous residues displayed by about 100 or 200 residues, respectively, from the diagonal (Walker *et al.* 1987; Saraste & Walker, 1982; Aquila *et al.* 1987). This signifies that similarities exist between the three domains of about 100 residues each within the AAC. With this indication, the sequence of the AAC can be divided up into three domains and aligned for maximizing the similarities (Fig. 1). The principle of this alignment is to search for a maximum of specific residues (acidic, basic, aromatic, glycine, proline) conserved through the three segments. For this reason, some insertions or deletions in the segments have to be taken into account. The result as shown in Fig. 1 is not perfect, since there are obviously more partial (twofold) similarities which could be aligned by more deletions. In the present alignment mainly the threefold conservations are stressed.

From this triplicated comparison analysis, a more convincing assignment for the localization of putative transmembrane α-structures can be made than by the hydrophobicity averaging plot. The comparative alignment actually helps to distinguish the more hydrophobic from the hydrophilic or polar region. There are stretches in the first section, from 10 to 27, 116 to 132, and 212 to 229, respectively, as well as near the end of each repeat, between 72 and 90, 175 and 194, 272 and 290, which are candidates for membrane-spanning α-helices. The threefold repeated occurrence strongly magnifies the significance of these regions.

A first glance at the distribution of conserved residues shows that they are clustered outside the putative hydrophobic α-ranges. Obviously for the α-helices the conservation of residues is not as stringent because of their mainly structural and less functional nature. The three, about equally long central hydrophilic regions appear to be constructed in a relatively conserved manner, as judged from the frequent repeats of the polar, aromatic and glycine residues. This alignment facilitates the discernment of residues conserved throughout the three domains. In two positions an acidic group and in six positions a basic group are retained throughout the three domains. Also important for structural reasons is the threefold repeat of α-structure breakers such as proline (positions 27, 132, 229) which obviously limits the first α-stretch, and the threefold conservation in three positions of glycines. Aromatic residues are threefold retained in positions 50 and 70. There are dual retentions in at least 10 additional positions. Also in this region one finds cysteine conserved, the blockage of which is known to impair the transport by the AAC (Leblanc & Clauser, 1972).

The few polar residues present within the putative membrane-spanning regions have exclusively positive charge. In the first α-regions only lysine 22 in one of the three domains stands out which has been shown to be the only lysine involved in the binding (Bogner *et al.*). In all the three second α-regions there is one positive residue, represented only by arginine. This agrees with the fact that arginine has a higher chance to exist within a hydrophobic region than lysine because of its greater polarizability. Lysine at position 22 seems to be an exception to this rule, but this can be rationalized in view of its unique accessibility and participation in the hydrophilic binding center, as shown below. Acid amino acids seem to be exclusively located outside the transmembrane α-helical regions. Threefold conserved acid residues are conspicuously located at the end of both transmembrane α-regions. Probably within ionic pairs, they seem to maintain the transmembrane segment "anchored" to the extramembranous domains.

Last not least, the primary structure and folding of the AAC also relates to the biosynthetic pathway, the import and sorting of the AAC from the cytosol into the inner mitochondrial membrane. Considerable information, particularly for the AAC from Neurospora, has accumulated (Pfanner & Neupert, 1987). Recently the occurrence of import-competent sequences within the last two domains of the AAC have been shown (Pfanner *et al.* 1987). Probably these sequences are present in each of the three domains.

PROBING RESIDUES

The most extensive data concerning the folding arrangement of the AAC come from probing the lysine distribution with a membrane-impermeant reagent such as pyridoxalphosphate (Bogner *et al.* 1986). The AAC structure affords a particular opportunity for this approach because of the wide distribution of 23 lysines throughout the AAC sequence. Applying pyridoxalphosphate to mitochondria should primarily label cytosolic (c-site) localized lysines, whereas in submitochondrial particles it will react with lysine on both the m- and c-side. Furthermore, the analysis of the accessibility can be refined by comparing the additional lysines labelled in the isolated protein with those in the membrane. These lysines are probably localized in the "collar" region of the protein, which in the membrane is covered by the phospholipid head groups. Also the difference in lysine reactivity in the two conformations of the AAC which can be induced by bongkrekate (m-state) and by carboxyatractylate (c-state) has been determined. Lysines which are particularly sensitive to these influences have been assigned to the translocation path. The results show that the lysines of the central hydrophilic segment in the first two domains are directed towards the m-side. Results on the third domain are weak and would point more to the c-side. However, for the overpowering reasons of symmetry and analogy we would conclude that also this polar region is at the m-side. In the connecting regions the lysines appear to be more accessible from the c-side. Therefore these two regions are concluded to be placed at the c-side.

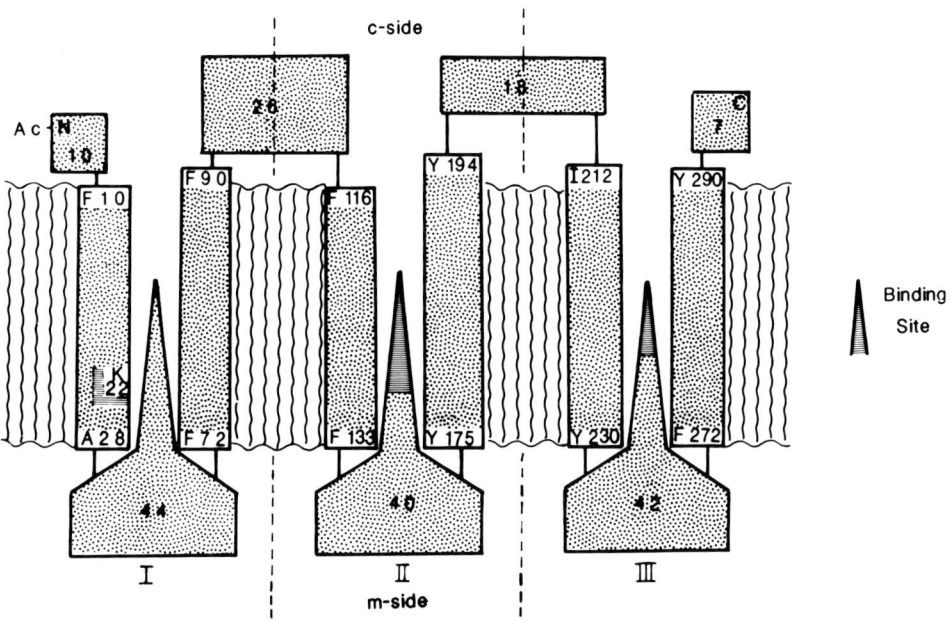

Fig. 2. A proposed secondary structure and folding arrangement of the AAC from bovine heart. The numbers refer to the length of the section or to the sequence location of the transmembrane α-helices. The triplicate structure is indicated by the numbering of the three repeat domains of Fig. 2 with I, II, III.

A folding of the AAC, as constructed within the framework so far discussed, is shown in Fig. 2, with (a) each domain containing two transmembrane α-helices, (b) the localization of the central hydrophilic region on the matrix side and (c) the localization of the connecting hydrophilic regions on the cytosolic side. This construct agrees with other evidence on the localization of the N- and C-termini. The uncoupling protein from brown adipose tissue mitochondria is homologous to the AAC with respect to the tripartite structure (5). Here the hydrophilic C-terminus stretch is longer by 9 amino acids than in the AAC. By proteolytic and crosslinking experiments it can be clearly localized on the c-side (Eckerson & Klingenberg, 1987). For the sake of homology, it can be concluded that also the C-terminus of the AAC is directed to the c-side. The N-terminus of the AAC is acetylated and therefore difficult to assay (Aquila et al. 1982). Recently, unpublished immunological evidence has been mentioned that the N-terminus is directed towards the c-side (Dalbon et al. 1988). It is clear that this simple and, because of its symmetry, aesthetically quite pleasing folding pattern can only be a superficial scheme which must be strongly skewed by the actual asymmetry in proteins so important for their function.

The search for functionally important residues is focused on those involved in the binding of ligands, *i.e.* ADP, ATP and the inhibitors. The pyridoxalphosphate approach has revealed at least two or three lysines which are connected with the translocation path and the binding (Bogner *et al.* 1986). Lysine 22 is specifically blocked by carboxyatractylate and at the same time accessible from the c-side, although this lysine is located within the first α-stretch more towards the m-side. Therefore it can be proposed to line the translocation path. Lysine 162 is primarily accessible from the cytosol, but also a member of the second domain hydrophilic center which is primarily directed to the m-side. Therefore the region around 165 is also assigned to the translocation path making it accessible to pyridoxalphosphate from the c-side and only in the c-state, when the c-gate is open. Lysine 106 is accessible both from the c- and m-side, but in a conformationally sensitive manner and is therefore also assigned to the translocation path. Lysines 42 and 48, being localized at the m-surface, are sensitive to conformation changes and thus visualized to be part of the translocation entrance.

This evidence indicates that the central hydrophilic sections in each domain contain parts of the translocation path with loops protruding into the inner membrane space. Recent evidence based on photoaffinity labelling of the ADP/ATP carrier with 2-azido-ADP also comes to the conclusion that at least part of the binding center is in the region round 162 and 165 (Dalbon *et al.* 1988). In yeast AAC, which can be cleaved in a more unequivocal manner than the beef heart AAC, 2-azido-ATP labelling is also found in the region between 160 to 170 (unpublished results). Since 2-azido-ATP is supposed to bind only from the c-side, this region was placed on the cytosolic surface (Dalbon *et al.* 1988). These conclusions, however, contrast to the lysine-probe findings. Moreover, such a localization on the c-surface would contradict the postulate that the AAC has only one binding center which therefore must be localized more within the central core of the carrier. This can be accommodated with our assignment of lysine 162 to the center of the translocation path (Bogner *et al.* 1986). One reason for the different folding pattern comes from allotting only 5 transmembrane helices (Saraste & Walker 1982; Dalbon *et al.* 1988). Here the potential region for the transmembrane second Á-helix is placed on the cytosolic surface, in contrast to our previous assignments based on the lysine probes. We conclude that the azido-labelling data, whether with arylazido-atractylate (Boulay *et al.* 1983) or 2-azido-ADP/ATP (Dalbon *et al.* 1988), are not contradictory to but in fact support and substantiate our folding model.

The partial disposition of the more conserved hydrophilic central sections in the translocation path emphasizes that the specificity and functional requirements of the translocation channel are not met by the relatively unconserved transmembrane α-structure, but by the lining with the more elaborate central sections. Besides the binding center, the protein also has to provide the gates, which are probably formed by ion bridges. These bridges are visualized to be released or locked by the interaction with neighboring polar groups, according to the change in the mutual distances resulting from the conformational change during the translocation process. The polar groups involved in the gating are still to be defined within these central sections. Finally, the disposition of these loops within the membrane is also in agreement with the hydrodynamic data, according to which the mass of protein within the membrane is larger than the amount of only about 40% of the six transmembrane α-helices (Hackenberg & Klingenberg, 1980).

THE AAC IN DIFFERENT SPECIES

Since the AAC is present in all aerobic eukaryotic cells it has been a target for primary structure analysis also in other cells. The molecular weights of the AACs from various species are quite similar, in the range between 32.5 to 35 kD. Because of its great abundance, cloning of the cDNA for the AAC from various species was achieved and in two cases even occurred inadvertently. The cDNA served to identify the primary structure and also the possible occurrence of leader sequences for the import of the AAC into the mitochondria. In no case so far has such a leader sequence been found for the AAC.

The comparison of the species of the AAC is confined here to AAC from mammalian, fungi and higher plant sources (Arends & Sebald, 1984; Adrian et al. 1986; Baker & Leaver, 1985). There are also isoforms in most cells, as will be discussed below. The overall amino acid composition as derived from the sequence shows that the AAC from mammalians has a higher excess of basic residues that from both fungi and plant. The cysteine content is somewhat higher in the mammalian AAC. Remarkable is the lack of histidine in the fungi AAC. This agrees with the non-conservation of histidine positions in the various AAC. Histidine is known as a ubiquitous and essential residue for enzymes involving nucleotides in phosphate transfer, hydrolysis, dehydrogenases, etc. In the functioning of the AAC, however, ADP/ATP are not chemically changed and therefore it is reasonable that the carrier to dismisses histidine as a non-essential residue. At least one cysteine appears to be essential and is conserved in all types of AAC. The total homology of the primary structures between the various species is remarkably high. It is even higher for those residues that are potentially important in structure and function, such as glycine *plus* proline, and the acidic residues. The highest homology is shown by the aromatic and basic residue.

The sequences for the four AAC species are aligned according to the triplicate principle which is found in the other AACs as well (Fig. 3). This alignment is used to obtain maximum homologies both between various species and between the three similar, about 100 residue long domains. The structural elements elucidated above for the beef heart AAC are present also in other species, such as the occurrence of two putative transmembrane A-stretches in each domain and the clustering of charged residues in the center region and to a lesser extent in the connecting regions. The reason for the high interspecies homology is strikingly illustrated.

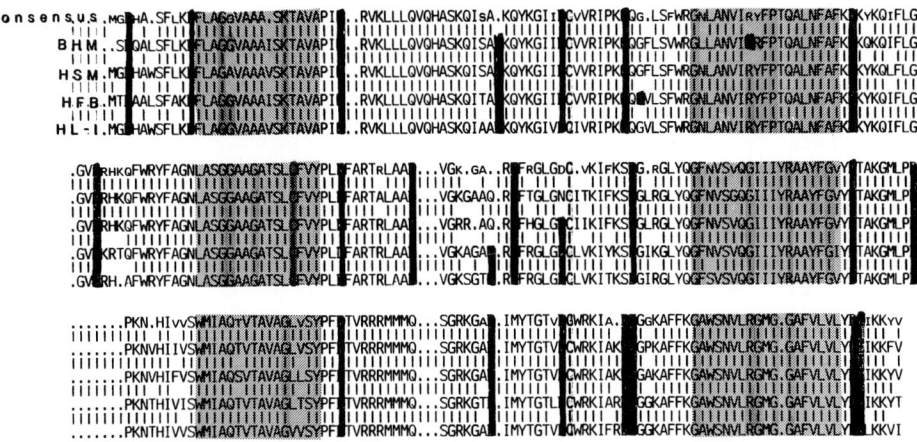

Fig. 3. Primary structure of the AAC from various cells. The tripartite arrangement is programmed for maximum interdomain homologies. The shaded area indicates the transmembrane α-regions. BHM = bovine heart muscle (Aquila *et al.* 1982), HSM = human skeletal muscle (Neckelmann *et al.* 1987), SC-1 = Saccharomyces cerevisiae (Adrian *et al.* 1986), NC = Neurospora crassa (Arends & Sebald, 1984), CM-1 = corn mitochondria (Baker & Leaver 1985).

A high proportion is retained of the polar, aromatic and glycine *plus* proline residues, not only in all these species, but also throughout the three repeated domains. We can conclude that the triplicate structure and the folding of these AACs is conserved in all these evolutionarily widely divergent eukaryotic cells. In fact, the origin of these domains from a common ancestral gene seems quite feasible.

What has been stated above for the occurrence of the binding centers, can also be extended to the AAC from other species. Recently, in yeast AAC, using photoaffinity labelling with 2-azido-ADP, residues were identified at positions homologous to those labelled in beef heart (unpublished results). However, 2-azido-ADP in yeast is not inserted in addition in the 230 to 250 region, as reported for the beef heart AAC (Dalbon *et al.* 1988). This may indicate that the latter insertion is not essential but rather a circumstantial unspecific photoaffinity labelling effect.

ISOFORMS OF THE AAC

Early immunological evidence suggested that there exists an organ specificity. For example, the antiserum raised against the intact AAC from beef heart reacted only poorly with the kidney and liver from beef, but strongly with the mitochondria from hearts of various other animals (Eiermann *et al.* 1977). There seemed to be a immunological organ specificity stronger than species specificity. Recently, in various human gene banks, isoforms of the AAC were found by their cDNA-defined sequences (Battini *et al.* 1987; Neckelmann *et al.* 1987; Houldsworth & Attardi, 1988). Thus in fibroblasts, an AAC sequence was found (Neckelmann *et al.* 1987) which differs significantly from that found in a cDNA library from skeletal muscle (Neckelman *et al.* 1987). In fact, Wallace established that the homology of the human skeletal muscle AAC to the beef heart muscle AAC is 94% and to the human fibroblast AAC only 88% (Neckelmann *et al.* 1987). Thus he speculated that the two human cDNAs must have diverged before or in the permean period. The occurrence of two different isoforms and additional isoforms was also demonstrated in a liver cDNA library (Houldsworth & Attardi, 1988).

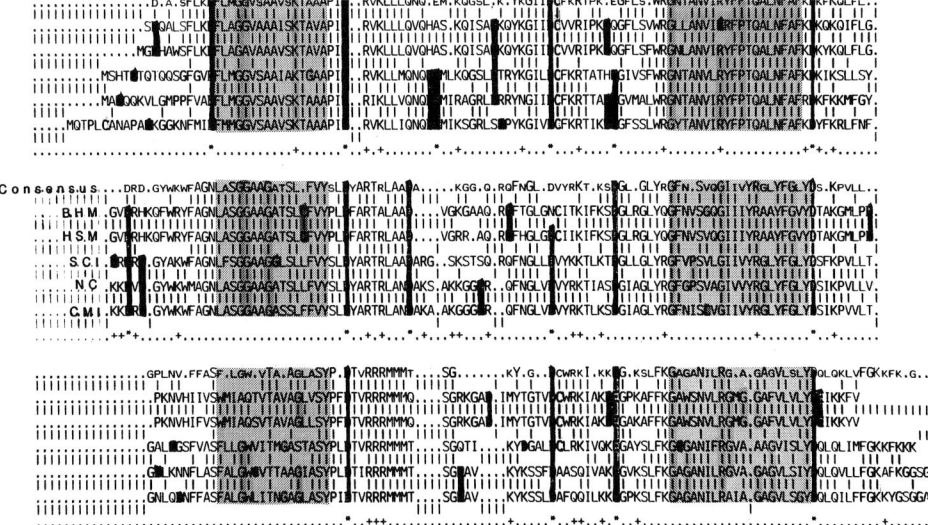

Fig. 4. The primary structure of the isoforms of the AAC from various human tissues. The tripartite arrangement of each sequence is programmed for maximum interdomain homologies. BHM = bovine heart muscle (Aquila *et al.* 1982), HSM = human skeletal muscle (Neckelmann *et al.* 1987), HFB = human fibroblasts (Battini *et al.* 1987), HL-1 = human liver (Houldsworth & Attardi, 1988).

One of these clones seems to be identical with that of fibroblasts and the other distinct, but still more similar to the skeletal muscle one. The existence of polymorphismic forms of the AAC in liver was made probable. It can be concluded that there exist at least three isoforms of AAC in human tissues (Fig. 4) which are differently expressed.

These studies support although do not yet prove the organ specificity indicated by the immunological studies (Eiermann *et al.* 1977; Schultheiss & Klingenberg, 1984). The reasons for these organ divergences are probably of a functional nature in that three could be subtle but important differences in the regulatory properties and responses of the AAC to the membrane potential and the ADP/ATP ratio, complying with organ specific requirements. It is known that in heart these ratios and possibly also the membrane potential are different than, for example, in liver (Soboll *et al.* 1978; Soboll & Bünger, 1981). It is not known whether the dimers in these organs are exclusively in the isoform (α_2) or may also occur in heteroforms ($\alpha\beta$). It can be expected that more isoforms will be found and that the occurrence of these isoforms in different organs will form a certain pattern. The degree of expression of these isoforms in various organs is not yet very well established. One possibility, also to be considered, is that particular isoforms of the AAC are present during the embryonic stage. It was earlier shown that the embryonic rat liver mitochondria have nearly no adenine nucleotides and are not able to exchange but only to import ADP and ATP, in contrast to adult mitochondria

CONCLUSIONS

The molecular mechanism, elucidated form the structure/function relationship, is far from being solved for the AAC or, for that matter, for any biomembrane carrier. Although the single binding center gated pore mechanism appears to be adequately accommodated within the available structural information, there is no unequivocal way of interpreting the structural information in terms of the mechanism. Site-directed mutagenesis has to a limited extent advanced the understanding of the lactose permease, however, the limitations of this approach are apparent (Püttner *et al.* 1986; Carrasco *et al.* 1986). The application of this method to the AAC in yeast is presently under way. One of the severest problems is the lack of information on the three-dimensional structure due to the absence of protein crystals.

The information obtainable with affinity probes for elucidating the binding center and other essential elements, is limited. Still, progress can be expected form probes concerning the bulk structure, in particular hydrophobic or hydrophilic surfaces and sidedness of the carrier. The increasing number of primary structures for the ADP/ATP carrier available from various species also provides a broad base of structural information which permits a differentiation of essential residues. These residues may be operative in the translocation or may serve to maintain the appropriate structure and folding. Eventually functional differences detected between AAC from various cells may be related to differences of certain residues which then can be assigned to the altered functions in the translocation process. Isoforms within one organism but differentially expressed in various tissues may also show functional differences related to specific residues. The role of the structural elements important for the sorting and import of the AAC will also have to be elucidated.

ACKNOWLEDGEMENTS

This work was supported by grants from the Deutsche Forschungsgemeinschaft (Kl 134/24) and from the Fonds der Chemischen Industrie.

REFERENCES

Adrian GS, McCammon MT, Montgomery DL, Douglas MG (1986) Sequences required for delivery and localization of the ADP/ATP translocator to the mitochondrial inner membrane. Mol Cell Biol 6:626-634

Aquila H, Link TA, Klingenberg M (1985) The uncoupling protein from brown fat mitochondria is related to the mitochondrial ADP/ATP carrier. Analysis of sequence homologies and of folding of the protein in the membrane. EMBO J 4:2369-2376

Aquila H, Link TA, Klingenberg M (1987) Solute carriers involved in energy transfer of mitochondria form a homologous protein family. FEBS Lett 212:1-9

Aquila H, Misra D, Eulitz M, Klingenberg M (1982) Complete amino acid sequence of the ADP/ATP carrier protein from beef mitochondria. Hoppe-Seyler's Z Physiol Chem 363:345-349

Arends H, Sebald W (1984) Nucleotide sequence of the cloned mRNA and gene of the ADP/ATP carrier from Neurospora crassa. EMBO J 3:377-382

Baker A, Leaver CJ (1985) Isolation and sequence analysis of a cDNA encoding the ADP/ATP translocator of *Zea mays L. Nucl Acids Res 13:5857-5867

Battini R, Ferrari S, Kaczmarek L, Calabretta B, Sing-Tsueng C, Baserga R (1987) Molecular cloning of a cDNA from a human ADP/ATP carrier which is growth-regulated. J Biol Chem 262:4355-4359

Bogner W, Aquila H, Klingenberg M (1986) The transmembrane arrangement of ADP/ATP carrier as elucidated by the lysine reagent pyridoxal 5-phosphate. Eur J Biochem 165:611-620

Boulay F, Lauquin GJM, Tsugita A, Vignais PV (1983) Photolabeling approach to the study of the topography of the atractyloside binding site in mitochondrial adenosine 5'-diphosphate/adenosine 5'-triphosphate carrier protein. Biochemistry 22:477-484

Büchel DE, Gronenborn B, Müller Hill B (1980) Sequence of the lactose permease gene. Nature 283:541-545

Carrasco N, Antes LM, Poonian MS, Kaback HR (1986) lac permease of Escherichia coli: histidine-322 and glutammic acid-325 may be components of a charge-relay system. Biochemistry 25:4486-4488

Dalbon P, Brandolin G, Boulay F, Hoppe J, Vignais PV (1988) Mapping of the nucleotide binding sites in the ADP/ATP carrier of beef heart mitochondria by photolabeling with 2-azido[α-^{32}p] adenosine diphosphate. Biochemistry (in press)

Eckerskorn C, Klingenberg M (1987) In the uncoupling protein from brown adipose tissue the C-terminus protrudes to the c-side of the membrane as shown by tryptic cleavage. FEBS Lett 226:166-170

Eiermann W, Aquila H, Klingenberg M (1977) Immunological characterization of the ADP, ATP translocator protein isolated from mitochondria of liver, heart and other organs. Evidence for an organ specificity. FEBS Lett 74:209-214

Hackenberg H, Klingenberg M (1980) Molecular weight and hydrodynamic parameters of the adenosine 5'-diphosphate-adenosine 5'-triphosphate carrier in Triton X-100. Biochemistry 19:548-555

Houldsworth J, Attardi G (1988) Two distinct genes for ADP/ATP translocase are expressed at the mRNA level in adult human liver. Proc Natl Acad Sci (USA) 85:377-381

Rao JKM, Argos P (1986) Conformational preference parameter to predict helices in integral membrane proteins. Biochim Biophys Acta 869 197-214

Kyte J, Doolittle RF (1982) A simple method for displaying the hydropathic character of a protein. J Mol Biol 157:105-132

Leblanc P, Caluser H (1972) ADP-dependent inhibition of sarcosomal adenine nucleotide translocase by N-ethylmaleimide. FEBS Lett 23:107-113

Neckelmann N, Li K, Wade RP, Shuster R, Wallace DC (1987) cDNA sequence of a human skeletal muscle ADP/ATP translocator: lack of a leader peptide, divergence from a fibroblast translocator cDNA, and co-evolution with mitochondrial DNA genes. Proc Natl Acad Sci (USA) 84:7580-7584

Pfanner N and Neupert W (1987) Biogenesis of energy transducing complexes In: Lee CP (ed) Current Topics in Bioenergetics, Vol 15, Academic Press, pp 177-220

Pfanner N, Hoeben P, Tropschug M, Neupert W (1987b) The carboxy-terminal two-thirds of the ADP/ATP carrier polypeptide contains sufficient information to direct translocation into mitochondria. J Biol Chem 262:14851-14854

Püttner IB, Sarkar HK, Poonian MS, Kaback HR (1986) lac permease of Escherichia coli: histidine-205 and histidine-322 play different roles in lactose/H+ symport. Biochemistry 25:4483-4485

Saraste M, Walker JE (1982) Internal sequence repeats and the path of polypeptide in mitochondrial ADP/ATP translocase. FEBS Lett 144:250-254

Schultheiss HP, Klingenberg M (1984) Immunochemical characterization of the adenine nucleotide translocator. Organ specificity and conformation specificity. Eur J Biochem 143:599-605

Soboll S, Bünger R (1981) Compartmentation of adenine nucleotides in the isolated working guinea pig heart stimulated by noradrenaline. Hoppe Seylers Z Physiol Chem 362:125-132

Soboll S, Scholz R, Heldt HW (1978) Subcellular metabolite concentrations. Dependence of mitochondrial and cytosolic ATP-system on the metabolic state of perfused rat liver. Eur J Biochem 87:377-390

Urry DW (1972) Protein conformation in biomembranes: optical rotation and absorption of membrane suspensions. Biochim Biophys Acta 265:115-168

Walker JE, Cozens Al, Dyer MR, Fearnley IM, Powell SJ, Runswick MJ (1987) Studies of the genes for ATP synthases in Eubacteria, chloroplasts and mitochondria: Implications for structure and function of the enzyme. Chem Scr 27B:97-105

Kinetic Mechanisms of the Adenylic and the Oxoglutaric Carriers: a Comparison

F.E. SLUSE, C.M. SLUSE-GOFFART, C. DUYCKAERTS

Unité de Bioénergétique, Université de Liège, Istitut de Chimie (B-6) Sart-Tilman, B-4000 Liège (Belgium)

Translocations by exchange through membranes must obey to one of the two general types of mechanisms: the ping-pong mechanism and the sequential mechanism.

If translocation follows a ping-pong mechanism the carrier possesses a single binding site for its substrates and it exists in two forms, one that can be loaded by the internal substrate. Since the transformation of one form into the other can only occur if the carrier is loaded with one substrate, the carrier can perform an obligatory one-to-one exchange by binding alternately both its substrates without formation of a ternary complex S_{ext}-carrier-S_{int}.

If the translocation follows a sequential mechanism the carrier possesses two binding sites, one accessible to the internal substrate and the other accessible to the external substrate, and the two substrates must bind to the translocator, forming the active ternary complex, before the exchange occurs. Studies that permit to discriminate between mechanisms of action are very important because they give at least direct informations on the minimum number of sites simultaneously accessible to substrates.

Study of the mechanism of transport of the mitochondrial carriers has been carried out in two ways: the kinetic approach and the molecular approach. The kinetic approach is the most powerful approach to discriminate between mechanisms of action. However, kinetic studies *in situ* have been underrated because the technical problems that make difficult the necessary initial-rate measurements at various concentrations of both internal and external exchangeable substrates.

A. Azzi et al. (Eds.)
Anion Carriers of Mitochondrial Membranes
© Springer-Verlag Berlin Heidelberg 1989

An initial-rate analysis permits to distinguish between the two types of mechanism: in Lineweaver-Burk plots, a ping-pong mechanism leads to a pattern of parallel straight lines, each of which corresponding to a fixed concentration of the second substrate while a sequential mechanism exhibits straight lines having a common point of intercept somewhere on the left of the origin.

The conclusions coming from such an analysis must be warranted by the measurements of actual initial rates and by the knowledge of the true free-substrate concentrations to which the binding sites are exposed, at least on one side of the membrane.

Direct accurate measurements of the actual initial rates of exchange may be extremely difficult to achieve even by inhibitor-stop technique. This can be reached if five conditions are fulfilled: (1) if during the steady-state rate measurement the product concentrations are negligible; (2) if the inhibitor-insensitive binding of the substrates to the membrane is negligible; (3) if the addition of the inhibitor does not modify the distribution of the substrates between the two compartments separated by the membrane; (4) if the inhibitor used to stop the exchange is very efficient, *i.e.* if it has a high rate of action and if it completely blocks the exchange; (5) if the exchange under study is catalysed by only one translocator species so that the rate is not the sum of different translocation activities and if there is only one exchangeable substrate on each side of the membrane.

In an ideal case the 5 conditions are fulfilled and the actual initial rate (v_o) is equal to the measured rate (v_m) that is the rate of accumulation in the pellet of the labelled substrate added to start the exchange. In other cases if only conditions (4) and (5) are fulfilled, the initial rate is equal to:

$$v_o = v_m \times F_1 \times F_2 \times F_3$$

where F_1, F_2 and F_3 are the correction factors corresponding to conditions (1), (2) and (3).

OXOGLUTARATE TRANSLOCATOR

An ideal situation was encountered with the oxoglutarate translocator in rat-heart mitochondria: all conditions to measure and use the initial rates properly are fulfilled (Sluse *et al.*, 1971, 1972, 1973a, 1973b, 1975, 1979) if its rate of exchange is measured at 2°C between 0.2 and 1 s using 20-mM phenyl-succinate as quenching inhibitor. As shown by Sluse *et al.*, 1979, Fig. 1), a small pre-steady state lag-time is observed, at low external oxoglutarate concentrations, followed by a steady state. This induction time diminishes as the substrate concentration increases.

The initial rate analysis allows us not only to eliminate the ping-pong mechanism but also to precise that the mechanism is rapid-equilibrium random and that the internal sites and the external sites are independent. These properties were demonstrated for 12 exchanges between internal oxoglutarate, malate or malonate and external oxoglutarate, malate, malonate or succinate.

Extensive studies of OG/Mal, OG/OG, Mal/Mal exchanges (Sluse, 1979, Sluse *et al.*, 1975, 1979); Sluse-Goffart *et al.*, 1983, 1986, 1988) have allowed us to show the presence of 3 intermediary plateaux in the saturation curve by external oxoglutarate (Fig. 3 in Sluse-Goffart *et al.*, 1983) contrary to external malate that is Michaelian (Fig. 1 in Sluse-Goffart *et al.*, 1983) whereas both are competitive for the same binding sites as we have proved it (Fig. 2 in Sluse-Goffart *et al.*, 1983). On the matricial side (Fig. 1) the internal oxoglutarate seems to be Michaelian for initial rate versus internal oxoglutarate concentration in intact mitochondria (apparent $K_m \approx 1$ mM) and for equilibrium binding for external oxoglutarate with inverted submitochondrial particles ($K_d \approx 800$ μM) while the internal-malate kinetic saturation curves reveal negative cooperatively modulated by the internal-aspartate concentrations (Fig. 2).

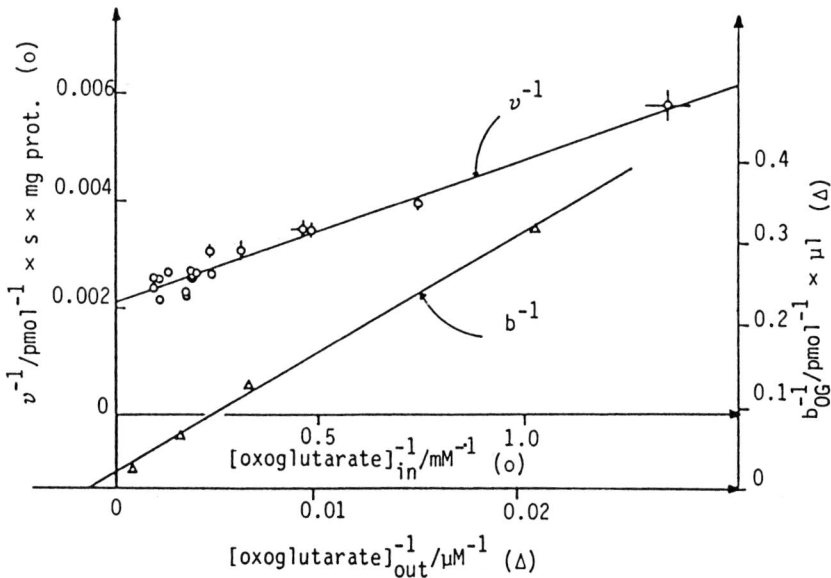

Fig. 1. Kinetics of 2-oxo [5-^{14}C]glutarate influx in exchange for internal oxoglutarate. v^{-1} as a function of $[OG]^{-1}_{in}$. 2-oxo[5-^{14}C]glutarate $= 11$ μM. Intact rat-heart mitochondria: 1.1 mg protein/ml.
Equilibrium binding of 2-oxo[5-^{14}C]glutarate to empty inverted submitochondrial particles. b^{-1} (bound oxoglutarate)$^{-1}$ as a function of OG^{-1}_{out}.

186

The exceptional behaviour of oxoglutarate translocator has been analysed and demonstrates that the translocator has an oligomeric structure (Sluse-Goffart *et al.*, 1983) and that several conformational changes are involved during its activity cycle (Sluse-Goffart & Sluse, 1986; Duyckaerts *et al.*, 1984). We have also shown that the oxoglutarate translocator is an allosteric enzyme since its activity may be regulated by effectors as internal aspartate (Sluse *et al.*, 1980) and external free ADP at micromolar range (Sluse *et al.*, 1985). Fig. 3 represents a functional subunit of the oligomeric translocator exchanging external oxoglutarate with internal malate and summarizes all the mechanistic properties of the oxoglutarate translocator we have evidenced.

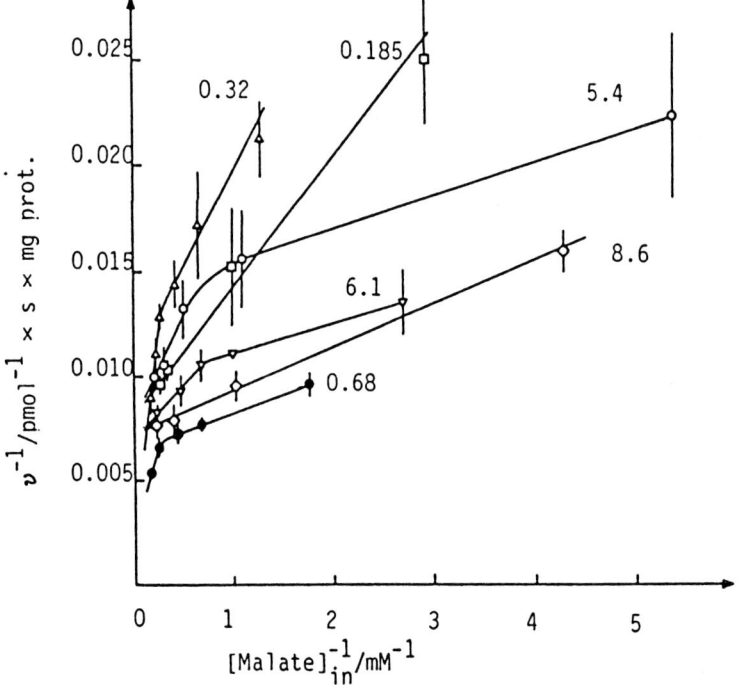

Fig. 2. Kinetics of ^{14}C-OG influx in exchange for internal malate at different internal-aspartate concentrations. v^{-1} as a function of $[Mal]^{-1}_{in}$ at internal aspartate concentrations between 0.18 and 8.6 mM. Intact rat-heart mitochondria: 0.90 mg protein/ml and 2-oxo $[5-^{14}C]$ glutarate = 5.1 μM.

ADENYLIC TRANSLOCATOR

The adenylic translocator has been intensively studied from a long time by the group of Klingenberg and by the group of Vignais mainly through the molecular approach. Klingenberg has claimed that the translocator follows a ping-pong mechanism. Two kinetic studies were made, a first with heart mitochondria by our group (Duyckaerts *et al.*, 1980) and a second with liver mitochondria by Barbour (Barbour & Chan, 1981).

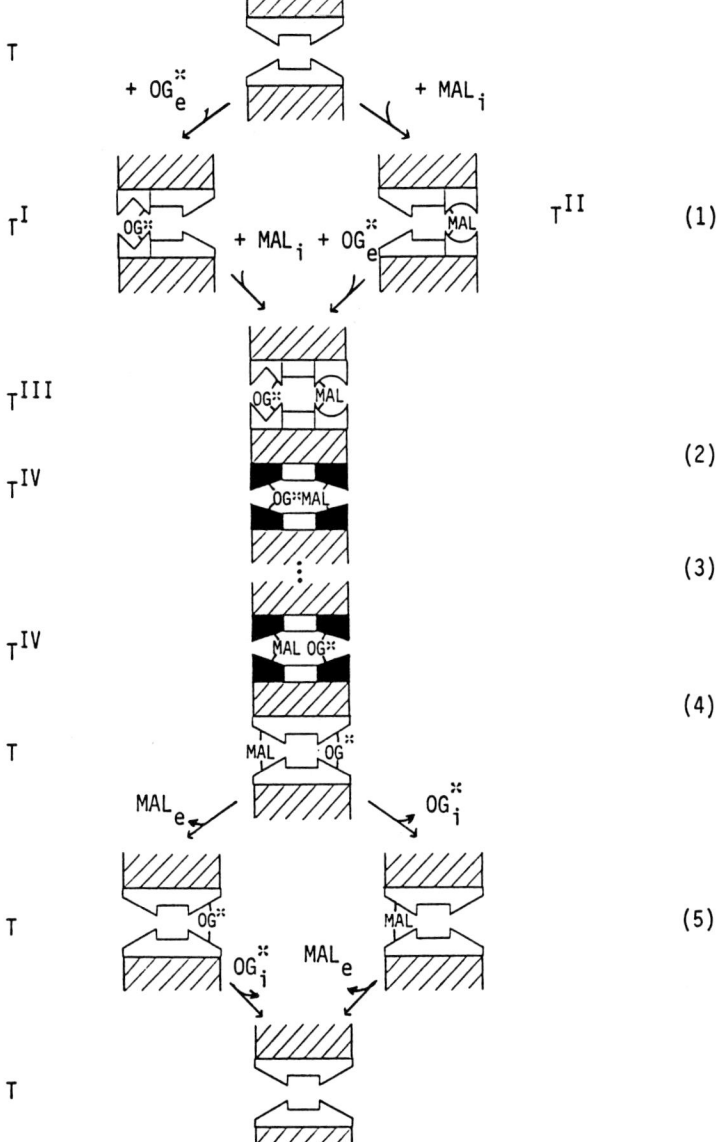

Fig. 3. Schematic representation of the mechanistic properties of a functional subunit of the oligomeric oxoglutarate translocator. (1) Random binding of the 2 substrates with local independent conformational changes induced by external OG and internal Mal. (2) transfer of the 2 substrates into a channel accompanied by conformational changes of the two surface parts. (3) Exchange. (4) Transfer to the surface sites. (5) Release of the products that is not, in this exchange, accompanied by conformational changes.

Both disagree with the ping-pong mechanism. However these kinetic studies, including ours, may be strongly criticized because the authors have supposed that the measured rate is the actual initial rate without taking into account all conditions we have stated above. Our first study was also criticizable for its narrow range of substrate concentrations. It is the reason why we have started a more careful study in order to reach sure conclusions from the kinetic analysis.

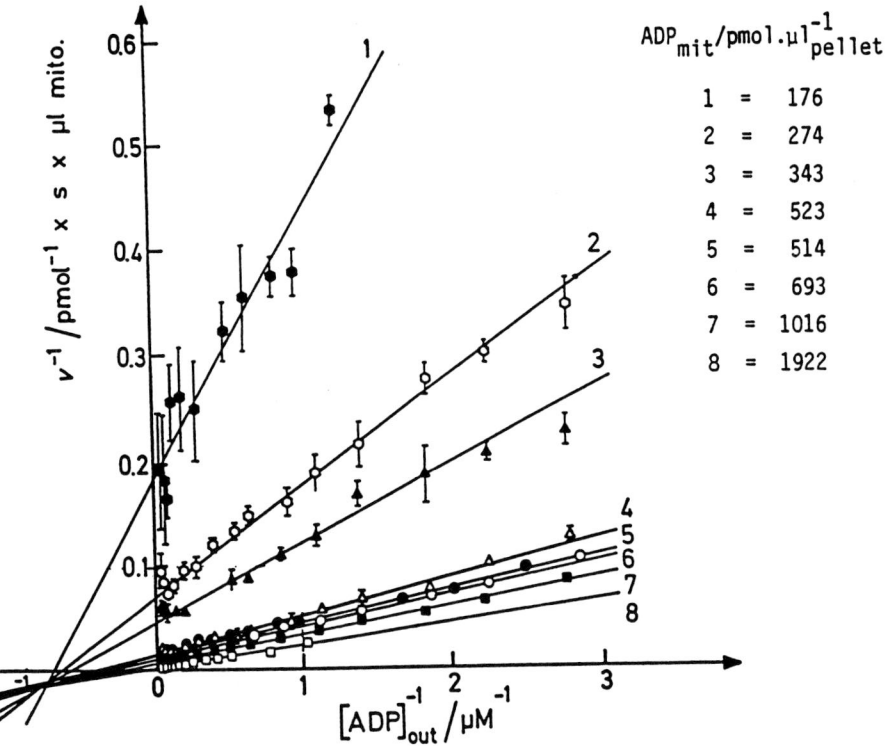

Fig. 4. Kinetics of ^{14}C-ADP influx in exchange for 8 mitochondrial ADP contents v^{-1} (crude) as a function of added $[ADP]^{-1}_{out}$. The highest ADP_{mit} was obtained by direct loading of ADP without pyrophosphate depletion.

This new attempt was also based on improvements (Sluse et al., 1987a, 1988a) of the mitochondrial preparation for the extension of the range of the internal concentrations of ADP that is, in our conditions, the only exchangeable internal substrate. This was obtained by a complete depletion by pyrophosphate of the endogenous nucleotides followed by an ADP reloading in the presence of various ADP-Mg concentrations (Mg constant).

In a first step, we have used these new preparations to study the influence of external concentrations (from 0.5 to 20 μM) and mitochondrial ADP contents on crude rates measured between 0.3 and 1 s using 10μm CAT to stop the exchange (Sluse *et al.*, 1987b, 1988a,b). Double reciprocal plots (Fig. 4) exhibit linear relationships and the straight lines obtained for the 8 different mitochondrial contents have a common intersection point below the abscissa, indicating that the translocator apparently follows a sequential mechanism. During this study, we have evidenced that the rate of action of CAT was too low when the concentration of the substrates increased, then we had to improve the efficiency of CAT inhibition.

In order to check the efficiency of the inhibitor we have tested at 30μm ^{14}C-ADP$_{out}$ different concentrations of CAT between 10 μM, that is largely enough to block completely the exchange, and 500 μM (Table 1). We have measured uptake at t=0 when CAT added 5 s before ADP and when CAT is added together with ADP and we have observed that a good efficiency is reached at 50μm CAT since the difference is equal to zero. We have also measured the crude rate that decreases from 10μm to 100μm CAT and remains constant up to 500μm CAT. Thus ultimately, we will use 100μm CAT to block the exchange.

Table 1. Efficiency of CAT inhibition

| CAT μM | ^{14}C-ADP uptake (pmol. μl^{-1}) at t = 0 | | | | Rate | Ampl. of burst |
	(1) CAT before	(2) CAT simult.	Δ (2)-(1)	(3) Extrapol. t = 0	pmol.μl^{-1}.s^{-1}	(3)-(1)
10	49 ± 1	59 ± 2	10 ± 3	100 ±3	65± 5	51 ± 4
20	49 ± 0.3	52 ± 2	3 ± 2	93 ± 5	61 ± 7	44 ± 5
50	50 ± 1	50 ± 1	0 ± 2	84 ± 2	52 ± 3	34 ± 3
100	46 ± 1	46 ± 1	0 ± 2	85 ± 3	50 ± 4	39 ± 4
200	38 ± 1	38 ± 2	0 ± 3	78 ± 2	55 ± 4	40 ± 3
500	32 ± 2	32 ± 2	0 ± 4	63 ± 4	52 ± 6	31 ± 6

The time course obtained with 100 μM CAT shows that the extrapolated ordinate (uptake of ^{14}C-ADP at time = 0) is much higher than the measured uptake at time = 0 (Table 1). This difference is not negligible compared with the amount of added ADP and this means that the rates are not measured in initial conditions, thus factor F_1 can not be neglected. Correction factor F_1 can be calculated easily if equilibrium data are known and is given by Eqn 1 in the case of the homologous exchange ^{14}C-ADP_{out}/ADP_{in} under study.

$$F_1 = \frac{1}{1 - \dfrac{ADP^*_{it}}{ADP^*_{ieq}}}$$

where ADP^*_{it}/ADP^*_{ieq} is equal to the ratio of the amount of ^{14}C-ADP in the pellet at time $= t$ *minus* ^{14}C-ADP in the pellet at time $= 0$ *verus* the amount of ^{14}C-ADP in the pellet as equilibrium *minus* ^{14}C-ADP in the pellet at time $t = 0$.

The second correction factor linked to the presence of ^{14}C-ADP bound externally in the presence of CAT is given by Eqn (2) and is almost equal to 1 (< 1.02).

$$F_2 = \frac{1}{1 - \dfrac{ADP^*_{pellet\ t=0}}{ADP^*_{added}}}$$

where the amount of ^{14}C-ADP_{pellet} is measured in the presence of CAT.

The correction factor F_3 takes into account the modifications of distribution that are not due to exchange : (1) those that would occur in the case of a ping-pong mechanism, (2) those that might be due to deleterious effects of $100\mu m$ CAT on membrane integrity. Moreover it contains a correction of specific radioactivity of ^{14}C-ADP required by the presence of cold-contaminant ADP already present in the external space when ^{14}C-ADP is added to start the exchange. F_3 is given by Eqn (3).

$$F_3 = \frac{(ADP_{tot} + ADP^*_{tot})_{out}}{ADP^*_{added}} \ X \ \frac{(ADP_{tot} + ADP^*_{tot})_{in}}{(ADP_{tot} + ADP^*_{tot})_{in(+\ CAT)}}$$

where "tot" means bound + free and where "out" and "in" mean extramatricial and intramatricial respectively. The intramatricial content in the presence of CAT $(in(+CAT))$ can be calculated rigorously. However in the absence of CAT (numerator of F_3), a non negligible extramatricial binding prevents a rigorous determination of the two ADP contents so that F_3 can only be evaluated between a maximal and a minimal calculated value.

Then in the next step we have studied the influence of 3 mitochondrial ADP contents and 6 external ADP concentrations on the actual initial rates by determining the correction factors in all conditions (Table 2).

Table 2. Example of determination of correction factors

Mitochondrial ADP = 235 pmol·μl pellet^{-1}	Determination of correction factors					
[ADP*] added (μM)	0.552	0.814	1.24	2.21	4.47	10.1
[ADP+ADP*]$^{free}_{out}$ measured (μM)	0.664	0.813	1.26	2.23	4.54	10.9
v_m/pmol s^{-1}·μl^{-1} pellet	1.63	2.02	2.53	2.99	4.17	4.38
F_1	1.034	1.029	1.034	1.034	1.041	1.042
F_2	1.016	1.016	1.015	1.014	1.013	1.011
F_3	2.01	1.76	1.57	1.42	1.32	1.26
Global corr. factor	2.11	1.84	1.65	1.49	1.39	1.33
v_o/pmol·s^{-1}·μl^{-1} pellet	3.45	3.72	4.17	4.44	5.80	5.81

The crude kinetic pattern obtained with 100μm CAT may be compared with the one obtained by blocking the exchange with 10μm CAT. The new data converge to the same common point and are situated as expected regarding the mitochondrial ADP content.

If we compare this pattern of crude rates with the pattern obtained with corrected rates, we can see that the convergence remains conclusive thanks to the large difference in mitochondrial contents tested (Fig. 5). The common point is still under the abscissa axis but moves to the left. We can conclude that steady-state kinetic results are really in contradiction with a ping-pong mechanism.

A complementary argument can be obtained from pre-steady state study. We have followed the kinetics from 35 ms by the quenching flow technique with a Durrum apparatus. Fig. 6 shows a burst that appears complex. Moreover the amplitude of the burst decreases when the internal-ADP content is decreased.

It must be noticed that if a ping-pong mechanism can exhibit a burst due to the reorientation of sites induced by the addition of the external substrate such a burst would have to increase when the internal-substrate content decreases contrary to our observation.

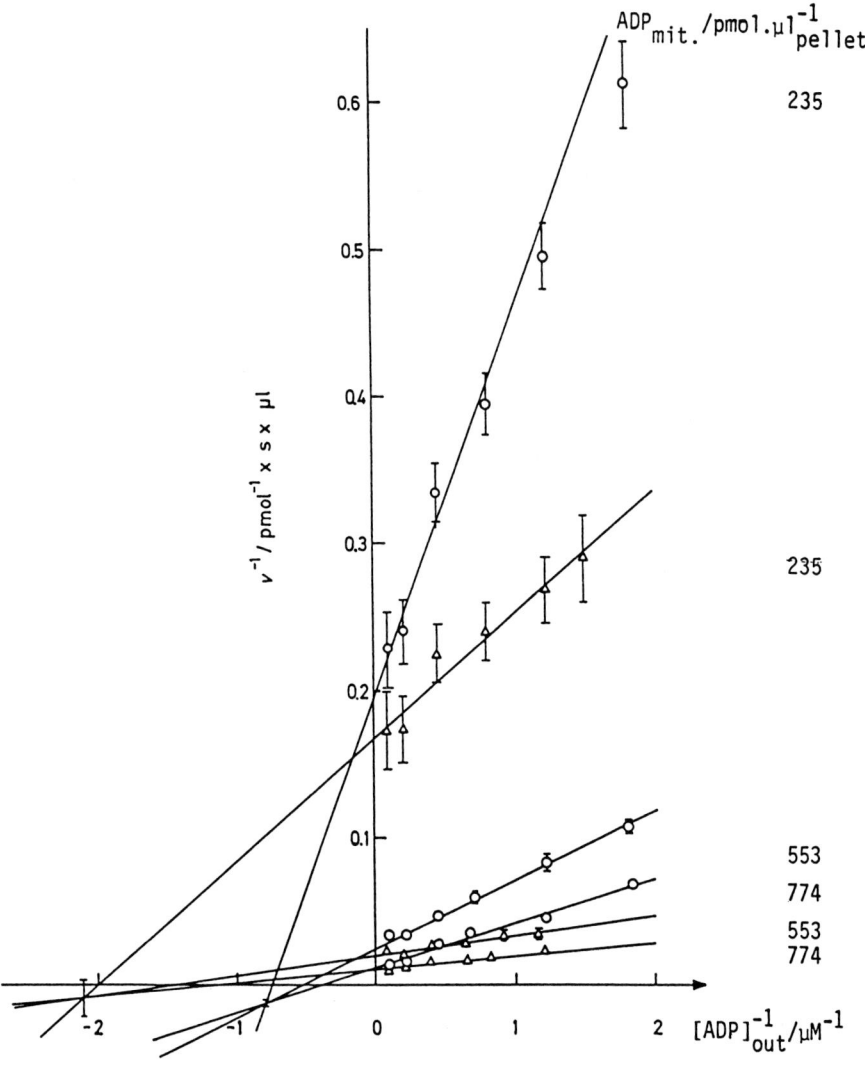

Fig. 5. Kinetic of ^{14}C-ADP influx in exchange for internal ADP, the exchange being blocked by 100μm CAT. Crude rate^{-1} (o) and corrected rate^{-1} (△) as a function of added [^{14}C-ADP]$^{-1}_{out}$ and measured free ADP $^{-1}$]$_{out}$ respectively. Coordinates of intercept (o) : - 0.78, - 0.012 ± 0.002; coordinates of intercept (△) : - 2.1, - 0.008 ± 0.012.

Thus this observation confirms the sequential mechanism. At high mitochondrial ADP content the amplitude of the burst decreases when the external-ADP concentration is decreased (not shown). When the mitochondrial ADP content is low, an initial lag is visible and when both external and internal ADP are low, the burst disappears, only a lag is observed (not shown).

This complex pre-steady state behaviour depending on both substrate concentrations means probably that several steps are implied possibly several conformational changes of the translocator-substrate complexes (Evens *et al.*, 1988).

Then it may be hoped that the pre-steady state studies together with steady state analysis will allow to reach more informations on the sequential mechanism followed by the adenylic translocator.

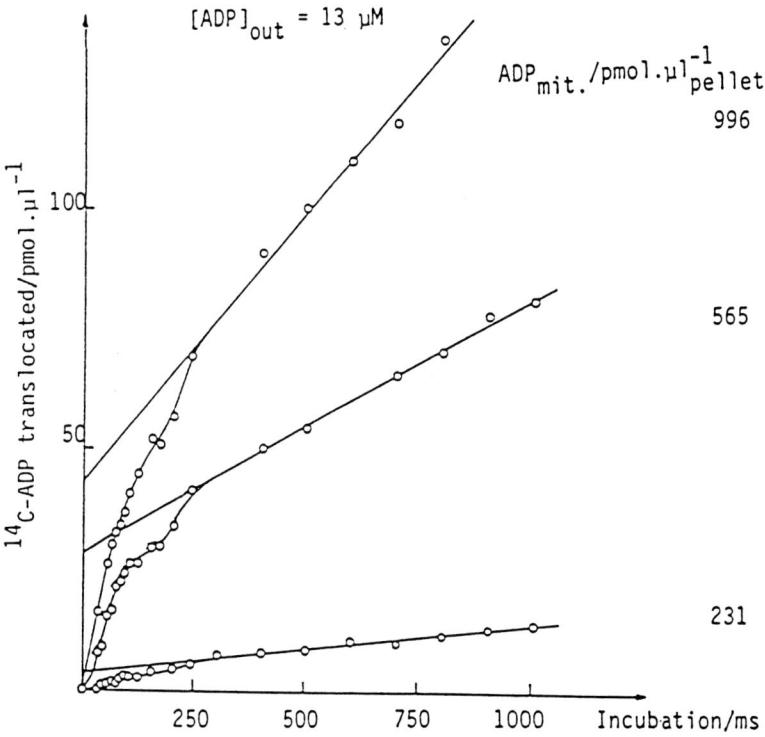

Fig. 6. Time course between 0 and 1000 ms : ^{14}C-ADP translocated *versus* time. The origin is an experimental point measured 5 or 10 times for each condition.

As a conclusion, it is remarkable that the kinetic approach has evidenced a sequential mechanism for the three mitochondrial translocators studied up to now in this way: for the oxoglutarate and the adenylic translocators in intact mitochondria (this paper) and for the aspartate-glutamate translocator in proteoliposomes (Dierks & Krämer, 1988).

REFERENCES

Barbour RL, Chan SHP (1981) Characterization of the kinetics and mechanism of the mitochondrial ADP-ATP carrier. J Biol Chem 256:1940-1948

Dierks T, Krämer R (1988) Reaction mechanism of the reconstituted aspartate/glutamate antiporter from mitochondria. Reversible switching to uniport function. In: Azzi A, Nałęcz KA, Nałęcz MJ, Wojtczak L (eds) The Anion Carriers of the Mitochondrial Membranes. Springer-Verlag Heidelberg, pp 99-110

Duyckaerts C, Sluse-Goffart CM, Fux JP, Sluse FE, Liébecq C (1980) Kinetic mechanism of the exchanges catalysed by the adenine-nucleotide carrier. Eur J Biochem 106:1-6

Duyckaerts C, Sluse-Goffart CM, Sluse FE, Gosselin-Rey C, Liébecq C (1984) Spontaneous modification of the oxoglutarate translocator *in vivo*. Eur J Biochem 142:203-208

Evens A, Sluse-Goffart CM, Duyckaerts C, Camus G, Sluse FE (1988) Etude de la cinétique transitoire du transporteur adénylique. Abstr Congrès d'Automne de la Société de chimie biologique sur "Systèmes Bioénergétiques : Structure, contrôle et évolution" Bombannes France

Sluse FE (1979) Alternating cooperativities in the kinetic and binding curves of the oxoglutarate translocator. Mathematical model. In: Quagliariello E, Palmieri F, Papa S, Klingenberg M (eds) Function and Molecular aspects of Biomembrane Transport. Elsevier, Amsterdam, pp 407-410

Sluse FE, Meijer AJ, Tager JM (1971) Anion translocators in rat-heart mitochondria. FEBS Lett 18:149-153

Sluse FE, Ranson M, Liébecq C (1972) Mechanism of the exchanges catalysed by the oxoglutarate translocator of rat-heart mitochondria. Kinetics of the exchange reactions between 2-oxoglutarate, malate and malonate. Eur J Biochem 25:207-217

Sluse FE, Goffart G, Liébecq C (1973a) Mechanism of the exchanges catalysed by the oxoglutarate translocator of rat-heart mitochondria. Kinetics of the external-product inhibition Eur J Biochem 32:283-291

Sluse FE, Liébecq C (1973b) Kinetics and mechanism of the exchange reactions catalysed by the oxoglutarate translocator of rat-heart mitochondria. Biochimie 55:747-754

Sluse FE, Sluse-Goffart CM, Duyckaerts C, Liébecq C (1975) Evidence for cooperative effects in the exchange reactions catalysed by the oxoglutarate translocator of rat-heart mitochondria. Eur J Biochem 56:1-14

Sluse FE, Duyckaerts C, Liébecq C, Sluse-Goffart CM (1979) Kinetic and binding properties of the oxoglutarate translocator of rat-heart mitochondria. Eur J Biochem 100:3-17

Sluse FE, Duyckaerts C, Sluse-Goffart CM, Fux JP, Liébecq C (1980) Oxoglutarate translocator of rat-heart mitochondria : regulation by aspartate. FEBS Lett 120:94-98

Sluse FE, Duyckaerts C, Bartolomé T, Sluse-Goffart CM, Liébecq C (1985) Regulation of the oxoglutarate translocator by adenine nucleotides. Abstr Special FEBS Meet on Metal Ions, Proteins and Membranes, Algarve, p 126

Sluse FE, Duyckaerts C, Sluse-Goffart CM (1987a) Kinetics of adenine nucleotide translocator revisited. Abstr Journée sur le Contrôle du Métabolisme Cellulaire, Bordeaux, France

Sluse FE, Sluse-Goffart CM (1987b) On kinetics of adenine nucleotide translocation. Lect on 9th Magdeburg Coll on "Mitochondrial Functions", Magdeburg

Sluse FE, Duyckaerts C, Sluse-Goffart CM (1988a) Kinetics of the adenine-nucleotide translocation in rat-heart mitochondria. Arch Int Physiol Biochim 96:B56

Sluse FE, Duyckaerts C, Evens A, Sluse-Goffart CM (1988b) New attempt to determine the kinetic mechanism of the adenylic translocator : methodological prerequisite. Abstr Int Symp on Molecular Basis of Biomembrane Transport, Ostuni, Italie, p 74

Sluse-Goffart CM, Sluse FE, Duyckaerts C, Richard M, Hengesch P, Liébecq C (1983) Conformational changes and possible structure of the oxoglutarate translocator of rat-heart mitochondria revealed by the kinetic study of malate and oxoglutarate uptake. Eur J Biochem 134:397-406

Sluse-Goffart CM, Sluse FE (1986) Kinetics as a tool for the study of transmembrane exchanges exemplified by the study of the oxoglutarate translocator. In: Damjanovich S, Keleti T, Tron T (eds) Dynamics of Biochemical Systems. Elsevier, Amsterdam, pp 521-535

Sluse-Goffart CM, Duyckaerts C, Bartholomé T, Sluse FE (1988) Study of the matricial face of the oxoglutarate translocator using inverted submitochondrial vesicles. Arch Int Physiol Biochim 96:B57

III. Porins

Porins from Mitochondrial and Bacterial Outer Membranes: Structural and Functional Aspects

ROLAND BENZ

Lehrstuhl für Biotechnologie, Universität Würzburg, Röntgenring 11,
D-8700 Würzburg, F.R.G.

The cell envelope of Gram-negative bacteria consists of three different layers, the outer membrane, the murein and the inner membrane (Beveridge, 1981). The inner membrane represents a real diffusion barrier and contains, similar to the mitochondrial inner membrane, the respiration chain and a large number of transport systems. The outer membrane of Gram-negative bacteria plays an important role in the physiology of these organisms. All nutrients or antibiotics either hydrophilic or hydrophobic have to cross this permeability barrier which means that it has special sieving properties. The active components of molecular sieving of the outer membrane are due to presence of a few major proteins called "porins" (Nakae, 1976). The porins are organized as trimers of three identical subunits and form transmembrane channels that have more general properties and sort according to the molecular weight of the solutes (Benz, 1985; Nikaido & Vaara, 1985). On the other hand, also specific porins have been identified in the outer membrane of Gram-negative bacteria which contain binding sites for substrates (Benz, 1988). The porin trimers form basically one channel in the outer membrane with three openings faced to the cell surface and one outlet into the periplasmic space (Engel et al., 1985; Lepault et al., 1988).

The origin of the first eukaryotic cell is still puzzling. The endosymbiontic theory proposes that several advanced prokaryotic cells formed the first eukaryotic cell. The analogy between the respiration chain of photosynthetic Gram-negative bacteria and that of mitochondria suggests that both could have common ancestors. Mitochondria

A. Azzi et al. (Eds.)
Anion Carriers of Mitochondrial Membranes
© Springer-Verlag Berlin Heidelberg 1989

contain also an outer membrane that is under full genetic control of the cell. The role of the mitochondrial outer membrane in physiology and metabolism of the cell organelle was underestimated in the past. More recent papers gave some insight into the function of the mitochondrial outer membrane and its role in mitochondrial metabolism (Bessman et al., 1985; Brdiczka et al., 1986; Ohlendieck et al., 1986; Gellerich et al. 1987; Benz et al., 1988). Similar to the bacterial outer membrane the outer membrane of mitochondria contains general diffusion pores which share some similarities to the bacterial porins (Colombini, 1979; Benz, 1985; Benz et al., 1985). The protein is arranged in the pore-forming complex in β-pleated sheet structure and no indication for α-helical regions (Mihara & Sato, 1985; Kleene et al., 1987). Mitochondrial porins characterized to date have molecular weights between 30 000 and 35 000 Dalton (Zalman et al., 1980; Lindén et al., 1982a; Freitag et al., 1982b; Benz, 1985; De Pinto et al., 1987a). They are encoded in the nucleus without N-terminal leader extension and are synthesized on free cytoplasmic ribosomes (Freitag et al., 1982a). The basic difference between both types of pores is the voltage-dependence. The voltage-dependence of bacterial porins, as observed in certain reconstitution experiments at transmembrane potentials larger than 100 mV (Schindler & Rosenbusch, 1978), has most likely no physiological significance (Sen et al., 1988). Mitochondrial porins switch already at 20 to 30 mV to substates of the open state (Schein et al., 1976; Freitag et al., 1982b; Roos et al., 1982). These substates have completely different properties than the open state and are probably not permeable for ATP and ADP (Benz et al., 1988; Ludwig et al., 1988). This means that the mitochondrial metabolism could be controlled by the outer membrane porin as discussed by Brdiczka et al. (1989).

PROPERTIES OF BACTERIAL PORINS

Isolation and reconstitution of bacterial porins

The porins of most Gram-negative bacteria are tightly associated with the peptidoglycan layer (Nikaido, 1983; Benz, 1985; Nikaido & Vaara, 1985). This allows the rapid isolation of the porins via the preparation of the murein-protein complex. The cells are harvested in the late logarithmic phase. They are washed and resuspended in a small volume before they are passed several times through a French pressure cell. The pellet of a subsequent centrifugation step contains the cell envelope. Most components of the cell envelope are soluble in detergents. The insoluble material is composed of murein and a few proteins either covalently bound to or associated with the peptidoglycan. The porins can be released by standard methods either by digestion of the murein or by the salt extraction method (Nikaido, 1983). After the release, pure porin may be obtained by column chromatography using gel filtration or affinity chromatography. Several different methods have successfully been used to reconstitute porin pores into model systems. Here only one method is described in detail. For the others the interested reader is referred to recent reviews (Hancock, 1986; Benz, 1988).

The addition of porins to the aqueous phase bathing a black lipid bilayer membrane is the simplest reconstitution method (Benz *et al.*, 1978). Fig. 1 shows a typical reconstitution experiment of this type. The phosphate-starvation inducible porin PhoE of *Escherichia coli* K-12 was added in final concentration of 4 ng/ml to the aqueous phase bathing a lipid bilayer membrane from phosphatidylcholine/n-decane. The addition of the porin resulted in a stepwise increase of the membrane conductance. The steps were specific for the presence of the porins and were absent when only detergents were present in the aqueous phase. A large variety of different ions were found to be permeable through the general diffusion pores of the bacterial outer membranes. Subsequently, the single channel conductance of many but not all porin pores was a linear function of the specific conductance σ of the bulk aqueous phase as is shown in Fig. 2 for OmpF of *E. coli* outer membrane.

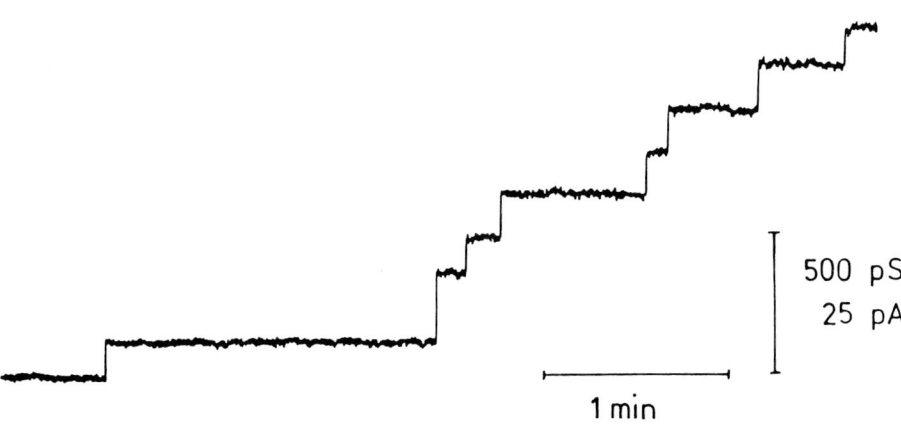

500 pS
25 pA

1 min

Fig. 1. Stepwise increase of membrane conductance (membrane current per unit voltage) in the presence of 4 ng/ml PhoE porin of *E. coli* K-12 in a 0.1 M KH_2PO_4 solution (pH 6). The membrane was formed from phosphatidylcholine/n-decane; T = 25°C, V_m = 50 mV.

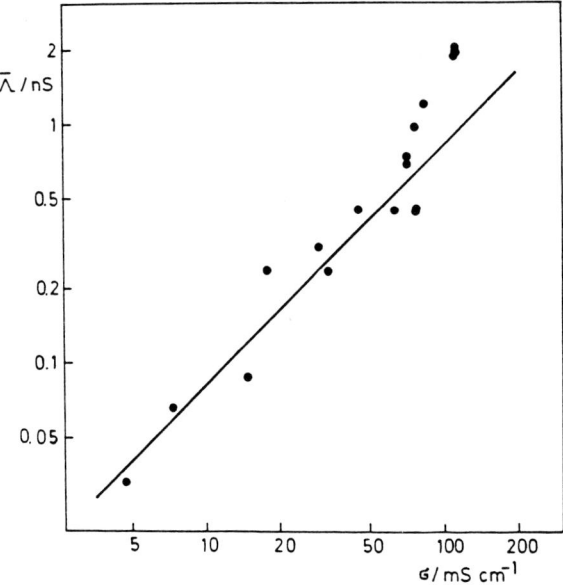

Fig. 2. Average pore conductance of OmpF of *E. coli* K-12 for different salts given as a function of the specific conductivity of the bulk aqueous phase. The data were taken from Benz *et al.* (1979).

Table 1. Zero-current membrane potentials V_m and effective diameter d of different *E.coli* porins in the presence of a ten-fold KCl gradient. The potential corresponds to the potential at the more dilute side of the membrane. P_c/P_a was calculated according to the Goldman-Hodgkin-Katz equation and d according to eqn (1) (σ = 110 mS/cm, 1 M KCl). The data was taken from Benz *et al.* (1985).

Porin	V_m (mV)	P_c/P_a	Λ(nS)	d (nm)
OmpF (B)	27	3.9	2.1	1.2
OmpF (K-12)	28	3.8	1.9	1.1
OmpC	50	26	1.5	1.0
PhoE	-24	0.30	1.8	1.1
K	46	7.6	1.5	1.0
NmpC	-26	0.27	1.3	1.0
LamB	-	-	0.16	-

This means that despite a large variation of the average single channel conductance the ratio Λ/σ varied only little (Benz *et al.*, 1979; Benz, 1985). The single-channel conductance of these general diffusion pores can be used to calculate the effective diameter of the porin pores. Assuming that the pores are filled with a solution of the same specific conductivity as the external solution and assuming a cylindrical pore with a length l of 6 nm, the effective pore diameter d (= 2r) can be calculated according to:

$$\Lambda = \sigma r^2 \pi/1 \qquad (1)$$

Where σ is the specific conductivity of the bulk aqueous phase. The diameters of the general diffusion pores using this method varied between 1.0 and 2.3 nm (Benz *et al.*, 1979, 1985; Benz, 1985). It has to be noted, that the effective pore diameters as calculated from the single channel data showed usually excellent agreement to the data derived from other methods including the vesicle permeability assay (Nikaido & Rosenberg, 1983; Hancock, 1986).

Specificity of porin channels

Ions move through the general diffusion pores similar to the way they move in the bulk aqueous phase (Benz, 1988). Nevertheless, the porin pores exhibit a certain specificity for charged solutes that can be detected *in vivo* and in *in vitro* experiments (Nikaido *et al.*, 1983; Nikaido & Rosenberg, 1983; Benz *et al.*, 1985). The lipid bilayer assay allows the evaluation of the ionic selectivity by measuring the membrane potential under zero-current conditions. From the measured V_m and the concentration gradient c"/c' across the membrane, the ratio P_c/P_a of the permeabilities (P_c for cations and P_a of anions) was calculated using the Goldman-Hodgkin-Katz equation (Benz *et al.*, 1979, 1985). Table 1 shows the zero-current membrane potentials for different porins of *E. coli*. OmpF and OmpC are present in the strain K-12. Their expression is regulated by the osmolarity of the growth media (van Alphen & Lugtenberg, 1978). PhoE is induced in *E. coli* under the conditions of phosphate-starvation (Tommassen & Lugtenberg, 1980). NmpC appears in revertants of porin-deficient mutants, whereas porin K was found in *E. coli* strains that form capsules (Benz *et al.*, 1985). It has to be noted that the selectivity of all these porins is not an absolute one. This means that the permeability ratio P_c/P_a is dependent on the mobility sequence of the ions in the aqueous phase and is generally larger for KCH_3COO than it is for KCl and even larger than it is for LiCl (Benz *et al.*, 1985).

Properties of specific porins

The general diffusion pores are wide water-filled channels. They sort between different substrates according to the molecular weight. The class of the molecules and their charge have only a small influence on their penetration through the pores. The outer membrane of certain Gram-negative bacteria contains, besides one or several general diffusion pores, channels that are specific for one class of solutes. These specific porins have completely different properties than the general pores because they contain a binding site for substrates inside the pore. It is obvious that in these cases, the effective channel diameter cannot be calculated using the formalism described above. Instead, a completely different procedure has to be used for the study of the channel properties, because the natural substrates for these specific porins are in many cases neutral solutes. Specific porins are induced in the outer membrane of *E. coli* (Luckey & Nikaido, 1980; Maier *et al.*,1988), *Pseudomonas aeruginosa* (Hancock & Carey 1980; Hancock *et al.*, 1982) and other Gram-negative bacteria when the organisms are grown under special growth conditions. Here only the properties of the sugar specific LamB porin of *E. coli* are discussed in detail.

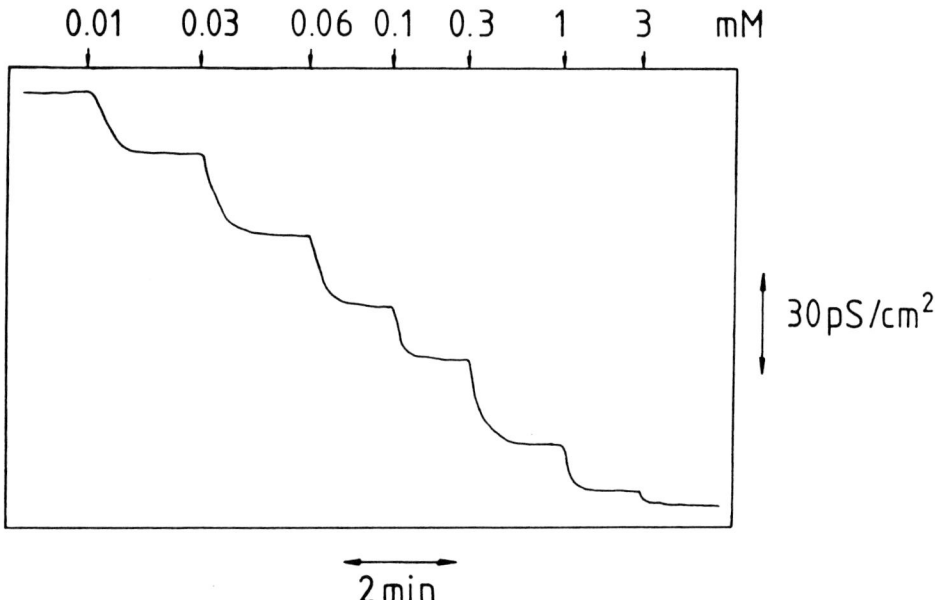

Fig. 3. Titration of LamB-induced membrane conductance with maltopentaose. The aqueous phase contained 50 ng/ml LamB, 1 M KCl, and maltopentaose at the concentrations shown at the top of the figure. The membrane was formed from phosphatidylcholine/n-decane; $T = 25^{\circ}C$, $V_m = 50$ mV.

The presence of maltose and oligosaccharides of the maltose series in the growth media of *E. coli* results in the expression of a number of different proteins located either in the inner membrane, the periplasmic space or the outer membrane (Szmelcman *et al.*, 1975). LamB, the protein induced in the outer membrane is the receptor for phage Lambda and plays an important role for the uptake of maltose into the cell. It forms small ion-permeable channels in lipid bilayer membranes (Benz *et al.*, 1987). The binding of sugars to LamB can be studied in reconstitution experiments using the following procedure. Lipid bilayer membranes are formed in an aqueous solution containing 1 M KCl and 50 ng/ml LamB protein. The membrane conductance increases during about 30 min due to the spontaneous insertion of the protein into the membrane. Then increasing concentrations of sugars are added to the aqueous phase as shown in Fig. 3 for the addition of maltopentose. The membrane conductance decreased as a function of the maltopentaose concentration. The data of Fig. 3 (and of similar experiments with other sugars) can be analyzed using a one-site two-barrier model for the sugar transport through LamB (Benz *et al.*, 1987). The conductance G(c) (max. conductance G_{max} in the absence of sugar) of the channel at a given sugar concentration c is given by the probability that the binding site is free:

$$(G_{max} - G(c))/G_{max} = Kc/(Kc + 1) \qquad (2)$$

where K is the stability constant for the binding of the sugar to the binding site. This means that the titration curve can be analyzed using a Lineweaver-Burk plot. For the data of Fig. 3 a stability constant K of 17,000 M^{-1} could be calculated corresponding to a half saturation constant of 0.06 mM. In similar experiments the binding of nucleosides to the nucleoside-specific Tsx protein of *E. coli* (Maier *et al.*, 1988) and the binding of phosphate to the phosphate-specific protein P of *P. aeruginosa* (Hancock & Benz, 1986) were determined. The binding sites inside the specific porins lead to a saturation of the substrate flow at high substrate concentration (Benz *et al.*, 1987).

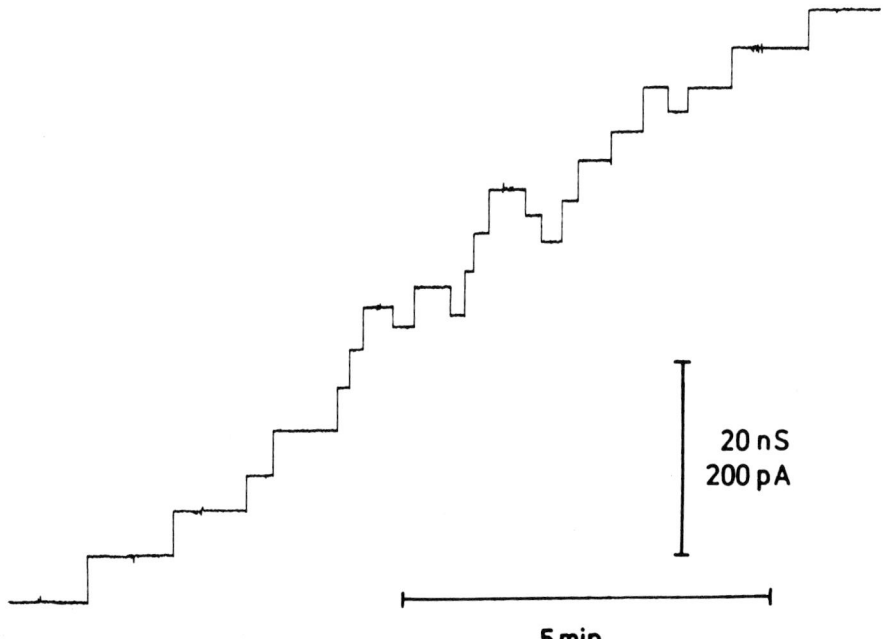

20 nS
200 pA

5 min

Fig. 4. Conductance fluctuations observed after the addition of 2 ng/ml yeast porin to 1 M KCl solution bathing a membrane of phosphatidylcholine/n-decane; T = 25°C, V_m = 10 mV.

On the other hand, the binding site facilitates the diffusion of the solutes at small concentrations which can be considered as the *in vivo* situation. This result demonstrates how transport *via* a binding site inside a channel is more efficient than transport through a general diffusion pore which is always a linear function of the driving force.

PROPERTIES OF MITOCHONDRIAL PORINS

Mitochondrial porins have been isolated using different methods (Zalman *et al.*, 1980; Freitag *et al.*, 1982b; Roos *et al.*, 1982; Lindén *et al.*, 1982; De Pinto *et al.*, 1987a). The most efficient method starts from whole mitochondria and uses a hydroxyapatite column as an essential step (De Pinto *et al.*, 1987b).

Reconstitution of mitochondrial porin

The reconstitution of mitochondrial porin is essentially the same as described above for the bacterial porins. The addition of small amounts of protein results in a conductance increase of many orders of magnitude. After an initial rapid increase for 15 to 20 min, the membrane conductance increased at a much slower rate. This slow conductance increase continued usually until membrane breakage. When the rate of conductance increase was relatively slow (as compared with the initial one) it was shown for different mitochondrial porins that the membrane conductance was a linear function of the protein concentration (Freitag *et al.*, 1982b; Roos *et al.*, 1982; Ludwig *et al.*, 1987).

Table 2. Single channel conductance of mitochondrial porins from different eukaryotic cells in 1 M KCl. The pore diameter d is calculated from eqn (1) assuming a length of the pore of 6 nm and a specific conductance σ of the pore interior of 110 mS/cm.

Porin from	Λ(nS)	d (nm)	Ref.
Neurospora crassa	4.5	1.8	Freitag *et al.*(1982b)
rat liver	4.3	1.7	Roos *et al.* (1982)
beef heart	4.0	1.7	Benz *et al.* (1985)
rabbit liver	4.0	1.7	Benz *et al.* (1985)
rat brain	4.0	1.7	De Pinto *et al.* (1987a)
rat kidney	4.0	1.7	De Pinto *et al.* (1987a)
pig heart	3.5	1.6	De Pinto *et al.* (1987a)
yeast	4.2	1.7	Ludwig *et al.*(1988)
Paramecium	2.4	1.3	Ludwig *et al.* (1988)

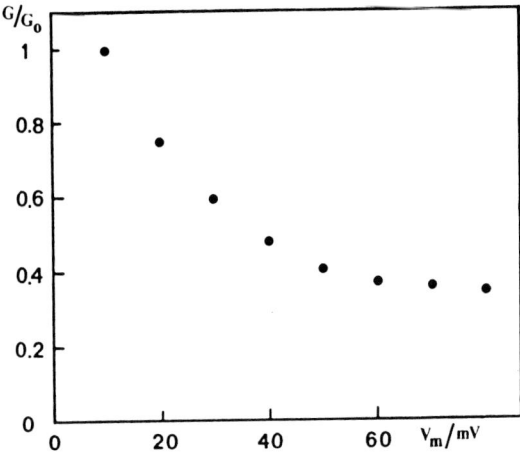

Fig. 5. Ratio of the conductance G divided by the conductance G_0 at zero potential as a function of V_m. The membrane was formed from phosphatidyl-choline/n-decane. The aqueous phase contained 1 M KCl and 12 ng/ml yeast porin (only at the cis-side). Only the positive branch (with respect to the cis-side) of the bell-shaped curve is shown; T = 25°C.

Single-channel analysis and pore diameter

The addition of smaller amounts of mitochondrial porin from different eukaryotic cells allowed the resolution of step increases in conductance as shown in Fig. 4 for yeast porin.

Most of the conductance steps were directed upwards and closing steps were only rarely observed at small transmembrane potentials of about 10 mV (see Fig. 4). The most frequent value for the single-channel conductance of yeast porin in 1 M KCl was about 4 nS. All mitochondrial porins (besides *Paramecium*) studied to date had the same single channel conductance under identical conditions (see Table 2).

Assuming that the same rough approximation as described above for the bacterial porins can also be used for the mitochondrial pore the pore diameter is calculated to be about 1.7 nm. Such a diameter is consistent with the results of research on isolated mitochondria (Wojtczak & Zaluska, 1969) and with those of research on the electron microscopy of mitochondrial outer membranes (Mannella & Frank, 1984a,b) but it disagrees with the diameter obtained by Colombini (1980) for the diffusion of polyethylene glycols through the mitochondrial pore. The reason for this discrepancy is not clear. It has to be noted, however, that there exists in the literature excellent agreement between the effective pore sizes of general diffusion pores calculated from the liposome swelling assay using hydrophilic solutes of different molecular weight and those estimated from the single-channel data (Nikaido & Rosenberg, 1983; Benz *et al.*, 1985).

Voltage-dependence of the mitochondrial porin

All mitochondrial porins studied to date were found to be voltage-dependent (Schein *et al.*, 1976; Freitag *et al.*, 1982b; De Pinto *et al.*, 1987a). The pore conductance is reduced at membrane potentials larger than 20 mV. The voltage-dependence of the mitochondrial porins can be demonstrated in experiments in which only a few channels were reconstituted into the membrane and in multi-channel experiments. In the multi-channel system the decay of the membrane current following a voltage step could be described by a single exponential decay. The steady-state conductance showed a bell-shaped curve as a function of the applied voltage as shown for yeast porin in Fig. 5. The data given in Fig. 5 could be analyzed as proposed by Schein *et al.* (1976):

$$N_o/N_c = \exp(nF(V_m - V_0)/RT) \quad (3)$$

where F, R and T have the usual meaning, n is the number of gating charges moving through the entire transmembrane potential gradient for channel gating and V_0 is the potential where 50% of the total number of channels are in the closed configuration. The open to closed ratio of the channels, N_o/N_c may be calculated from the data given in Fig. 5 according to:

$$N_o/N_c = (G - G_{min})/(G_0 - G) \quad (4)$$

G in this equation is the conductance at a given membrane potential V_m, G_0 and G_{min} are the conductances at zero voltage and very high potentials, respectively.

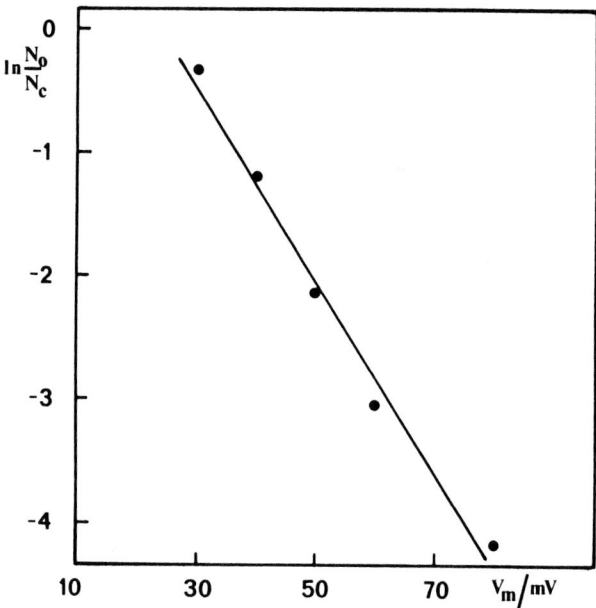

Fig. 6. Semilogarithmic plot of the ratio N_o/N_c as a function of the applied potential V_m. The data were calculated from the results of Fig. 5. The slope of the line was about 13 mV for an e-fold change of V_m; $V_0 = 16$ mV.

A semilogarithmic plot of the data given in Fig. 5 showed that they could be fitted to a straight line with a slope of 13 mV for an e-fold change of V_m. This result suggested that the number of charges involved in the gating process was approximately two in the case of yeast porin (see Fig. 6). For other mitochondrial porins n varied between 1 and 3.5 (De Pinto et al., 1987a).

Ionic selectivity of the closed state

The open state of all mitochondrial porins characterized so far is slightly anion-selective for salts with equally mobile cations and anions such as KCl. This means that the ionic selectivity is dependent on the mobility of the ions in the aqueous phase. Such a behavior is expected for a general diffusion pores because ions move through these pathways similar to the way they move through the aqueous phase. The zero-current membrane potentials were of the order of -10 mV at the more dilute side of a tenfold KCl gradient (Benz, 1985). This corresponds to a twofold higher permeability for Cl⁻ over K⁺ according to the Goldman-Hodgkin-Katz equation (Benz et al., 1979).

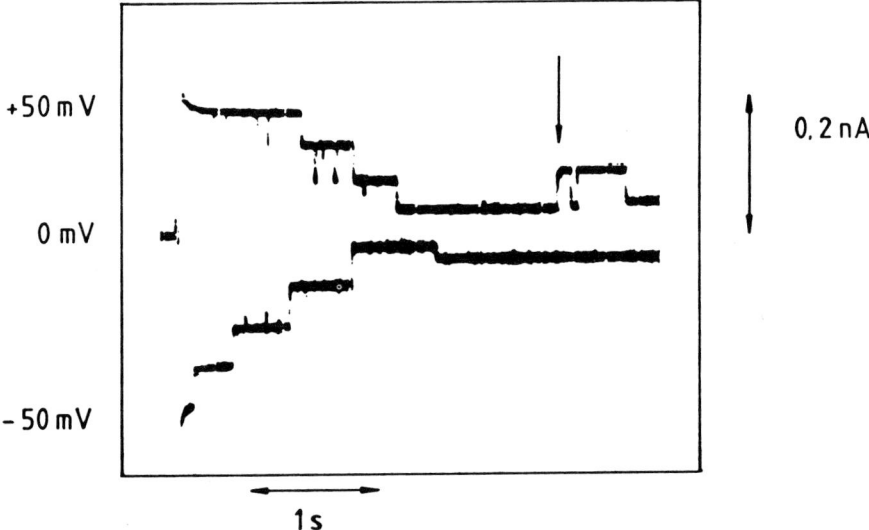

Fig. 7. Relaxation of the membrane current in the presence of the *Paramecium* porin. The membrane potential was first switched to +50 mV (three pores, upper trace) and then to -50 mV (four pores, lower trace). The arrow indicates the insertion of the fourth pore. The membrane was formed of phosphatidylcholine/n-decane in 0.5 M Tris-Cl (pH 7.2); T = 25°C.

As described above, the mitochondrial porins switched to closed states if the transmembrane voltages exceeded 15-20 mV. This states definitely have a reduced permeability towards ions. Their ionic selectivity is difficult to measure, because it is impossible to perform zero-current membrane potential measurements under the conditions in which the pores are in the closed state. To get some insight into the ionic selectivity of the closed state the following measurements were performed. The decay of the membrane current was studied as a function of time after the application of a voltage in KCl and in salts other than KCl. For these experiments a highly mobile cation was combined with a less mobile anion, and *vice versa*. Fig. 7 shows an experiment in which a voltage of 50 mV was applied to a membrane bathed in 0.5 M Tris-Cl (pH 7.2) containing 3 *Paramecium* porins (upper trace) and 4 pores (lower trace). Surprisingly, the current through the pores decayed to less than 20% of the initial value (as compared to about 55% in KCl alone).

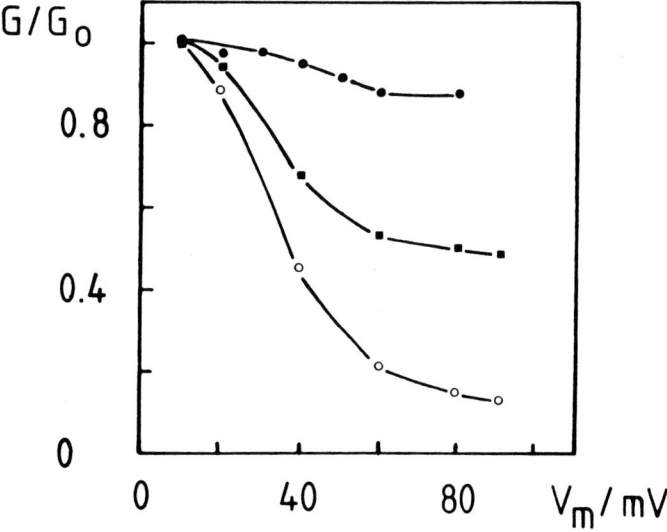

Fig. 8. Ratio of the conductance G divided by the conductance G_0 at 5 mV as a function of V_m. The membranes were formed from phosphatidylcholine/n-decane. The aqueous phase contained either 0.5 M KCl (pH 7.2) (full squares), 0.5 M K-MES (pH 7.2) (closed circles), or 0.5 M Tris-Cl (pH 7.2) (open circles) and 50 ng/ml *Paramecium* porin at the cis-side; T = 25°C.

Fig. 8 shows the reduced data for G/G_0 derived from similar experiments with *Paramecium* porin as a function of the transmembrane voltage for 0.5 M KCl (pH 7.2), 0.5 M Tris-Cl (pH 7.2), and 0.5 M K-MES (pH 7.2). It is obvious that the voltage dependence was in all cases similar. However, G/G_0 is dependent on the type of the cation and anion present in the aqueous solution. G/G_0 is considerably larger (at high voltages) in K-MES than in KCl or in Tris-Cl. This result could only be explained by the assumption that the substances of the *Paramecium* porin (and those of other mitochondrial pores) have a much lower permeability towards anions than to cations. Otherwise, the results of Fig. 8 cannot be explained. The exact value for the permeability ratio of potassium and chloride was difficult to obtain because the mobility of Tris and MES inside the pore is not known. However, because of the comparably small mobility of Tris and MES in the aqueous phase we are sure that P_K/P_{Cl} of the closed state is at least 10. The value may be even larger if the closed state is impermeable to anions. On the other hand, it is also evident from Fig. 8 that potassium is almost equally mobile through the open and the closed state which also suggested that the latter state has completely different properties. In fact, recent studies of rat liver porin with a polyanion has shown that the low conductance state of this porin is not permeable for ATP and ADP (Benz et al., 1988). This result is consistent with the cation- selectivity of the closed state.

CONCLUSION

The outer membranes of Gram-negative bacteria and mitochondria have special sieving properties as compared with normal cell membranes. Based on the presence of one or only a few proteins, these membranes have an extremely high permeability for a variety of different solutes. However, whereas the bacterial outer membranes have only passive properties which means their permeability is not voltage-regulated but still under full genetic control, the permeability of the mitochondrial membrane appears to be regulated by a transmembrane potential. In particular ADP and ATP cannot pass through the closed states of the mitochondrial porin. This could mean that a potential across the mitochondrial outer membrane controls the mitochondrial metabolism and the formation of compartments within the intermembrane space.

ACKNOWLEDGEMENTS

I would like to thank my colleagues for the excellent and fruitful collaboration in the fields of research into the permeability of bacterial and mitochondrial outer membranes. My own research was supported by the Deutsche Forschungsgemeinschaft (project B9 of the Sonderforschungsbereich 176).

REFERENCES

Benz R (1985) Porins from bacterial and mitochondrial outer membranes. CRC Crit Rev Biochem 19:145-190

Benz R (1988) Structure and function of porins from Gram-negative bacteria. Ann Rev Microbiol 42:359-393

Benz R, Janko K, Boos W, Läuger P (1978) Formation of large, ion-permeable membrane channels by the matrix protein (porin) of Escherichia coli. Biochim Biophys Acta 511:305-319

Benz R, Janko K, Läuger P (1979) Ionic selectivity of pores formed by the matrix protein (porin) of Escherichia coli. Biochim Biophys Acta 551:238-247

Benz R, Ludwig O, De Pinto V, Palmieri F (1985) Permeability properties of mitochondrial porins of different eukaryotic cells. In: Quagliariello E *et al.* (eds) Achievements and Perspectives of Mitochondrial Research, vol 1. Elsevier, Amsterdam, pp 317-327

Benz R, Schmid A, Hancock REW (1985) Ion selectivity of Gram-negative bacterial porins. J Bacteriol 162:722-727

Benz R, Schmid A, Vos-Scheperkeuter GH (1987) Mechanism of sugar transport through the sugar-specific LamB channel of *Escherichia coli* outer membrane. J Membrane Biol 100:12-29

Benz R, Wojtczak L, Bosch W, Brdiczka D (1988) Inhibition of adenine nucleotide transport through the mitochondrial porin by a synthetic polyanion. FEBS Lett 210:75-80

Bessman SP, Carpenter CL, (1985) The creatine-phosphate energy shuttle. Ann Rev Cytochem 54:831-865

Beveridge TJ (1981) Ultrastructure, chemistry and function of the bacterial wall. Int Rev Cytol 72:229-317

Brdiczka D, Knoll G, Riesinger I, Weiler U, Klug G, Benz R, Krause J (1986) Microcompartmentation at the mitochondrial surface: its function in metabolic regulation. Myocardial and skeletal muscle bioenergetics. Brautbar N (ed) Plenum Press, New York, pp 55-69

Brdiczka D, Adams V, Kottke M, Benz R (1989) Topology of peripheral kinases: Its importance in transmission of mitochondrial energy. In: Azzi A, Naɫecz KA, Naɫecz MJ, Wojtczak L (eds) The Anion Carriers of the Mitochondrial Membranes. Springer-Verlag Heidelberg, pp 361-372

Brooks SPJ, Suelter CH (1987) Compartmented coupling of chicken heart creatine kinase to the nucleotide translocase requires the outer membrane. Arch Biochem Biophys 257:144-153

Colombini M (1979) A candidate for the permeability pathway of the outer mitochondrial membrane. Nature 279:643-645

Colombini M (1980) Pore size and properties of channels from mitochondria isolated from *Neurospora crassa*. J Membrane Biol 53:79-84

De Pinto V, Ludwig O, Krause J, Benz R, Palmieri F (1987a) Porin pores of mitochondrial outer membranes from high and low eukaryotic cells: biochemical and biophysical characterization. Biochim Biophys Acta 894:109-119

De Pinto V, Prezioso G, Palmieri F (1987b) A simple and rapid method for the purification of the mitochondrial porin from mammalian sources. Biochim Biophys Acta 905:499-502

De Pinto V (1989) Purification of mammalian porins In: Azzi A, Naɫecz KA, Naɫecz MJ, Wojtczak L (eds) The Anion Carriers of the Mitochondrial Membranes. Springer-Verlag Heidelberg, pp 237-248

Engel A, Massalski A, Schindler M, Dorset DL, Rosenbusch, JP (1985) Porin channels merge into single outlets in *Escherichia coli* outer membranes. Nature 317:643-645

Fiek C, Benz R, Roos N, Brdiczka D (1982) Evidence for identity between the hexokinase-binding protein and the mitochondrial porin in the outer membrane of rat liver mitochondria. Biochim Biophys Acta 688:429-440

Freitag H, Janes M, Neupert W (1982a) Biosynthesis of mitochondrial porin and insertion into the outer mitochondrial membrane of *Neurospora crassa*. Eur J Biochem 126:197-202

Freitag H, Neupert W, Benz R (1982b) Purification and characterization of a pore protein of the outer mitochondrial membrane from *Neurospora crassa*. Eur J Biochem 123:629-636

Gellerich FN, Schlame M, Bohnensack R, Kunz W (1987) Dynamic compartmentation of adenine nucleotides in the mitochondrial intermembrane space of rat-heart mitochondria. Biochim Biophys Acta 890:117-126

Hancock REW (1986) Model membrane studies of porin function. In: Inouye M (ed) Bacterial outer Membranes as Model Systems. Wiley and Sons, New York, pp 187-225

Hancock REW, Benz R (1986) Demonstration and chemical modification of a specific phosphate binding site in the phosphate-starvation-inducible outer membrane porin protein P of *Pseudomonas aeruginosa*. Biochim Biophys Acta 860:699-707

Hancock REW, Poole K, Benz R (1982) Outer membrane protein P of *Pseudomonas aeruginosa*: regulation by phosphate deficiency and formation of small anion-specific channels in lipid bilayer membranes. J Bacteriol 150:730-738

Kleene R, Pfanner N, Pfaller R, Link TA, Sebald W, Neupert W, Tropschug M (1987) Mitochondrial porin of *Neurospora crassa*: cDNA cloning, *in vitro* expression and import into mitochondria. EMBO J 9:2627-2633

Lepault J, Dargent B, Tichelaar W, Rosenbusch JP, Leonard K, Pattus F (1988) Three dimensional reconstitution of maltoporin from electron microscopy and image processing. EMBO J 7:261-268

Lindén M, Gellerfors P, Nelson BD (1982a) Purification of a protein having pore forming activity from the rat liver mitochondrial outer membrane. Biochem J 208:77-82

Lindén M, Gellerfors P, Nelson BD (1982b) Pore protein and hexokinase-binding protein from the outer membrane of rat liver mitochondria are identical. FEBS Lett 141:189-192

Luckey M, Nikaido H (1980) Specificity of diffusion channels produced by -phage receptor protein of *Escherichia coli*. Proc Natl Acad Sci (USA) 77:167-171

Ludwig O, Krause J, Hay R, Benz R (1988a) Purification and characterization of the pore forming protein of yeast mitochondrial outer membrane. Eur Biophys J 15:269-276

Ludwig O, Benz R, Schultz IE (1988b) Porin of *Paramecium* mitochondria: Isolation characterization and ion selectivity of the closed state. Eur J Biochem (submitted)

Maier C, Bremer E, Schmid A, Benz R (1987) Pore-forming activity of the Tsx protein from the outer membrane of *Escherichia coli*. Demonstration of a nucleoside-specific binding site. J Biol Chem 263:2493-2499

Mannella CA, Frank J (1984a) Negative staining characteristics of arrays of mitochondrial pore protein: Use of correspondence analysis to classify different staining patterns. Ultramicroscopy 13:93-102

Mannella CA, Frank J (1984b) Electron microscopic stains as probes of the surface charge of mitochondrial outer membrane channels. Biophys J 45:139-141

Mihara K, Sato R (1985) Molecular cloning and sequencing of cDNA of yeast porin, an outer mitochondrial membrane protein: a search for targeting signal in the primary structure. EMBO J 4:769-774

Nakae T (1976) Identification of the major outer membrane protein of *Escherichia coli* that produces transmembrane channels in reconstituted vesicle membranes. Biochem Biophys Res Commun 71:877-884

Nikaido H (1983) Proteins forming large channels from bacterial and mitochondrial outer membranes: Porins and phage lambda receptor protein. Methods Enzymol 97:85-100

Nikaido H, Rosenberg EY (1983) Porin channels in *Escherichia coli*: studies with liposomes reconstituted from purified proteins. J Bacteriol 153:241-252

Nikaido H, Rosenberg EY, Foulds J (1983) Porin channels in *Escherichia coli*: studies with β-lactams in intact cells. J Bacteriol 153:232-240

Nikaido H, Vaara M (1985) Molecular basis of bacterial outer membrane permeability. Microbiol Rev 49:1-32

Ohlendieck K, Riesinger I, Adams V, Krause J, Brdiczka D (1986) Enrichment and biochemical characterization of boundary membrane contact sites in rat-liver mitochondria. Biochim Biophys Acta 860:672-689

Roos AK, Benz R, Brdiczka D (1982) Identification and characterization of the pore-forming protein in the outer membrane of rat liver mitochondria. Biochim Biophys Acta 686:204-214

Schein SJ, Colombini M, Finkelstein A (1976) Reconstitution in planar lipid bilayers of a voltage-dependent anion-selective channel obtained from *Paramecium* mitochondria. J Membrane Biol 30:99-120

Schindler H, Rosenbusch JP (1978) Matrix protein from *Escherichia coli* outer membranes forms voltage-controlled channels in lipid bilayers. Proc Natl Acad Sci (USA) 75:3751-3755

Sen K, Hellman J, Nikaido H (1988) Porin channels in intact cell of *Escherichia coli* are not affected by Donnan potentials across the outer membrane. J Biol Chem 263:1182-1187

Szmelcman S, Hofnung M (1975) Maltose transport in *Escherichia coli* K-12: Involvement of the bacteriophage lambda receptor. J Bacteriol 124:112-118

Tommassen J, Lugtenberg B (1980) Outer membrane protein e of *Escherichia coli* K-12 is coregulated with alkaline phosphatase. J Bacteriol 143:151-157

Wojtczak L, Zaluska H (1969) On the impermeability of the outer mitochondrial membrane to cytochrome c: I. Studies on whole mitochondria. Biochim Biophys Acta 193:64-72

Zalman LS, Nikaido H, Kagawa Y (1980) Mitochondrial outer membrane contains a protein producing nonspecific diffusion channels. J Biol Chem 255:1771-1774

Modulation of the Mitochondrial Channel VDAC by a Variety of Agents

Marco Colombini, Marcia J. Holden, and Patrick S. Mangan

Department of Zoology, University of Maryland, College Park, MD 20742, United States of America

A major difference between prokaryotes and eukaryotes is the extensive use of intracellular membranous organelles, by eukaryotes, to compartmentalize metabolic processes. The full implications of this compartmentation will probably not be appreciated for some time. However, it is clear that this compartmentation poses barriers to the movement of metabolites from one compartment to another. Transport systems both overcome these barriers and provide a means for regulating the flux of metabolites and thus the rate and direction of metabolic processes.

The mitochondrial outer membrane is one such barrier. The major permeability pathway for metabolite flux through this membrane is the channel-forming protein called VDAC (see Mini- Review Series, 1987). This protein, with essentially unchanged properties has been isolated from all mitochondria tested including all 4 eukaryotic kingdoms (Schein *et al.*, 1976; Colombini, 1979; Smack & Colombini, 1985). Like other transport systems, VDAC probably acts not merely as a conduit for the flow of metabolites between the cytoplasm and the intermembrane space but also as a means of regulating this flux. By studying the properties of VDAC channels inserted into phospholipid membranes, we have been able to demonstrate many ways of regulating the permeability of these channels to ions. Although we have used almost exclusively the ions of simple salts, the same (if not larger) effects should be observed with

A. Azzi et al. (Eds.)
Anion Carriers of Mitochondrial Membranes
© Springer-Verlag Berlin Heidelberg 1989

physiologically important ions such as succinate, citrate, ATP, P_i, *etc.* (in a few cases where this was examined, this expectation proved correct).

Since some confusion exists in the literature, a disclaimer must be expressed at this point. None of the results that will be presented are to be misconstrued as applying to the bacterial channel called porin. Since VDAC (referred to by some as mitochondrial porin) and porin are unrelated, as judged from electrophysiological properties, structural properties (Mannella & Tedeschi, 1987), and primary sequence (Forte *et al.*, 1987; Kleene *et al.*, 1987), there is no reason to expect that VDAC's characteristics can be generalized to porin.

VDAC IS A REGULATABLE CHANNEL

In order for a protein to be regulatable, it must fulfill some rather rigid criteria. It must exist in different conformational states whose properties are very different. The energy level of the different states must be extremely close so that relatively small perturbations will induce a change in conformational state. Finally, there must be ways of inducing these conformational transitions. These conditions are so stringent that if a protein were to fulfill them then it is very likely that the regulatory property was selected and is maintained by evolutionary pressure. It will be demonstrated that VDAC does fulfill the above conditions and therefore it will be argued that the regulation of VDAC must be important physiologically.

QUASI-DEGENERATE CONFORMATIONAL STATES

VDAC can occupy conformational states that differ in their ability to allow ions and small molecules to cross membranes. The state which induces the highest permeability is referred to as the open state. The variety of states with lower permeability are referred to collectively as closed states. When a VDAC channel is induced to close, its ion permeability, as measured by the single-channel conductance can become one of a set of discrete conductance levels. Since molecular processes are probable, the closed channel can open and then close again into the same or a different conformational state. When many channels are under investigation, it is, at present, impossible to determine which closed configuration they will assume. Therefore, the closed conformations are considered as a group. The open state differs from the closed states in the following ways: 1) The closed state's permeability to salts of alkali metal ions is about 40% of that of the open state (Colombini, 1986). For larger ions the permeability change is greater: 10% for succinate. 2) The effective radius of the aqueous pore formed by the open state is about 1.5 nm while that of the closed state has been estimated at about 0.9 nm (Colombini et al., 1987). 3) The volume of the channel's aqueous pore is reduced by about 30 nm^3 upon channel closure (Zimmerberg & Parsegian, 1986). 4) The weak selectivity of the open state for anions is lost upon channel closure (Colombini, 1980).

These differences in the properties of the open and closed states indicate substantial changes in the protein's structure. However, the energy difference between the open and closed states is only about 8 kJ/mole (Doring & Colombini, 1985), about half of the energy of the average hydrogen bond. Thus, despite the relatively large change in properties of the different conformational states, their structure is such that their energy levels are almost the same.

REGULATION BY TRANSMEMBRANE POTENTIAL

The most straight-forward way of inducing conformational changes in VDAC is to apply a transmembrane potential. In the absence of a potential, the channel exists, almost exclusively, in an open conformation (Schein *et al.*, 1976). If a small potential is applied (about 20 mV), VDAC occupies closed conformations with increasing probability. At higher potentials (40 mV and above), the channels are almost always closed. The channels close when either positive and negative potentials are applied but the closed states achieved at positive potentials differ from those achieved at negative potentials (Colombini, 1986).

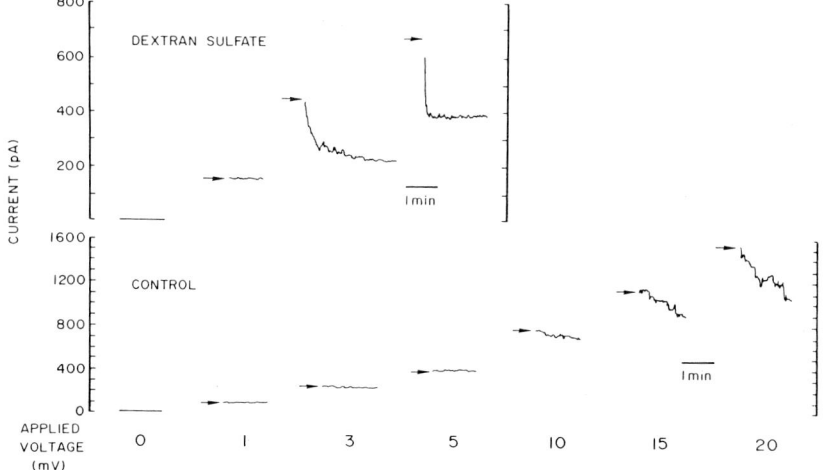

Fig.1. The voltage dependence of VDAC is amplified by dextran sulfate. The current flow through a channel-containing membrane was recorded with time after the application of various membrane potentials (numbers at the bottom refer to the voltage applied to obtain the traces above them). The lower panel (the control) shows recordings made prior to the addition of dextran sulfate. Those in the upper panel were made after dextran sulfate addition (from Mangan & Colombini, 1987).

The lower panel (control) in Fig.1 illustrates the behavior of VDAC channels isolated from *N. crassa* mitochondria. About 25 channels had inserted into a phospholipid membrane. The current flow through the membrane (essentially all this current is a result of ion flow through these channels) was monitored for the time shown after the application of a membrane potential (indicated by the numbers below the recordings). When the potential was first applied the resulting current represents the ion flow through open channels (the magnitude of the current is proportional to the applied voltage). At the higher potentials the current decreased with time as VDAC channels underwent transitions from the open to the closed conformation. Channel closure occurred more often and more rapidly as the membrane potential was increased. In this figure (1, control), closure was not very pronounced because small voltages were applied.

Could a change in potential across the outer membrane be used to control VDAC's conformation and thus the flux of metabolites across the outer membrane? No information is available regarding a potential across the outer membrane. It is unlikely that potentials based on ion gradients or ion pumps exist across the outer membrane because of its high permeability. However, potentials maintained by non-dissipative mechanisms, such as Donnan potentials or surface potentials, could exist. Such a potential could be changed by 1) changing the charge of the potential-inducing structure (the impermeant ion for Donnan potential; the charged entity on the surface) by protein phosphorylation or Ca^{++} binding (by changing cytoplasmic levels); 2) changing the concentration of this charged structure by controlling its synthesis or degradation. All these mechanism of altering the membrane potential are slow but the regulation of mitochondrial energy production should probably be a slow process responding to changes in cellular function or environmental conditions.

REGULATION BY POLYANIONS

The presence of highly negatively-charged polymers can dramatically increase VDAC's probability of being closed (Mangan & Colombini, 1987; Colombini *et al.*, 1987). The action of dextran sulfate is shown in the upper panel of Fig. 1. In the presence of dextran sulfate, VDAC channels closed completely with an applied potential of 3 mV. This contrasts with the control done on the same channels prior to dextran sulfate addition. In the control, no closure was observed at 3 mV, and closure was slow and incomplete even at 20 mV. Results obtained on a variety of polyanions are summarized in Table 1. The switching voltage, V_o (right column), is the voltage needed to close 1/2 the channels.

The polyanions reduce V_o resulting in channel closure at lower potentials. The steepness of the voltage dependence (2nd column from right) is a measure of the voltage range over which the channels go from essentially all open to all closed (the larger the number the narrower the range). The polyanions greatly increase this steepness resulting in a conformational change in a large proportion of the protein population with a very small change in membrane potential (as in Fig. 1). The synthetic polymers (such as dextran sulfate, Konig's polyanion, and polyaspartate) have a higher potency than natural, highly-charged polymers (pepsin and RNA). Perhaps this is due to the defined structure of the natural polymers which is probably less deformable than that of the synthetic polymers. Their deformability may allow the synthetic polymers to interact more effectively with the sensor that controls VDAC's conformation.

TABLE 1. Augmentation of VDAC's Voltage Dependence by Polyanions

Agent	Polyanion Concentration μM	(mg/ml)	Steepness of Voltage Dependence	Switching Voltage, V_o mV
Control			3.5	23.
Dextran Sulfate (8kDa)	6.2	(0.05)	10	6.7
	62	(0.5)	19	3.3
	120	(1.0)	25	3.2
	620	(5.0)	31	2.5
Dextran Sulfate (500kDa)	2	(1.0)	53	2.7
Polyaspartate (15kDa)	67	(1.0)	22	4.3
Konig's polyanion(10kDa)	0.3	(0.003)	16	4.2
RNA (25kDa)	40	(1.0)	6.9	14
tRNA (25kDa)	40	(1.0)	8.5	11
Pepsin(34.5kDa)	320	(10)	7.1	8.0

The polyanion-induced channel closure is intimately linked to the voltage-gating process. These represent two ways of inducing the channels to undergo transitions among the available conformations built into the native protein. Indeed, the polyanions markedly increase the voltage-dependence of the channels. It has been proposed that these polymers interact with the sensor to induce VDAC closure (Mangan & Colombini, 1987). The electric field increases the probability cf finding the polyanion at the mouth of the channel where it interacts with the sensor to induce channel closure. The closed states achieved with the aid of the polyanions appear to be the same ones as those achieved with membrane potential alone. Before dextran sulfate addition the average closed-state conductance was $38\pm1\%$ of the open-state conductance. As much as 10 mg/ml dextran sulfate had no significant effect on this value ($39\pm1\%$).

The characteristics of the polyanion effects are not consistent with a non-specific/artifactual phenomenon, rather these polymers seem to be triggering a process that is programmed into the structure of VDAC. While the effects are pronounced, a

220

basic characteristic is left unaltered. The energy difference between the open and closed states of VDAC, in the absence of an electric field is essentially unchanged by the addition of dextran sulfate.

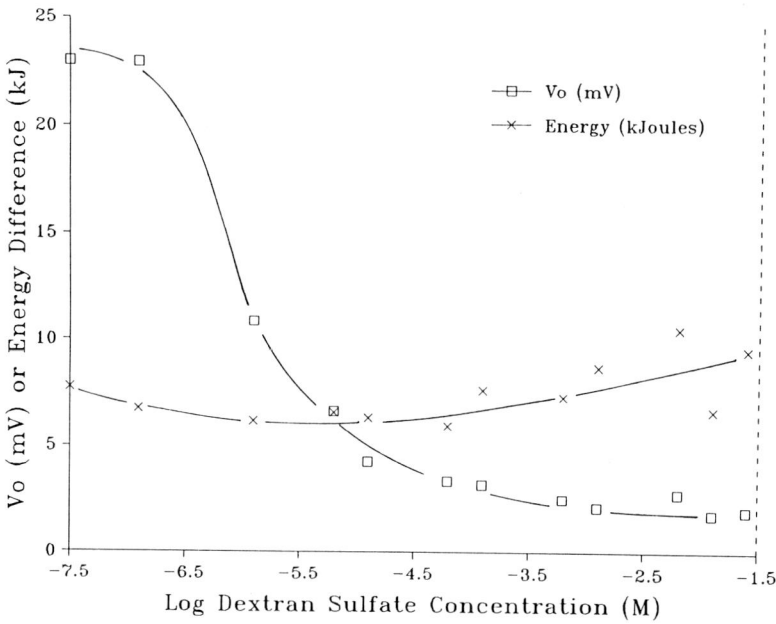

Fig.2. Dextran sulfate (8 kDa) greatly reduces the voltage needed to close half the channels (V_0, open squares) while leaving the energy difference between the open and closed states (X) essentially unchanged (see Mangan & Colombini, 1987). Each point is the mean of 1 to 5 estimations made by analyzing different VDAC - containing membranes with the indicated amount of dextran sulfate added to the aqueous phase on each side of the membrane.

If dextran sulfate were perturbing VDAC's structure, it is likely that it would perturb the open conformation differently than the closed conformation. This is especially true in view that dextran sulfate induces channels to close at much lower potentials (Fig. 1). Fig. 2 shows that while the voltage needed to close half the channels (V_0) is markedly reduced by increasing concentrations of dextran sulfate, the energy difference between the open and closed conformations changes very little.

If the polyanions are acting on a regulatory system that is programmed into VDAC's structure, there must be a physiological role for the system. There may be a substance in cells that acts on this regulatory system in a way that is mimicked by the polyanions. A substance has been found in mitochondrial fractions from yeast and *N. crassa* that induces VDAC to close. This substance has been dubbed the VDAC modulator (Holden & Colombini, 1988).

REGULATION BY THE MODULATOR

A soluble factor can be extracted from mitochondrial fractions that greatly increases VDAC's tendency to close when a potential is applied. Whether this substance is the endogenous counterpart to the polyanions used *in vitro* is unclear. Preliminary experiments indicate that it does not have a great deal of negative charge, although the presence of a highly negatively charged domain is possible. In detail, the modulator's effects differ somewhat from those observed with the polyanions.

The only assay currently available to detect modulator activity is to insert VDAC channels into the planar membranes and examine how the addition to the aqueous phase of a sample alters VDAC's voltage dependence. It is more convenient to perform the assay by using phospholipid membranes (made with diphytanoyl phosphatidyl choline) in which VDAC closes at higher than normal potentials. The resulting effect is more dramatic.

An example of the action of the modulator is shown in Fig. 3. Prior to modulator addition, the channels closed poorly at 60 mV. The addition of an aliquot of a modulator-containing fraction caused a rapid increase in the rate of channel closure which became even more pronounced with time. Essentially all the channels closed. The rate of opening of some of the channels is often markedly reduced (not shown). The addition of pronase (36 μg/ml final) caused a return to pre-modulator behavior within a minute. Similar results were obtained with trypsin indicating that the modulator is a soluble protein.

Does the modulator facilitate channel closure or act by physically blocking of the channel? This question was addressed by examining the modulator's activity at the single channel level. VDAC's voltage-induced closed conformation is still capable of allowing small ions to flow through the membrane, the permeability to these ions is only reduced by channel closure. The conductance decrement averaged 2.3 nS with voltage alone (pronase was added to insure that modulator was absent) and 2.4 nS with modulator present. In the presence of modulator one often observed a second conductance decrement of 1.0 nS. This was rarely observed without modulator but when it did occur its average conductance was 0.9 nS. Thus, the ion flow through the modulator-facilitated closed state is essentially the same as that through the voltage-induced closed state. However, with modulator present, the closed states were occupied more often and there was a greater preference for states of lower conductance. These findings are consistent with modulator-induced channel closure rather than blockage.

The modulator is extremely potent. From the activity exhibited by a partially purified preparation, the modulator can act at the low nanomolar range in concentration. This is a high estimate since the fraction was only partially purified.

Not enough is known about the modulator and its cellular distribution to make firm statements about how it may regulate the permeability of the outer membrane. Its effects on VDAC differ somewhat from those observed with the polyanions. From experiments mentioned here and others, the modulator is a soluble protein. Thus it is probably located in an aqueous compartment although it is possible that it might be adsorbed to the surface of a membrane. Since it is found in the mitochondrial fraction, two likely compartments are the intermembrane space and the matrix space.

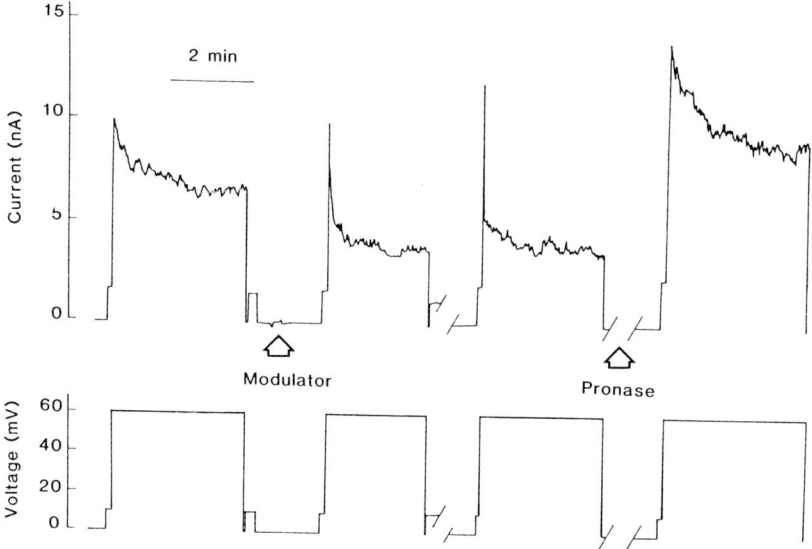

Fig. 3. A soluble protein modulator induces VDAC closure. To a phosphatidyl choline (diphytanoyl) membrane (in 1M KCl, 5mM CaCl$_2$) containing VDAC channels was added an aliquot of a sample containing the modulator (100 μl sample to 5 ml). The upper record shows the current flowing through the channels resulting from the applied voltage (lower tracing). The indicated breaks in the record were only a few minutes. At the point indicated, pronase was added to a final conc. of 37 μg/ml. During the experiment channels inserted into the membrane resulting in an increase in the current reading when the 10 mV test pulse was applied.

If the modulator were located in the intermembrane space, its concentration might be controlled so as to regulate the permeability of the outer membrane. If a potential exists across the outer membrane, raising or lowering the active modulator concentration could close or open the channels. The modulator could do this by changing V_0, the voltage at which the channels undergo their conformational transition, below or above the level of the transmembrane potential. It is possible that at high enough concentration the modulator alone (i.e. no potential) could induce channel closure (this has not been demonstrated).

Physiological Regulatory Role for VDAC

Direct evidence for the regulation of mitochondrial function by VDAC is not yet available. However there are a number of indications that such regulation exists. Most importantly, VDAC's structure and its built-in regulatory mechanisms. Secondly, the existence of a protein modulator that can change VDAC's response to an applied electric field. Finally, it has been demonstrated that the outer membrane can limit the rate of mitochondrial metabolism (Gellerich & Kunz, 1987) and access to drugs (Mannella et al., 1986). If such regulation exists, it will have great impact on our understanding of cellular energy metabolism.

Conclusions

The mitochondrial outer membrane channel, VDAC, is a pathway which allows molecules and ions to travel between the cytoplasm and the mitochondrial intermembrane space. When purified and reconstituted into phospholipid membranes, VDAC channels can undergo conformational transitions between quasi-degenerate states (overall energy differences are less than a hydrogen bond) with different permeabilities to ions. Small changes in the transmembrane potential will convert the channels from states of high permeability to ones with low permeability. These conformational changes can also be induced by negatively-charged polymers and by a soluble protein modulator found in the mitochondrial fraction. The negative polymers dramatically increase the voltage dependence of VDAC channels. Under appropriate conditions, the polymers can close VDAC channels in the absence of an externally applied electric field. The mode of action of these polymers is probably through the same process as occurs with an applied membrane potential. The modulator greatly increases the rate of closure of VDAC and reduces the voltage that must be applied to close the channels. Since the modulator is greatly diluted in the assay procedure, it is likely that the in vivo concentrations are sufficient to close the channels in the absence of an applied potential. VDAC therefore has the structural and behavioral characteristics of a regulated transport process. Perhaps cells use these channels to regulate mitochondrial functions.

References

Colombini M (1979) A candidate for the permeability pathway of the outer mitochondrial membrane. Nature 279:643-645

Colombini M (1980) Structure and mode of action of a voltage dependent anion-selective channel (VDAC) located in the outer mitochondrial membrane. Ann N Y Acad Sci (USA) 341:552-563

Colombini M (1986) Voltage gating in VDAC: toward a molecular mechanism. In: Miller C (ed) Ion Channel Reconstitution. Plenum Press, New York, p 533

Colombini M, Choh LY, Tung J, Konig T (1987) The mitochondrial outer membrane channel, VDAC, is regulated by a synthetic polyanion. Biochim Biophys Acta 905:279-286

Doring C, Colombini M (1985) Voltage dependence and ion selectivity of the mitochondrial channel, VDAC, are modified by succinic anhydride. J Membr Biol 83:81-86

Forte M, Guy HR, Mannella CA (1987) Molecular genetics of the VDAC ion channel: structural model and sequence analysis. J Bioenerg Biomemb 19:341-350

Gellerich FN, Kunz W (1987) Cause and consequences of dynamic compartmentation of adenine nucleotides in the mitochondrial intermembrane space in respect to exchange of energy rich phosphates between cytosol and mitochondria. Biomed Biochim Acta 46:S545-S548

Holden MJ, Colombini M (1988) A cellular protein modulates the behavior of the mitochondrial outer membrane channel, VDAC. Biophys J 53:266a

Kleene R, Pfanner N, Link TA, Sebald W, Neupert W, Tropschug M (1987) Mitochondrial porin of *N. crassa*: cDNA cloning, *in vitro* expression and import into mitochondria. EMBO J 6:2627-2633

Mangan PS, Colombini M (1987) Ultrasteep voltage dependence in a membrane channel. Proc Natl Acad Sci (USA) 84:4896-4900

Mannella CA, Capolongo N, Berkowitz R (1986) Correlation between outer-membrane lysis and susceptibility of mitochondria to inhibition by adriamycin and polyamines. Biochim Biophys Acta 848:312-316

Mannella CA, Tedeschi H (1987) Importance of the mitochondrial outer membrane channel as a model biological channel. J Bioenerg Biomemb 19:305-308

Mini-Review Series (1987) Biological channels: focus on the outer mitochondrial membrane. J Bioenerg Biomemb 19:305-358

Smack D, Colombini M (1985) Voltage-dependent channels found in the membrane fraction of corn mitochondria. Plant Physiol 79:1094-1097

Schein SJ, Colombini M, Finkelstein A (1976) Reconstitution in planar lipid bilayers of a voltage-dependent anion-selective channel obtained from *Paramecium* mitochondria. J Membr Biol 30:99-120

Zimmerberg J, Parsegian VA (1986) Polymer inaccessible volume changes during opening and closing of a voltage-dependent ionic channel. Nature 323:36-39

Bioenergetic Consequences of the Lack of Mitochondrial Porin: Identification of a Putative New Pore

J. MICHEJDA, X. J. GUO AND G. J.-M. LAUQUIN

Laboratoire de Physiologie Cellulaire Faculté des Sciences de Luminy Université Aix-Marseille II 70 Rte L. Lachamp 13288 Marseille-cedex 9 France
and
Department of Bioenergetics, University of Poznan, Fredry 10, 61701 Poznan, Poland

The mitochondrial porin, encoded in the nucleus and synthesized at the cytoplasmic ribosomes, is the major integral protein of the outer mitochondrial membrane, where it forms an oligomeric general diffusion pore or voltage-dependent anion channel (VDAC), allowing the non-specific passage of low molecular-weight hydrophilic solutes in its open state (Benz, 1985). An estimated diameter of approximately 2 nm and a defined exclusion limit of M_r 4,000 to 6,000 should make the channel as a crucial device allowing the free passage of substrates and metabolites, vital for the energetic and metabolic functions of mitochondria, through the outer membrane. Nevertheless, the possible role of the porin in controlling the metabolism and physiology of mitochondria is not fully understood, although several regulatory properties of porin-formed channels, especially the dependence of the conductivity of channels formed by porins isolated from various organisms on membrane voltage have been described in the reconstituted systems (De Pinto et al., 1987). In addition evidence was recently provided to prove that the closed state of the pore in mitochondria is impermeable for ATP and ADP (Benz et al., 1988).One would expect the absence of porin from mitochondria to largely limit the uptake of charged molecules e.g. ADP, ATP and the respiratory substrates, and would impose severe constraints on the

A. Azzi et al. (Eds.)
Anion Carriers of Mitochondrial Membranes
© Springer-Verlag Berlin Heidelberg 1989

226

```
           .-1300                                      .-1250
GACCCAGGTA GGTAGGAATT GAGAGGGAGC AAATATTTGA AAAGTGTTAT ACAAGAAGGT GAGTTCATCA TCAGTGATCT GACGTACGGG TTTTCCGTTT ACTGTCGGAA
        .-1200                             .1150
TAATACCACC GAGATTCGAG TTTTGTAAAA ATTTGATGTA TGTTTTAACC CAATTTAGAA AGTCATTAGG TGAGGTTAAC ATTGGTGGTG GTCTGACATA TTTTTTAGTG
 .-1100                              .-1050                                         .-1000
GATGTCATAT CAGAGTCCGC TGAGGATGAA TCAGTAAATG TATTACCTGA CTCAGGTGAT GGAGTGCTCA GAGGCGTTCC AACTGATGAT GGATACTGCG GAAACTGTGA
ATGTGGCCCA GGTGGAAAGT ACATAGGCGA CATTTGATAA GGTGTATACG GAATCATACA TGGGCCGTCCG TAAAATGACC AATCAGATGG ATTGGCTTGG TTTTGGGTCA
             .-950                                      .-900
TCATGCACTG CTGTGGGTAC GGCCCATTCT GTGGTGAATG TGACTGAGCA GTTTGAGGAG AGGCATGATG GGGGTTCTCT GGAACAGCTG ATGAAGCAGG TGTTGTTGTC
             .-850                             .-800
TGTTGAGAGT TAGCCTTAGT GGAAGCCTTA TCATATTCTT GAATTTTGGA AGCTGAAACG TCTAACGGAT CTTGATTTGT GTGGCACTTCC TTAGAAGTAA CCGAAGCACA
             .-750                             .-700
▶-650
GGCGCTACCA TCAGAAATGG GTGAATGTTG AGATAATTGT TGGGATTCCA TTGTTGATAA AGGCTATAAT ATTAGGTATA CAGAATATAC TAGAAGTTCT TCCTCGAGGA
 .-550                             .-500                          XhoI    .-450
TTTAGGAATC CATAAAAGGA AATCTGTAAT TCTACACAAT TCTATAAATA TTATTATCAT CGTTTTATAT GTTAATATTC ATTGATCCTA TTACATTATC AATCCTTGCG
                  .-400                                      .-350
TTTCAGCTTC CACTAATTTA GATGACTATT TCTCATCATT TGCGTCATCT TCTAACACCG TATATGATAA TATACTAGTA AGGTAAATAC TAGTTAGTAG ATGATAGTTG
                          ▶          .-300                     SpeI    .-250
ATTTTTATTC CAACAAGTTT AATGGTCAGA ATGCCCTTGT CGCGTGCCAG ATCGGGGTTC AATTCCCCGT CGCCGGAGATT TTTTGCCGCA CGTTTCCGCG TCAGACGAGC
             .-200                             .-150
TTTTTTTCACG GTTCCGGTGC TTTTTGCCGG CTGCGGCCCA ATCAAACACC GCCATTTCGG CCGTTCCTGA AGAAAGAACC CCTTTTATAG CCAGCAGAGC ACGAGTTGAT
                                         EagI
     .-100                             .-50                          .-10
CTACTATATA ACTACCCCCA ACTCGTTCCA CTACTCATTA GTGCTACGGA TTCTCCCAAC ACGAAACAGC CAAGCCGTACC CAAAGCAAAA ATCAAACCAA CCTCTCAACA
 .+1                                      .+50
ATG TCT CCT CCA GTT TAC AGC GAT ATC TCC AGA AAT ATC AAT GAC CTA TTG AAC AAG GAT TTC TAT CAT GCT ACC CCA GCT GCC TTT GAT
Met Ser Pro Pro Val Tyr Ser Asp Ile Ser Arg Asn Ile Asn Asp Leu Leu Asn Lys Asp Phe Tyr His Ala Thr Pro Ala Ala Phe Asp
                        EcoRV
        .+100                                      .+150
GTG CAA ACA ACA GCC AAT GGC ATT AAG TTC TCA TTG AAG GCT AAA CAG CCT GTC AAA GAC GGT CCA CTG TCT ACT AAC GTG GAA GCA
Val Gln Thr Thr Thr Ala Asn Gly Ile Lys Phe Ser Leu Lys Ala Lys Gln Pro Val Lys Asp Gly Pro Leu Ser Thr Asn Val Glu Ala
        .+200                                      .+250
AAG TTG AAT GAC AAG CAA ACC GGC TTG GGT CTA ACT CAA GGC TGG TCT AAC ACA AAC AAC TTG CAA ACC AAA TTA GAG TTT GCC AAC TTG
Lys Leu Asn Asp Lys Gln Thr Gly Leu Gly Leu Thr Gln Gly Trp Ser Asn Thr Asn Asn Leu Gln Thr Lys Leu Glu Phe Ala Asn Leu
                   .+300                                .+350
ACC CCT GGT CTA AAG AAC GAA TTG ATC ACT TCT TTG ACT CCA GGC GTC GCC AAG TCC GCC GTC TTA AAC ACT ACG TTC ACA GAA CCT TTC
Thr Pro Gly Leu Lys Asn Glu Leu Ile Thr Ser Leu Thr Pro Gly Val Ala Lys Ser Ala Val Leu Asn Thr Thr Phe Thr Glu Pro Phe
                        BclI                   .+400                                  .+450
TTC ACC GCA AGA GGT GCC TTT GAC TTG TGT TTG AAG TCA CCA ACA TTT GTT GGT GAC TTA ACT ATG GCC CAC GAA GGT ATT GTT GGT GGC
Phe Thr Ala Arg Gly Ala Phe Asp Leu Cys Leu Lys Ser Pro Thr Phe Val Gly Asp Leu Thr Met Ala His Glu Gly Ile Val Gly Gly
                                            .+500
GCA GAG TTT GGT TAC GAT ATC AGC GCC GGT TCC ATT TCT CGT TAT GCC ATG GCT TTA AGT TAT TTC GCC AAA GAC TAC TCC TTG GGC GCT
Ala Glu Phe Gly Tyr Asp Ile Ser Ala Gly Ser Ile Ser Arg Tyr Ala Met Ala Leu Ser Tyr Phe Ala Lys Asp Tyr Ser Leu Gly Ala
              EcoRV                            NcoI
        .+550                                      .+600
ACA TTG AAC AAC GAG CAA ATA ACT ACC GTT GAC TTC TTC CAA AAC GTC AAC GCC TTT TTA CAG GTC GGT GCT AAG GCT ACA ATG AAC TGC
Thr Leu Asn Asn Glu Gln Ile Thr Thr Val Asp Phe Phe Gln Asn Val Asn Ala Phe Leu Gln Val Gly Ala Lys Ala Thr Met Asn Cys
        .+650                                      .+700
AAA CTA CCT AAC TCC AAT GTC AAC ATC GAA TTC GCC ACT AGA TAT TTG CCT GAT GCA TCT TCC CAA GTT AAG GCT AAG GTG TCC GAT TCC
Lys Leu Pro Asn Ser Asn Val Asn Ile Glu Phe Ala Thr Arg Tyr Leu Pro Asp Ala Ser Ser Gln Val Lys Ala Lys Val Ser Asp Ser
                             EcoRI
                                  .+750                                  .+800
GGT ATT GTC ACT TTG GCT TAC AAG CAA TTG TTA AGA CCT GGC GTC ACT CTG GGT GTC GGT TCC TCT TTC GAT GCT TTG AAG TTG TCT GAA
Gly Ile Val Thr Leu Ala Tyr Lys Gln Leu Leu Arg Pro Gly Val Thr Leu Gly Val Gly Ser Ser Phe Asp Ala Leu Lys Leu Ser Glu
                                  .+850                                  .+900
CCT GTT CAC AAG CTA GGT TGG TCT TTG TCC TTC GAC GCT TGA ACGTATAT CTAATATATA TATGTTCACT ATATACCATA TATGTGCTCG TTCTTGTTTC
Pro Val His Lys Leu Gly Trp Ser Leu Ser Phe Asp Ala ***
                                  .+950                                  .+1000
CATTTTTCTA CTTGATCTTT AGACTTGTTG TTTTATTTAT ACAAATATAT TAATGCACGA AAATATTCTT TGTAACGGCT GGTTCCCATA ATCATATTGG TTGCTTGGCA
        .+1050                                      .+1100
GTGAGAATAG ACTCACTTCC TGACTCATCA TCACTTCCTT CCCGCAGCCTC TAGCTCATTC AAATACAAGC AGTGTTTAGC AAAGCCAGGG ATAACGTTTC TCGGTATTTT
        .+1150                                      .+1200
CAGATTGGTT TCCGAACATTG TGATAAACTC CAAATCTACT CCATCTTTCA TCCCACCAGA AAATAACCCG TACTCTTGGC ATATACCTGC AGTGGAACCA TCCCTGTAAC
        .+1250                                      .+1300
AACTTTCTTA TGATTCCCAC AACTACACCA ACACGGTGCT TGCCCTTATT GGAATGCACT AAGATGGGAT AGTTTCTGAC AACCAGCACC AAGTGCAAGA CCTGCTTCAT
```

Fig. 1. DNA sequence of the porin gene. Position +1 corresponds to the first nucleotide of the ATG initiation codon.

efficiency of energy generation and especially of ATP formation in these organelles with a large impact on cell growth and physiology.

Therefore several yeast mutants characterized by the phenotypic absence of the mitochondrial porin were constructed by disruption-deletion of the cloned porin gene (Guo & Lauquin, 1986, Guo, 1988) and the energy coupling was examined in the cells and the isolated mitochondria of these mutants.

RESULTS

Analysis of the porin gene

The yeast porin gene was isolated (Lauquin *et al.*, 1984) from a genomic expression library constructed in the λgt11 phage vector after screening with anti-porin antibodies raised against the porin protein isolated as the lower molecular weight band of an enriched preparation subjected to polyacrylamide electrophoresis (Boulay *et al.*, 1979). The DNA sequence was determined (Fig. 1) and indicated that the transcribed region was identical to the cDNA sequence reported by Mihara and Sato (1985). The presence of a single copy of the gene within the yeast genome was revealed after Southern blot using an intragenic porin probe. In addition the DNA sequence of the 5' flanking region revealed the presence of a Ty1 element flanked with two direct repeats (AGTTT) reflecting the insertion point of that element at the -315 position. The orientation of the Ty1 element is opposite to that of the porin transcript thus indicating that a putative alteration of the adjacent gene expression is possible. Further analysis did not show significant modulation of the porin gene expression. It is possible, however that the disruption of the porin gene could result in a strong activation of the Ty1 expression because it is accompanied by a huge accumulation within the cells of a 86 kDa protein the size of which is very similar to that of the major product of Ty1 element.

Construction of mutants lacking porin

Several different deletion/disruptions of the porin gene have been constructed *in vitro* (Fig. 2) and introduced at the chromosomal porin locus. All the resultant porin-deficient mutants were viable whatever the carbon source used for the growth medium. However the porin- mutants exhibited a marked thermosensitivity at 37°C when respirarory substrates (glycerol, ethanol or lactate) were the major carbon source of the culture medium. All the constructions made lead to the lack of porin within the cells, although some transcripts shorter than the parental could be detected with a porin probe in Northern experiments (Fig. 3) with the exception of the B5 mutant which did not exhibit any porin transcript. A suppressor mutant strain R1 has been isolated from the D9 porin- mutant by looking for cells able to grow at 37°C on medium supplemented with glycerol. When subjected to polyacrylamide gel electrophoresis cellular extracts from the R1 mutant did not exhibit any accumulation of the 86 kDa protein. In addition control experiment ascertained that the deletion /disruption of the D9 mutant was kept unchanged within the R1 strain.

Energy coupling of cell respiration

Cells were grown on glycerol as a nonfermentable carbon source; the endogenous respiration of cells of the wild strains starved in the absence of oligomycin was significantly increased by the uncoupler FCCP; 1-5 μM FCCP was most effective, depending on the strain. The cells starved in the presence of oligomycin revealed 20-40 % less O_2 uptake with ethanol, but this respiration was stimulated to threefold by FCCP. There was no significant difference in this pattern of respiratory response between the mutant and wild strains, except for about 1.5 times lower maximal (fully uncoupled) respiration in the former.

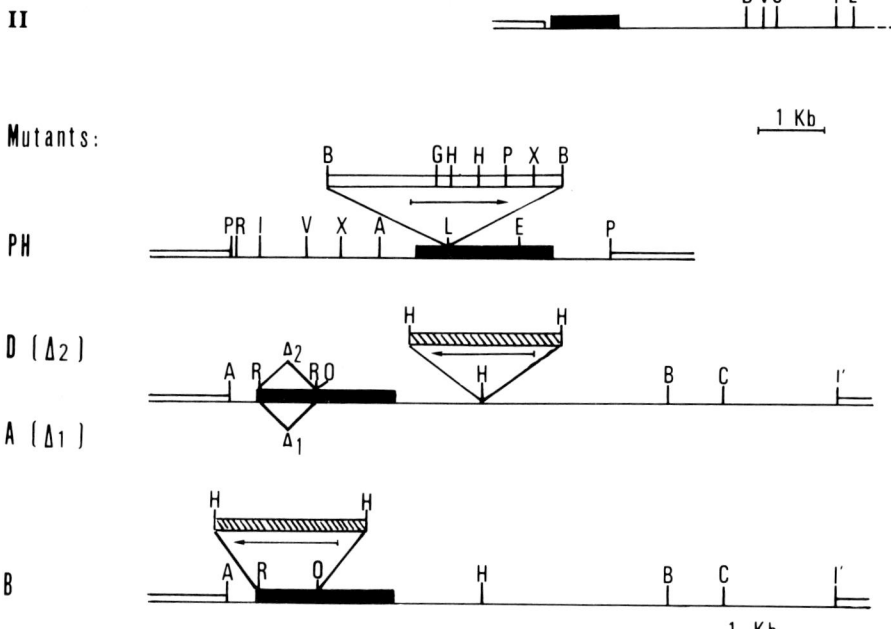

Fig. 2. Restriction maps of the cloned porin gene and of the different gene disruptions constructed. Open, hatched and filled bars are HIS3, URA3 and porin genes, respectively. A= EagI; B= BamHI; C= ClaI; E= EcoRI; G= BglII; H= HindIII; I= HpaI; L= BclI; O= NcoI; P= PstI; R= EcoRV; S= SalI; V= PvuI; X= XhoI; I and II are the two DNA fragments carrying the porin gene isolated using a probe from the isolated λgt11 porin plaque.

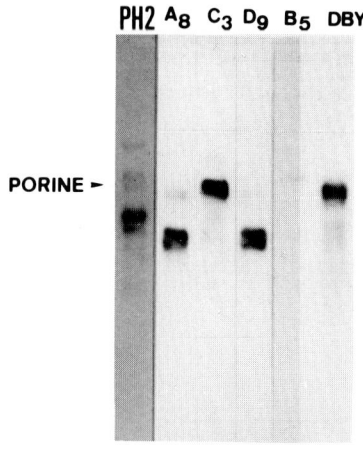

PORINE ►

Fig. 3. Northern blot of porin deficient mutant (PH2, A8, D9, B5) and control (DBY747, C3) strains. The probe was the ApaI-BclI DNA fragment from the 5' region of the porin gene.

The same was shown for wild and mutant cells grown on a glucose limited (0.3 %) medium but harvested after full derepression. The results suggest an *in vivo* efficient coupling in the cells lacking mitochondrial porin.

Energy coupling in isolated mitochondria

The outer mitochondrial membrane integrity was measured according to Douce *et al.*, (1984) as the difference between the KCN-sensitive oxygen uptake in the presence of ascorbate and exogenous cytochrome c in untreated mitochondria and that in mitochondria incubated with Lubrol or treated hypotonically. No significant difference in this respect was found between the mitochondria of mutants and those of the wild strains, both being 85-90 % intact.

In the absence of exogenous substrates the respiration of isolated mitochondria was negligible (less than 10 nAt O_2 x min-1 x mg-1 prot.) and the level of $\Delta\psi$ was less than 100 mV (near the limit of resolution under the experimental conditions applied). In the presence of 8 mM ethanol neither the state 4 respiration nor the fully uncoupled (in the presence of valinomycin 0.1 μg/mg) respiration, measured in mitochondria from mutants lacking porin, were significantly different from those in mitochondria from parental wild strains. In addition, there were no differences either in the rate of $\Delta\psi$ generation and its final level between the two types of mitochondria. In contrast, however, there was a significant difference in the dependence of the initial state 3 oxygen uptake and of the extension of $\Delta\psi$ depression on the concentration of ADP added (Table 1). While the curves of this dependence in the control (wild strain) mitochondria revealed a hyperbolic shape with an app. K_m about 50 μM ADP those in mitochondria from porin lacking mutants were more linear with an app. K_m about 160 μM ADP (Fig. 4). K_m symbolizes here the ADP concentration supporting the half-maximal respiratory control of the respiration ($V_{state\ 3}$ - $V_{state\ 4})_{0.5}$.

Correspondingly, the couse of return of state 3 to state 4 reflecting the decrease in external ADP concentration was sluggish on both, oxygen and $\Delta\psi$ tracings in mutant mitochondria in contrast to a much sharper state 3 - state 4 transition in mitochondria of the wild strains. This again reflected the increased K_m ADP for respiratory control in mutant mitochondria.

Table 1. Coupling properties of isolated mitochondria. Ethanol (8mM) supported O_2 uptake in nAt. min^{-1}.mg^{-1} protein; Val: valinomycin, 0.1 μg. mg^{-1} mitochondrial protein; Mg^{2+}: required for induction of the phosphorylating state. K_m ADP (μM) in the presence of 4mM MgCl$_2$. CATR (μM) to inhibit state 3 within 0.5 min in the presence of 200 μM ADP.

strain	O$_2$ uptake			K_mADP	CATR
	state 4	+ Val	Mg^{2+}		
GRF18	76	310	-	42	1
DBY747	72	250	-	47	1
PH2	80	225	+	145	15
D9	76	270	+	165	14
R1	83	230	+	130	10

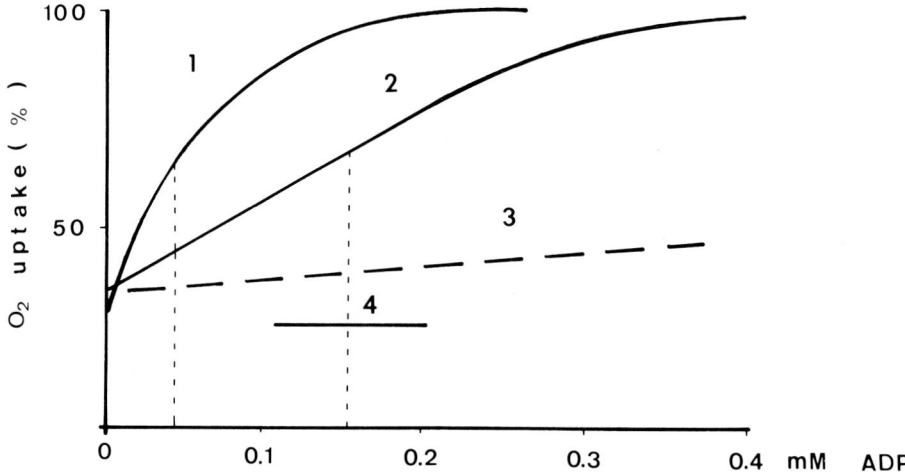

Fig. 4. Dependence of ethanol supported state 3 respiration on ADP concentration. Wild type (1); mutant +Mg (2) and -Mg (3); wild type and mutant +Mg +CATR (4). Ethanol: 8 mM; MgCl$_2$: 4 mM; Carboxyatractylate (CATR) 1 and 15 μM in wild type and mutant mitochondria, respectively.

This pattern of response caused serious difficulties in measuring the ADP:O ratio with small pulses of ADP in mutant mitochondria. However, with pulses of higher than 300 μM the respiratory control index (3.0-3.5) and the ADP:O ratio could be measured in mutant mitochondria, the values being close to those in mitochondria from the wild strains.

While the omission of Mg^{2+} from the incubation medium did not change the response of oxygen uptake and $\Delta\psi$ level to the pulses of ADP and did not increase the apparent K_m ADP for inducing state 3 respiration in wild strain mitochondria, the presence of external Mg^{2+} was necessary for such a response in mutant mitochondria.

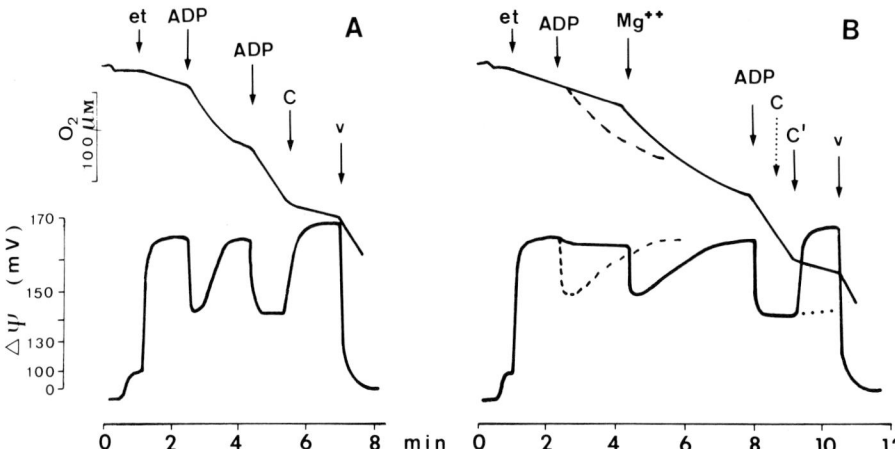

Fig. 5. Transitions of respiratory states driven by ADP and CATR in control (A) and mutant (B) mitochondria. A: ± Mg^{2+}. B: + 4 mM $MgCl_2$ (dashed line), - Mg^{2+} (full line). Pulses of ADP: first 150 μM, second 400 μM. et: ethanol 8 mM; Mg^{++} : 4 μM; CATR: (C) 1 μM, (C') 15 μM; v: valinomycin 0.1 μg.

In fact, in spite of the presence of 200-400 μM ADP, only after the addition of Mg^{2+} could the reversible transition from the resting state to the phosphorylating one be observed both on oxygen and$\Delta\psi$tracings and the ADP:O ratio could be measured in mutant mitochondria (Fig. 5B). Ca^{2+} was similarly efficient in the restoration of state 4-state 3-state 4 transitions following the pulse of ADP. A 4 mM concentration of these bivalent cations was necessary to make these transitions most effective (saturate these effects). It has to be mentioned, that Ca^{2+} is not accumulated by *S. cerevisiae* mitochondria and does not by itself stimulate oxygen uptake nor decrease the level of $\Delta\psi$.

In wild strain mitochondria 1 μM carboxyatractylate, an uncompetitive inhibitor of the adenine translocase, is efficient in the transition of state 3 into state 4 within half a minute in the presence of 300 μM external ADP, as proved by the inhibition oxygen uptake and generation of $\Delta\psi$ level a few millivolts higher than that at the previous state 4 (Fig. 5A). On the other hand, 10-15 times higher concentration of this inhibitor is needed to drive such a transition within half a minute in mitochondria of mutants lacking porin (Fig. 5B).

Similar differences between mitochondria from mutant and wild strains with respect to ADP, carboxyatractylate and Mg^{2+} concentrations necessary for transitions of the respiratory states, were found with succinate and external NADH as the substrates (though in this case some limitation in substrate transport seemed also to be involved).

There was no difference in effective concentrations of FCCP (1-2 μM) or valinomycin (0.1 μg per mg mitochondrial protein) necessary to fully uncouple mitochondria of wild strains and of porin lacking mutants, i.e. to induce the highest rate of oxygen uptake and a total collapse of $\Delta\psi$ within a few seconds.

All the above mentioned features, characteristic for the mitochondria of the strains lacking porin and exhibiting the positive temperature sensitivity (ts$^+$) paralleled by the accumulation of the 86 kDa protein (thus the strains PH2 and D9) were also found in the mitochondria of the suppressor strain R1 (derived from D9), which reveals no temperature sensitivity (ts$^-$) and no accumulation of the 86 kDa protein (Table 1).

DISCUSSION

Several yeast mutants lacking the porin gene have been constructed by chromosomal integration at the porin locus of several different disruption/deletions of the cloned porin gene. All the mutants were viable and stable. Analysis of the mitochondrial membranes of these mutants with anti-porin antibodies has shown the lack of porin protein. Although the porin- yeast mutants showed an unexpected viability whether the cells were grown on glucose or on nonfermentable carbon sources, the growth rate of the mutants on respiratory substrates was significantly reduced when compared to the parental strain. In addition, when grown on nonfermentable carbon sources the mutants exhibited a clear ts phenotype at 37°C and accumulated within the cells a huge amount of a 86 kDa protein. The latter result was previously reported by Dihanich et al., (1987) who made a similar porin-free mutant although their mutant needs a three days adaptation period before growth on glycerol could be possible for about 25% of the initially plated cells.

Ethanol is an efficient respiratory substrate in yeast cells and isolated mitochondria and its access to the mitochondrial matrix does not require any pore or carrier. In Saccharomyces cerevisiae oxidation of ethanol is coupled only at the II and III phosphorylating sites, like the oxidation of succinate and external NADH. Both state 4 and the uncoupled state respiration supported by ethanol were in the present studies approximately the same in mutant and control mitochondria. Thus the efficiency of the respiratory chain does not seem to be strongly influenced by the lack of porin. The results proved that valinomycin and FCCP are freely accessible to the inner mitochondrial membrane of porin-free mutants, totally collapsing its $\Delta\psi$.

A much higher increase in external CATR concentration required in mutant mitochondria to drive an effective state 3-state 4 transition (thus to inhibit the adenine nucleotide translocase) than that in the ADP concentration to allow the maximal state 3 respiration might be due to CATR having a 1.8 times higher molecular weight (Mw 760) and 1 more negative charge than ADP (Mw 480). As it is well known that the CATR inhibition of the translocase is not competitive with respect to ADP, one can assume some competition between both molecules to pass through the outer membrane. With isolated mitochondria this passage is apparently dependent on the presence of bivalent cations and this might be interpreted either in terms of the screening of the negative charges of these molecules to allow them to pass through another pathway of less anionic selectivity than that of the porin or in terms of controlling the open state of this pathway by cations. Apparently, the concentration of Mg^{2+} or other multivalent cations in the cytoplasm of mutant cells is favorable enough *in vivo* to enable the access of ADP to mitochondria as suggested by an efficient growth of the cells on a nonfermentable carbon source and by the same pattern of respiratory coupling in mutant and wild cells. In both types of cells the respiration seems to proceed at a similar intermediary state as suggested by the finding that the stimulation of the ethanol supported cell respiration by the uncoupler FCCP is significantly higher in the presence of oligomycin (state 4), than in its absence (attenuated state 3). Finding that the mitochondria of a porin-free suppressor strain (R1), not accumulating *in vivo* the 86 kDa protein reveal the same apparent limitation of the accessibility of ADP, CATR and NADH as the mitochondria of those porin-free mutants (D9, PH2 and B5), which accumulate the 86 kDa protein, suggests the absence of any relationship between these two features. The possibility of the relation between the accumulation of that protein and the efficiency of the energy metabolism (respiration and growth) has been suggested by Dihanich *et al.*, (1987) who proposed that the 86 kDa protein, abundant in porin lacking yeast cells adapted to grow on glycerol, might loosely bind to the mitochondrial surface *in vivo* and thereby increase the efficiency of a minor, outer membrane pore (perhaps by concentrating the transported molecules at the transport site).

Although much higher concentrations of ADP and CATR (and NADH) are required to allow their access to the inner membrane, these molecules can obviously pass through the outer membrane in the absence of the major pore. Though unlikely, the direct passive diffusion through the lipid bilayer of the outer membrane can not be excluded, in presence of Mg^{2+} the fluxes of ADP^{3-} through the porin-free membrane being similar to those of the wild type membrane, *e.g.* very high indeed, the existence of another channel-forming porin might be expected.

CONCLUSIONS

The ethanol supported respiration of porin lacking mutant and wild cells of yeast *S. cerevisiae* appears to be similarly coupled *in vivo*, as indicated by the stimulation of respiration by the uncoupler FCCP.

The absence of the major porin does not seem to influence the electron flux and $\Delta\psi$generation efficiency of isolated mitochondria but to increase the requirement for external concentration of ADP and CATR to drive the effective state 4-state 3 and state 3-state 4 transitions, respectively.

The results prove the penetration of ADP, NADH and CATR through the outer membrane deficient in the major porin. As the passive transport of these large charged molecules through the lipid bilayer of the outer membrane seems rather improbable, the existence of another porin species, possibly controlled by bivalent cations, is postulated. Alternatively, the necessity of divalent cations for the respiratory control might be interpreted in terms of screening the negative charges of ADP and CATR which seem to hinder their penetration, in the absence of the major porin, through the postulated new pore. Thus this channel would exhibit some cation selectivity in contrast to the major porin which is slightly anion selective.

The accumulation of the 86 kDa protein (and a parallel temperature sensitivity effect) observed in most of the porin deficient mutants does not seem to be related to the bioenergetic effects of the absence of the major porin.

REFERENCES

Benz R (1985) Porin from bacterial and mitochondrial outer membranes. CRC Crit Rev in Biochem 19:145-190

Benz R, Wojtczak L, Bosch W, Brdiczka D (1988) Inhibition of the adenine nucleotide transport through the mitochondrial porin by a synthetic polyanion. FEBS Lett 231:75-80

Boulay F, Brandolin G, Lauquin GJM, Jolles J, Jolles P, Vignais PV (1979) An ADP and atractyloside-binding protein involved in ADP/ATP transport in yeast mitochondria. FEBS Lett 98:161-164

Colombini M, Yeung CL, Tung J, König T (1987) The mitochondrial outer membrane channel, VDAC, is regulated by a synthetic polyanion. Biochim Biophys Acta 905:279-286

De Pinto V, Ludwig O, Krause J, Benz R, Palmieri F (1987) Porin pores of mitochondrial outer membranes from high and low eukaryotic cells: biochemical and biophysical characterization. Biochim Biophys Acta 894:109-119

Dihanich M, Suda K, Schatz G (1987) A yeast mutant lacking mitochondrial porin is respiratory-deficient but can recover respiration with simultaneous accumulation of an 86 kDa extramitochondrial protein. EMBO J 6:723-728

Douce R, Bourguignon J, Brouquisse R, Neuberger M (1984) Isolation of plant mitochondria. General principles and criteria of integrity. Methods Enzymol 148:403-415

Guo XJ (1988) Isolement et caractérisation du gène de la porine mitochondriale de *Saccharomyces cerevisiae*. Etude de l'expression du gène et construction de mutants déficients en porine.D.Sc. thesis, University of Aix-Marseille II.

Guo XJ, Lauquin GJM (1986) Mitochondrial porin-deficient mutant of *Saccharomyces cerevisiae: in vitro* construction and preliminary properties. In: EBEC Reports, Prague, 4:292

Lauquin GJM, Buhler JM, Marguet D (1984) Cloning of yeast mitochondrial porin gene with immunological screening. 12th Int Conf Yeast Genet Mol Biol abstact F.48.

Mihara K, Sato R (1985) Molecular cloning and sequencing of cDNA for yeast porin, an outer mitochondrial membrane protein: a search for targeting signal in the primary structure. EMBO J 4:769-774

Oestlund AK, Göhring U, Krause J, Brdiczka D (1983) The binding of glycerol kinase to the outer membrane of rat liver mitochondria. Its importance in metabolic regulation. Biochem Med, 30:231-245

Roos N, Benz R, Brdiczka D (1982) Identification and characterization of pore-forming protein in the outer membrane of rat liver mitochondria. Biochim Biophys Acta 686:204-241

Purification of Mammalian Porins

Vito de Pinto, Laura Gaballo, Roland Benz and Ferdinando
Palmieri

Department of Pharmaco-Biology, Laboratory of Biochemistry, University of Bari,
C.N.R. Unit for the Study of Mitochondria and Bioenergetics, trav. 200 v. Re David, 4
I-70125 Bari, Italy
and
Lehrstuhl für Biotechnologie, Universität Würzburg, Roentgenring 11, D-8700
Würzburg, Federal Republic of Germany

The matrix space of mitochondria is surrounded by two unit membranes. Whereas
the role of the inner membrane in oxidative phosphorylation was studied in full detail
in recent years, the role of the mitochondrial outer membrane in the mitochondrial
metabolism has been neglected because of its apparently high permeability for small
hydrophilic solutes. The mitochondrial outer membrane contains indeed general diffusion pores which explain its high permeability. The component of the membrane responsible for its molecular filter properties is a protein, called porin or VDAC (Benz,
1985). Mitochondrial porins were identified and characterized from a variety of eukaryotic cells by reconstitution experiments with planar lipid bilayers and liposomes
(Colombini, 1979, De Pinto et al., 1987). The mitochondrial pore has a diameter of
about 2 nm in the open state and is slightly anion selective for low transmembrane po-

Abbreviations: CHAPS: 3-(3-cholamidopropyl)-dimethyl-ammonio-1-propane-sul-
fonate; DCCD: dicyclohexylcarbodiimide; EDTA: ethylenediaminetetraacetic acid;
LDAO: lauryl-(dimethyl)-amineoxide; Lyso PC: lysophosphatidylcholine; octyl-POE:
octyl-polydisperse-oligooxyethylene; SDS-PAGE: sodium dodecyl sulfate polyacrylamide gel electrophoresis; TLC: thin-layer chromatography; zwittergent:
3(alkyldimethylammonio)-1-propanesulfonate.

A. Azzi et al. (Eds.)
Anion Carriers of Mitochondrial Membranes
© Springer-Verlag Berlin Heidelberg 1989

tentials. Voltages larger than 20 mV cause the shift of the pore into closed states with reduced permeability towards substrates and a changed selectivity.

More recent investigations gave some insight into the role and structural relationships between the outer mitochondrial membrane and the porin-pore. Porin has been identified to be identical with the hexokinase-binding protein. Furthermore, the subfractionation of mitochondrial membranes by density gradient centrifugation allowed the identification of contact sites between inner and outer membranes: these sites were especially enriched in porin and hexokinase-binding capacity (Ohlendiek *et al.*, 1986). This indicated that the outer membrane pore could represent a crucial point for adenine nucleotide transport from and to mitochondria. In yeast, the disruption of the gene encoding for the pore-forming protein unexpectedly resulted in a living mutant, which was able, after some days of adaptation, to grow and save the most important mitochondrial functions (Dihanich *et al.*, 1987). This result, together with other evidence obtained by patch-clamping experiments (Kinnally *et al.*, 1987), suggested the existence of a second porin. This means that the view of the outer mitochondrial membrane as a leaky barrier is replaced in favor of its important role in the mitochondrial metabolism.

In this paper we report the isolation of mitochondrial porin from bovine heart using different nonionic detergents. We screened their power for solubilizing the protein, their effect on chromatographic purification process and their influence on the functional activity of the protein. Two new porin purification procedures are presented. The former, in the presence of Triton X-100, allows the fast purification of high amounts of mitochondrial porins. The latter, in the presence of LDAO, allowed us to analyze the composition of the pore-forming complex. Our results give new information about the structure of the mitochondrial pore and are of general interest for the application of different detergents to the purification of membrane proteins.

METHODS

Analysis of porin solubilization

Bovine heart mitochondrial membranes were labelled with ^{14}C-DCCD (2 nmol/mg). This was the starting material for further solubilization experiments. The protein solubilization was carried out in a buffer containing 10 mM Tris/HCl, 1 mM EDTA, pH 7.5, 5 mg protein/ml and detergent in concentration between 0.1 and 4% (w/v) at 4^0 C. 30 min after the start of the solubilization, the nonsolubilized material was spun down at 147,000 x g for 30 min and the supernatant was analyzed. The total solubilization of protein was measured by the Lowry method modified for the presence of detergents. The solubilization of porin was obtained from a comparison of the radioactivity of the gel traces revealed by fluorography (De Pinto *et al.*, 1988).

Fast purification of the porin in the presence of Triton X-100

After the solubilization of the total mitochondrial membranes (40 mg) by 3% Triton X-100 in the same conditions as described above, the solubilized fraction was applied onto a dry, well packed hydroxyapatite/celite (6 g, ratio . 2 : 1) column. Elution was performed with the solubilization buffer. The first 12 ml were usually collected and porin was the only protein present in the eluate (De Pinto *et al.*, 1987b).

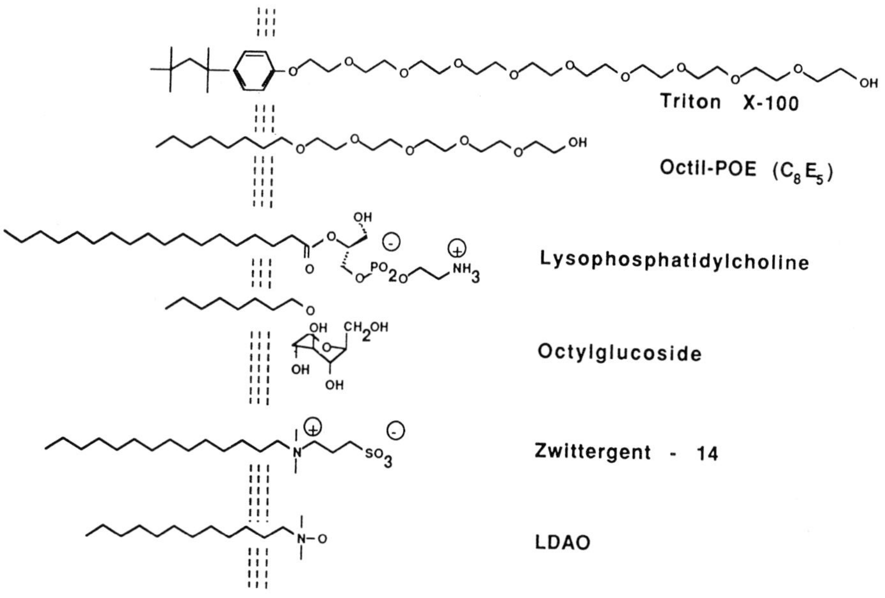

Fig. 1: Comparison of structural formulas of some detergents used in this paper.

Purification of the porin in the presence of LDAO

Mitochondrial membranes (40 mg/ml) were solubilized as described above by 2 % LDAO. After centrifugation, the supernatant was applied onto a dry hydroxyapatite/celite (6 g, ratio 2 : 1) column. The elution was performed first by the application of 18 ml of the solubilization buffer; and then by applying 27 ml of the solubilization buffer containing in addition 5 mM KP_i and 50 mM KCl. Fractions of 4.5 ml were collected. The 8th and 9th fractions contained pure porin (De Pinto *et al.*, 1988).

Lipid bilayer experiments

The methods used for the black lipid bilayer experiments were described previously (Benz *et al.*, 1978). The membranes were formed across circular holes (surface area about 0.1 mm^2 for the macroscopic conductance and the selectivity measurements) in the thin wall of a Teflon cell separating two aqueous compartments. The membranes were formed by painting onto the holes a 1 % (w/v) solution of diphytanoyl phosphatidylcholine (Avanti Biochemicals, Birmingham AL) dissolved in n-decane.

Other methods

SDS-polyacrylamide slab gel electrophoresis, fluorography, protein determination in the presence of detergents, lipid and sugar analysis were performed as described in De Pinto *et al.* (1988).

RESULTS

Porin solubilization by various detergents

Various nonionic, ionic and dipole or zwitterionic detergents were used to solubilize the mitochondrial membrane proteins and, particularly, the porin.

Nonionic detergents such as Tritons or Brijs have been already extensively used and their solubilization power has been studied. Much less is known about relatively new detergents, nonionic or zwitterionic, which may be of considerable interest for the use in crystallization of membrane proteins (Fig. 1) (Michel, 1983).

The solubilization buffers used in this study contained no salts and they were applied to mitochondrial membranes as starting material. Among the various physical and chemical properties of detergents, the HLB (hydrophobic lipophilic balance) number was used as a measure of the relative hydrophobicity of the nonionic detergents. The most hydrophobic detergents have an HLB number around zero, whereas it is for the most hydrophilic around 20. Almost all the detergents effective in membrane solubilization have HLB numbers in the range between 12 and 15. Fig. 2 shows the plot of the solubilization of protein from mitochondrial membranes as a function of the HLB number of different nonionic detergents. The total protein solubilization had a maximum around HLB 13. In comparison, porin showed better solubilization for smaller HLB number, *i.e.* for more hydrophobic detergents.

The solubilization of porin and of total membrane proteins was also studied for families of different detergents. These detergents series are available for fixed hydrophobic moiety and a variable length of hydrophilic part or, *vice-versa*, with a fixed hydrophilic part and a variable hydrophobic region. Among the same series, the detergent concentration and all the other parameters were kept constant, so that the degree of protein solubilization on the variable region of the tenside molecule could be compared.

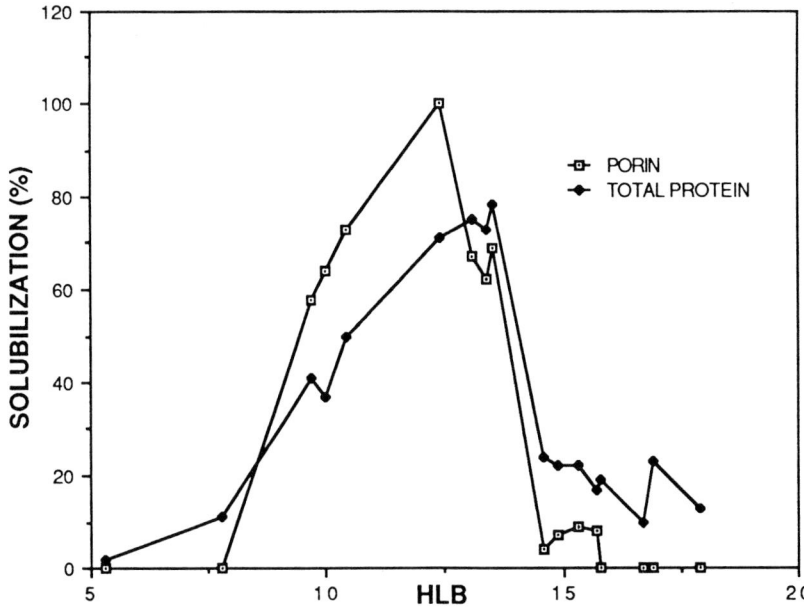

Figure 2. Porin and mitochondrial membrane proteins solubilization dependence as a function of the HLB number of different nonionic detergents. ^{14}C-DCCD labelled mitochondrial membranes were solubilized for 30 min at 0°C in solubilization buffers containing 2% (w/v) detergent. Each point represents a solubilization with a different detergent.

The Triton-X series has a common hydrophobic t-octyl-phenoxy group and a different number of added ethylene oxide units. The most effective detergents were Triton X-100 and Triton X-114. The whole curve was shifted towards the more hydrophobic part of the plot for the porin. In similar experiments, the series of dimethylamino N-oxides was employed: the polar head of this small detergent was fixed and the length of the hydrocarbon chain varied from 9 to 12 carbon units. In the case of total mitochondrial membrane proteins a plateau around 50-60% solubilization of total protein was achieved by detergents with alkyl chain lengths between 10 and 12. Porin was completely solubilized with LDAO, the detergent with the longest alkyl chain used (number of C = 12). The zwittergents are a homologous series of zwitterionic detergents with a sulfobetaine polar head. The polar head was again fixed and the length of hydrophilic alkyl tail ranged between 8 and 16 carbon units. Zwittergents 8 and 10 appeared to be very poor detergents, but zwittergent 14 solubilized 60% of total membrane proteins. A complete solubilization of porin was achieved with zwittergent 16 (data not shown).

With Tritons, alkyl dimethyl amino oxides and zwittergents, porin solubilization was best accomplished with the more hydrophobic detergents of the series with respect to the average behaviour of mitochondrial membrane proteins.

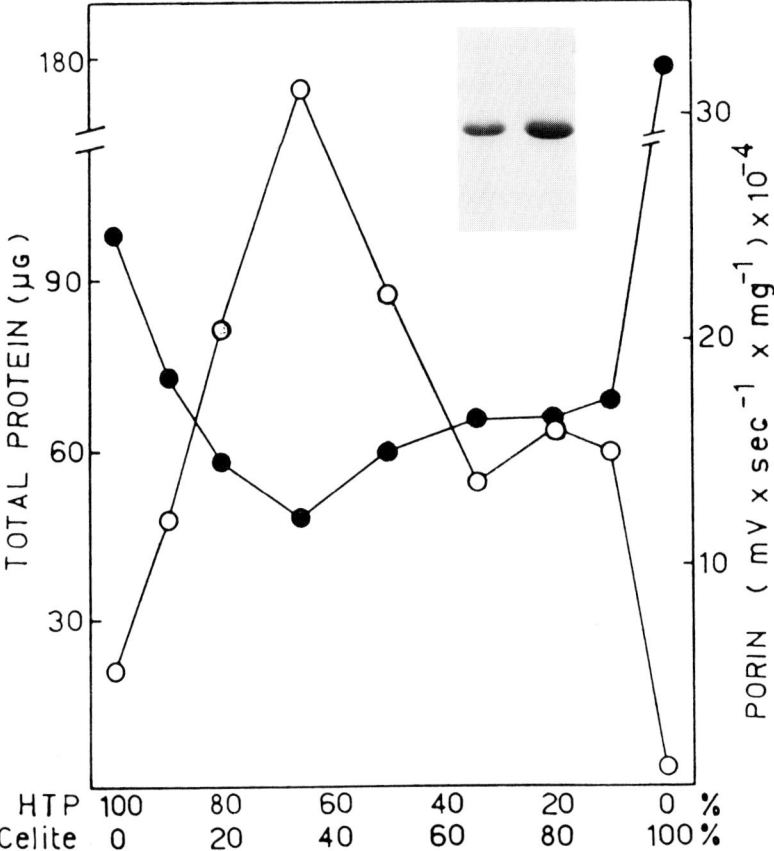

Fig. 3. Dependence of the total protein (●) and the relative amount of porin (o), present in the eluates, upon the ratio between hydroxyapatite and celite in the column. The relative amount of porin is expressed as area of porin/mg protein. The inset shows the SDS-PAGE of 10 and 20 μg of porin purified from bovine heart mitochondria by one step HTP/celite chromatography.

Fast purification of the mitochondrial porin

A new, simple and rapid procedure for the purification of high amounts of mitochondrial porin has recently been developed (De Pinto et al., 1987b). This procedure requires only a single chromatographic step on hydroxyapatite/celite following the solubilization of the mitochondrial membranes in the non-ionic detergent

Triton X-100. In this study, the appropriate conditions of the ionic strength in the solubilization and elution media were investigated in detail. Furthermore, the protein/detergent ratio and the hydroxyapatite/celite ratio were varied to obtain pure porin. Relatively few very hydrophobic proteins are eluted from hydroxyapatite columns when Triton-solubilized mitochondria are applied. The presence and the relative amount of proteins in the pass-through, however, largely depend on the composition of the media in which the proteins were solubilized and applied to dry hydroxyapatite/celite columns. The addition of increasing concentrations of KCl in the solubilization buffer resulted in a progressive increase of the number and of the total amount of proteins which were eluted from the column. In contrast to the majority of proteins, porin was eluted from hydroxyapatite/celite columns without the need of salt, indicating only minor or no interactions between this protein and the adsorption materials.

Fig. 3 shows the effect of different hydroxyapatite/celite ratios on the purification of porin. The protein content and the relative amount of porin in the eluates are given as a function of hydroxyapatite/celite ratios. With hydroxyapatite or celite alone a relatively large amount of protein passed through the column. Any combination of hydroxyapatite and celite decreased the elution of total protein with a minimum at a hydroxyapatite/celite ratio 2:1.

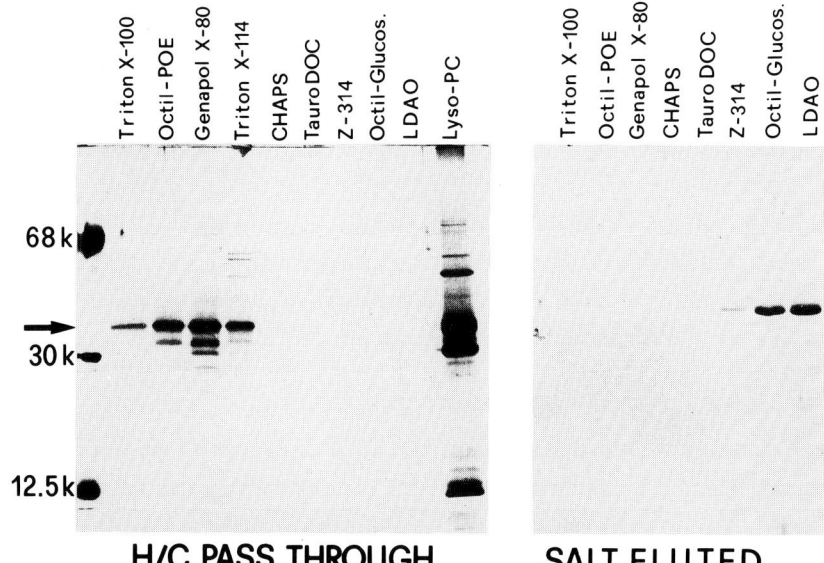

Fig. 4. SDS-gel electrophoresis pattern of HTP/celite eluates of mitochondrial membranes using different detergents (2% w/v). Protein elution on the left-hand side was performed with the solubilization buffer. The proteins on the right-hand side were then eluted by loading the same buffer containing in addition 5 mM P_i and 50 mM KCl.

In contrast to this, the relative amount of porin, as calculated from densitometric profiles, showed a maximum at the same 2:1 ratio. Endogenous cardiolipin prevents the elution of porin from hydroxyapatite (De Pinto *et al.*, 1985). Thus a low protein/detergent ratio was used to dilute endogenous cardiolipin. This procedure, together with the 2:1 HTP/celite column and the use of low ionic strength buffers in the solubilization and elution media allowed the purification of mitochondrial porin from several mammalian tissues to homogeneity (De Pinto *et al.*, 1987b).

Chromatographic behaviour of the mitochondrial porin and its purification in the presence of different detergents

Fig. 4 shows the behaviour of solubilized mitochondrial membranes on HTP/celite (2:1) columns when Triton X-100 was substituted with the other detergents listed in the figure. Doing this the interaction between the chromatographic material and the detergent-lipid-protein micelles was only influenced by the chemical nature of the detergents. Hydroxyapatite essentially acts as a mixed ion exchange material. The addition of celite increases the adsorption power of the column.

The left panel of Fig. 4 shows the SDS-PAGE of the HTP/celite pass-through fractions, namely the protein not adsorbed under the given experimental conditions. In the presence of the classical nonionic detergents like Triton X-100, Triton X-114 and Genapol X-80, the porin (indicated by the arrow) was obtained alone or with only few other bands. A very similar pattern was obtained by using octyl-POE, a nonionic detergent with a chemical structure similar to the Brij series but much shorter in both the alkyl and the polyoxyethylene chains. No protein at all was detected when the nonionic octylglucoside, the bile salt taurodeoxycholate, the zwitterionic detergents CHAPS or Z-314, or lauryldimethylamino N-oxide were used as a detergent. In the case of the zwitterionic detergent lysophosphatidylcholine, a natural compound, the pass-through contained several protein bands, especially in the region 28-37 kDa. Most metabolite carrier have such a molecular weight. The elution was performed extensively, until the eluates did not contain any protein. Then we added to the buffer phosphate (5 mM) and increased also the ionic strength (KCl 50 mM) and eluted again. The right side of Fig. 4 shows the protein eluted under these conditions with different detergents. With Triton X-100, Genapol X-80, octyl-POE, CHAPS and taurodeoxycholate no protein was found in the eluates. This means that the proteins were so tightly adsorbed in the presence of the ionic detergents CHAPS and tauro DOC that they could not be eluted. In the presence of octylglucoside, LDAO and Z-314 essentially one band, namely the porin, was present, even at variable concentrations. The porin, obtained in the presence of different detergents, was tested for its functional activity. Table 1 shows a comparison of specific activities of porin preparations obtained as HTP/celite pass-through or after salt addition in the elution buffer. The porin purified in the presence of LDAO showed a higher specific activity than in the presence of Triton X-100. Also the single channel analysis (not shown) indicated a better preserved function in the presence of LDAO. Other detergents like lysophosphatidylcholine or CHAPS were not able to save the activity of the porin. The possibility to have pure and still active porin adsorbed to a chromatographic column, allowed us to study the lipid and sugar content of the functionally active pore unit.

In the case of HTP/celite chromatography of Triton solubilized mitochondrial membranes, the eluate contains all components of the membrane including lipids and sugars beside those eventually bound to the protein.

Table 1. Chromatographic behaviour and functional activity of mitochondrial porin purified in the presence of different classes of detergents.

Detergent	HTP/Celite Elution		Specific Activity
	Pass-through	+5mM KP$_i$	(s/mm^2 mg)
Triton X-100	YES	NO	8×10^{-7}
LDAO	NO	YES	1×10^{-4}
LysoP	YES	NO	—
CHAPS	NO	NO	-(a)

(a) When porin prepared in the presence of LDAO was adsorbed to an ion-exchange column, and then eluted with CHAPS, it had no activity.

However, when mitochondrial membranes were solubilized by LDAO, lipid-detergent micelles or pure detergent micelles were found in the void volume of HTP/celite chromatography but protein-lipid-detergent micelles were adsorbed to the column. The extensive washing of the protein-lipid-detergent micelles, adsorbed to the column, by detergent buffer, resulted, through a fast equilibrium, in the "cleaning" or delipidation of every component not tightly bound to the proteic part of the micelle (Helenius & Simons, 1975).

Our analysis of the bound material showed that the cholesterol content was 5 moles cholesterol per mole 35 kDa porin polypeptide. Cholesterol is specifically present in the outer mitochondrial membrane, while it is practically absent in the inner membrane. No phospholipid specifically involved in the active porin-lipid-detergent micelle was found: nor by direct measure of the phosphorus content neither by TLC chromatography (data not shown).

The possible sugar content of the active protein was investigated by α-naphthol staining, specific for glycolipids, of TLC plates of LDAO-purified porin extracts without revealing any bluish spot. Furthermore, SDS-PAGE overloaded with LDAO-purified porin was stained for sugars by the PAS staining procedure (not shown). No glucide band was found on this gel, neither in the 30-35 kDa region, nor close to the electrophoretic front.

CONCLUSIONS

In the simplest and most naive approach, porin pore in the open state can be assumed to form a hollow cylinder spanning the outer mitochondrial membrane. The pore interior is probably filled with the same aqueous solution bathing the two sides of the membrane. This picture is consistent with the experimental results. In fact the porin has typical features of membrane proteins.

In the solubilization experiments reported here, the porin seems to have a more hydrophobic behaviour than the bulk of other mitochondrial membrane proteins despite its relatively polar primary structure (Kleene *et al.*, 1988). The solubilization dependence from the HLB number of detergents used indicates that the curve for the porin is shifted towards more hydrophobic values. In all the cases tested here the porin is better solubilized by the more hydrophobic members of detergent classes. When mitochondrial membrane proteins, solubilized by various detergents, were subjected to HTP/celite chromatography, we obtained three main different results: i) porin was eluted in the pass-through; ii) porin was eluted only after the application of a moderate ionic strength step; iii) porin was not eluted at all. Considering the results shown in Fig. 4 and in Table 1, the following conclusions could be drawn about the influence of detergents on the chromatographic behaviour of the porin: a) the hydrophobic moiety has no special influence on the elution of porin. b) Most influence on the chromatographic behaviour of porin is exerted by the hydrophilic head of different detergents. In the presence of detergents with very large hydrophilic heads, like Tritons or octyl-POE or lyso-PC, the porin and other membrane proteins cannot interact with the chromatographic material. But with small hydrophilic head detergents, like octyl glucoside or LDAO or zwittergent Z-314, the proteins can interact (see also Fig. 1 for a comparison). Other detergents like, for example CHAPS and taurodeoxycholate denature the porin.

For an integral membrane protein, deeply embedded in the phospholipid bilayer, only a very small part of its surface is exposed to the aqueous solutions. Only this part can interact with chromatographic material and the detergents effect on the elution can be explained as a shielding/de-shielding by the detergent of the water exposed surfaces of the membrane protein. With "small" detergents, porin and all the other mitochondrial membrane proteins remain attached to HTP/celite column; and only the addition of an increasing ionic strength can remove them (Fig. 4). Surprisingly, the porin is the first protein removed, demonstrating again to be one of the least hydrophilic, or one of the least provided with exposed charges, among the mitochondrial membrane proteins.

The use of the new, "small" detergents is indeed favorable to the purification of porin, and presumably of other membrane proteins. The integrity of functional activity was always tested. Several of them were not-denaturing in the case of porin. Their chemical structure allows the easier interaction of the protein-detergent micelles with the ion-exchange chromatographic material, thus improving the purification procedures. The exchange of detergents after adsorption of the protein to columns, and the use of some suitable detergent in the whole purification procedure, would probably increase the possibility of crystallizing the mammalian porins and other membrane proteins, through which their three dimensional structure can be studied in more detail.

ACKNOWLEDGEMENTS

This work was supported by the Consiglio Nazionale delle Ricerche and by the Deutsche Forschungsgemeinschaft (Project BR of SFB 176).

REFERENCES

Benz R, Janko K, Boos W, Läuger P (1978) Formation of large ion-permeable membrane channels by matrix porin (porin) of *Escherichia coli*. Biochim Biophys Acta 511:305-319
Benz R (1985) Porin from bacterial and mitochondrial outer membranes. CRC Critical Review in Biochemistry 19:145-190
Colombini M (1979) A candidate for the permeability pathway of the outer mitochondrial membrane. Nature (London) 279:643-645
De Pinto V, Tommasino M, Benz R, Palmieri F (1985) The 35 kDa DCCD-binding protein from pig heart mitochondria is the mitochondrial porin. Biochim Biophys Acta 813:230-242
De Pinto V, Ludwig O, Krause J, Benz R, Palmieri F (1987a) Porin pores of mitochondrial outer membranes from high and low eukaryotic cells: biochemical and biophysical characterization. Biochim Biophys Acta 894:109-119
De Pinto V, Prezioso G, Palmieri F (1987b) A simple and rapid method for the purification of the mitochondrial porin from mammalian tissues. Biochim Biophys Acta 905:499-502
De Pinto V, Benz R, Palmieri F (1988) The interaction of detergents with the mitochondrial porin, in preparation
Dihanich M, Suda K, Schatz G (1987) A yeast mutant lacking mitochondrial porin is respiratory deficient, but can recover respiration with simultaneous accumulation of an 86-kd extramitochondrial protein. EMBO J 6:723-728
Helenius A, Simons K (1975) Solubilization of membranes by detergents. Biochim Biophys Acta 415:29-79
Kinnally KW, Tedeschi H, Mannella CA (1987) Evidence for a novel voltage-activated channel in the outer mitochondrial membrane. FEBS Lett 226:83-87
Kleene R, Pfanner N, Pfaller R, Link TA, Sebald W, Neupert W, Tropschung M (1987) Mitochondrial porin in *Neurospora crassa*: cDNA cloning *in vitro* expression and import in mitochondria. EMBO J 6:2627-2633
Michel H (1983) Crystallization of membrane proteins. Trends Biochem Sci 8:56-59

Ohlendieck K, Riesinger I, Adams V, Krause J, Brdiczka D (1986) Enrichment and biochemical characterization of boundary membrane contact sites from rat-liver mitochondria. Biochim Biophys Acta 860:672-689

IV. Uncoupling Protein of Brown Adipose Tissue

A Molecular Biology Study of the Uncoupling Protein of Brown Fat Mitochondria. A Contribution to the Analysis of Genes of Mitochondrial Carriers

FREDERIC BOUILLAUD, SERGE RAIMBAULT, LOUIS CASTEILLA, ANNE-MARIE CASSARD AND DANIEL RICQUIER

Centre National de la Recherche Scientifique
Centre de Recherche sur la Nutrition, 9 rue Jules Hetzel
F-92190 Meudon-Bellevue, France

The uncoupling protein (UCP) is a specialized mitochondrial carrier unique to brown adipose tissue mitochondria. It acts as a proton carrier able to dissipate the proton gradient, to bypass ATP synthesis and to dissipate energy as heat (Nicholls & Locke, 1984). The proton translocating activity of UCP is inhibited by di- and triphosphate purine nucleotides and activated by free fatty acids (Nicholls et al., 1986). Moreover, UCP can be used as a marker to identify thermogenic adipocytes.

Besides UCP, brown fat mitochondria contain ubiquitous proteins such as ADP/ATP carrier (AAC) and phosphate carrier (PCP), of which the main function is to provide the cytosol with ATP synthesized by oxidative phosphorylation.

Interestingly, several reports from different laboratories have recently described sequence and structure homologies between AAC, PCP and UCP (Aquila et al., 1985, 1987; Battini et al., 1987; Bouillaud et al., 1986; Casteilla et al., 1988a; Neckelman et al., 1987; Ridley et al., 1986; Runswick et al., 1987).

Abbreviations: UCP, uncoupling protein; PCP, phosphate carrier; AAC, adenine nucleotide carrier

---

A. Azzi et al. (Eds.)
Anion Carriers of Mitochondrial Membranes
© Springer-Verlag Berlin Heidelberg 1989

UNCOUPLING PROTEIN, ADENINENUCLEOTIDE CARRIER AND
PHOSPHATE CARRIER SHARE COMMON PROPERTIES

Table 1 is a comparison of the three mitochondrial carriers. UCP, AAC and PCP monomer have almost the same molecular weight. The three proteins have been purified through filtration on hydroxyapatite. They share also a similar organization corresponding to a triplicated sequence with repeats in the three proteins related to each other (Aquila *et al.*, 1985, 1987; Casteilla *et al.*, 1988a; Runswick *et al.*, 1987). Particularly, AAC and UCP share additional properties such as the capacity to bind nucleotides and the absence of N-terminal targeting sequence in their neo-synthesized form (Bouillaud *et al.*, 1986; Freeman *et al.*, 1983; Ricquier *et al.*, 1983; Ridley *et al.*, 1986).

Table 1. Comparison of mammalian uncoupling protein (UP), ADP/ATP carrier (AAC) and phosphate carrier (PCP). Data were obtained from references quoted in the text. AA : amino acids, BAT : brown adipose tissue.

	UP	AAC	PCP
Organs	brown fat	all tissues	all tissues
Subcellular localization	inner mitochondrial membrane		
Function	$H^+ \longrightarrow$ matrix	$ADP \rightleftharpoons ATP$	$H_3PO_4^- \rightleftharpoons OH^-$
Mol. wt	33 kD	30 kD	35 kD
AA residues	306	297	313
N-terminal signal	no	no	yes
Nucleotide binding	ADP, ATP GDP, GTP 1 per dimer	ADP, ATP 1 per dimer	- -
mRNA	1.6 & 1.9 kb rodents	1.2 & 1.5 kb bovines	1.4 kb bovines
	1.9 kb, man, bovines,ovines	1.3, 1.4 & 1.6, man	
Gene(s)	1	2	-

ISOLATION OF MOLECULAR PROBES FOR UCP

UCP is not only a proton carrier and a major component of brown fat mitochondria but also an inducible component. UCP level is low in non-thermogenic brown adipose tissue (warm-adapted rats, adult guinea-pigs) whereas it is high in activated brown fat (cold-exposed rats, newborn guinea-pig) (Nicholls & Locke, 1984; Ricquier & Mory, 1984).

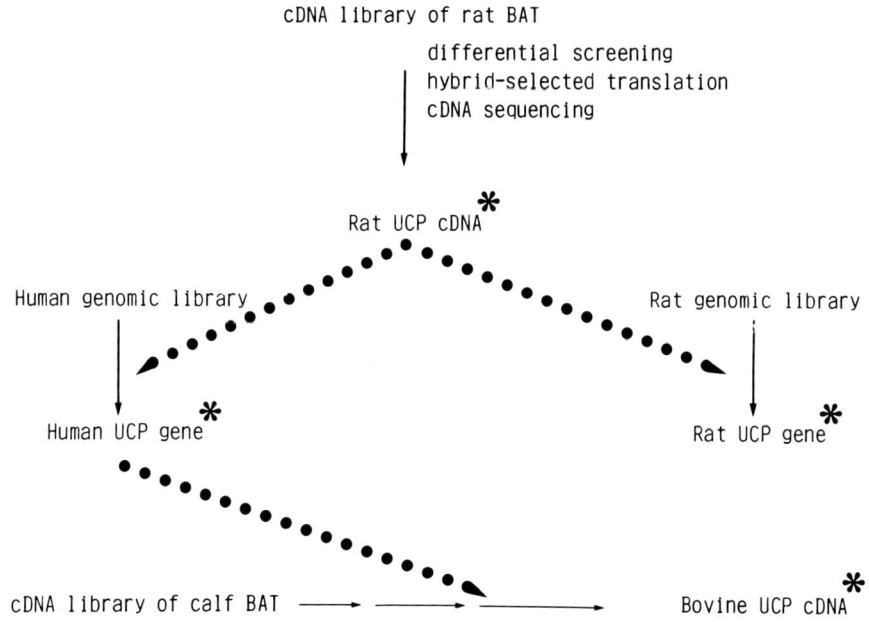

Fig. 1. Strategy of molecular cloning of rat, human and bovine uncoupling protein.

It is generally considered that the thermogenic capacity of brown adipose tissue is related to its UCP content. Moreover, UCP synthesis in brown adipose tissue is essentially controlled by neural factors and norepinephrine released at the surface of brown adipocytes (Ricquier & Mory, 1984). Thus, UCP expression is completely turn-off in non-brown fat cells while it can be rapidly modulated in brown adipocytes.These various features of UCP led us to undertake a molecular approach of thermogenesis in brown fat cells and to use recombinant DNA technology to study UCP. The strategy used is shown in Fig. 1.

A cDNA library of rat brown adipose tissue was constructed from poly(A$^+$) RNA isolated from brown adipose tissue of cold-exposed rats. This cDNA library was analysed by differential screening with single strand cDNAs constructed from poly(A$^+$)RNA of activated or non-activated tissue. Candidate clones were chosen and then used in hybrid-selected translation experiments combined with immuno-precipitation of translated product. Several rat UCP cDNA were then identified (Bouillaud *et al.*, 1985) and further characterized by DNA sequencing (Bouillaud *et al.*, 1986). Two other teams have also isolated cDNAs for mouse UCP (Jacobson *et al.*, 1985) and rat UCP (Ridley *et al.*, 1986).

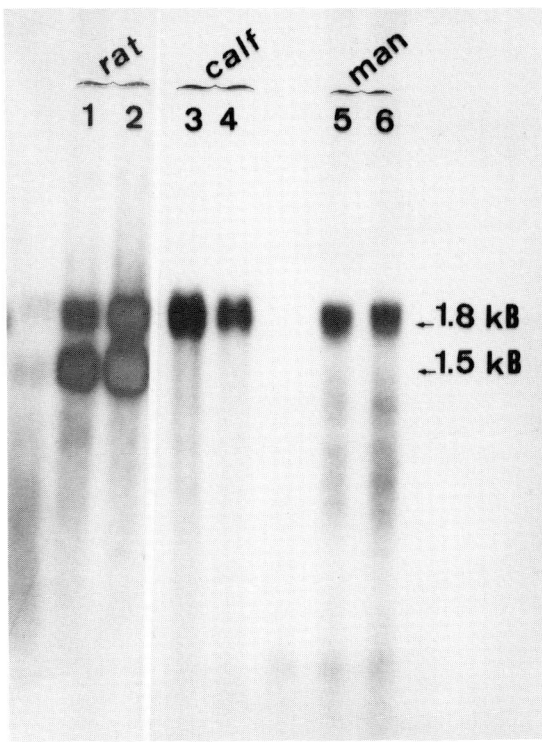

Fig. 2. Northern analysis of UCP mRNA in rat, calf and man. UCP mRNA was detected using rat cDNA, bovine cDNA and human genomic probe, respectively. RNA correspond to rat interscapular brown adipose tissue (1, 2), newborn calf perirenal tissue (3,4) and human hibernoma (5,6), respectively.

We have used rat UCP cDNA to isolate the rat UCP gene (Fig. 1) and to analyze rat UCP mRNA. However, we rapidly observed that the cross-hybridization between rat UCP cDNA and human, bovine or ovine RNA was weak. A strategy leading to the isolation of human UCP gene and bovine UCP cDNA was then set up (Fig. 1, Bouillaud *et al.*, 1988a; Casteilla *et al.*, 1988a).

ANALYSIS OF UCP MRNA

Rat UCP cDNA, bovine UCP cDNA and human UCP genomic fragments were used in Northern analysis to hybridize to UCP mRNA in rodent, bovine and human brown adipose tissue, respectively. No mRNA for UCP could be detected in white adipose tissue or liver of rats, calf or man. In rodents, UCP exhibits two mRNAs (Fig. 2) whereas in calf (Fig. 2), lamb (non shown) and man (Fig. 2) only one mRNA was observed. The presence of two mRNAs in rodents is due to two polyadenylation sites in UCP gene (Bouillaud *et al.*, 1988b). In rat, UCP mRNA level is rapidly and strong increased following exposure of animals in cold (Bouillaud *et al.*, 1985; Jacobson *et al.*, 1985; Ricquier *et al.*, 1984, 1986) and this induction could be mimicked by injection of norepinephrine or β-adrenoceptor agonist to animals (Mory *et al.*, 1984;

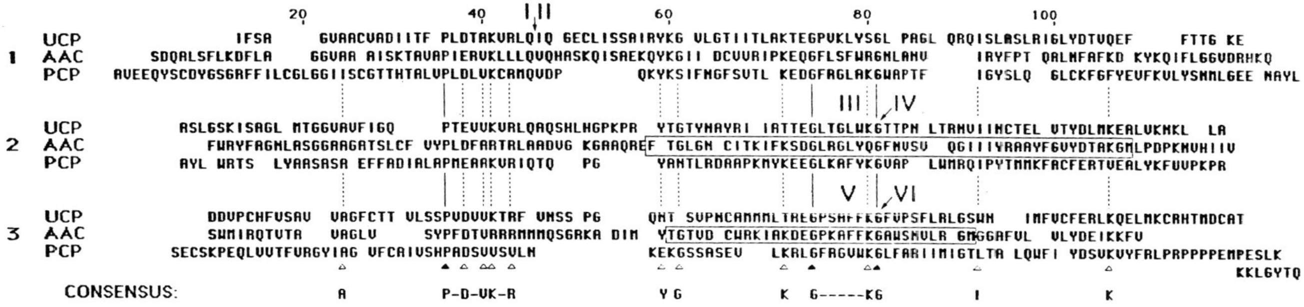

Fig. 3. Alignment of three repeated sequences in the three bovine carriers. Uncoupling protein (UCP, Casteilla *et al.*, 1988), ADP/ATP carrier (AAC, Aquila *et al.*, 1982) and phosphate carrier (PCP, Runswick *et al.*, 1987). Open and closed triangles, solid and dotted lines indicate identities. The AAC sequences identified as cytosolic-sided ADP binding sites by Dalbon *et al.* (1988) are boxed. Arrows indicate exon/intron limit and exons are numbered using Roman figures.

Ricquier et al., 1986). Moreover, run-on transcription experiments with isolated nuclei of brown adipose tissue have demonstrated that the gene of UCP is regulated at the transcriptional level (Ricquier et al., 1986). In calf, UCP mRNA is present in foetuses and is significantly increased at birth. However, it disappears during the first days of life (Casteilla et al., 1988b).

In man, UCP mRNA was detected in perirenal adipose tissue of newborn infants or adult patients with pheochromocytoma. Moreover, high level of UCP mRNA were recorded in RNA extracted from an hibernoma which is a brown fat tumor (Bouillaud et al., 1988a).

CDNA-DERIVED AMINO-ACID SEQUENCE OF UCP

The full-length rat UCP cDNA isolated by Bouillaud et al. (1985) was sequenced and rat UCP primary structure was identified (Bouillaud et al., 1986). This sequence is almost entirely similar to that obtained through sequencing of the purified hamster protein by Aquila et al., (1985). Both rat and hamster proteins have 306 amino-acid residues. More recently, 95% of bovine UCP sequence was obtained by sequencing of calf UCP cDNA (Casteilla et al., 1988a) and 80% of human UCP sequence was identified from sequencing of several human genomic probes (Cassard et al., unpublished data). Rat, hamster, bovine and human UCP are highly homologous.

Partial or complete sequences of bovine AAC and PCP have been determined by several laboratories (Aquila et al., 1982, 1987; Kolbe & Wohlrab, 1985; Rasmussen & Wohlrab, 1986; Runswick et al., 1987). An optimized alignment of the three repeats of the three bovine proteins was made (Fig. 3). As previously observed by Aquila et al. (1985, 1987) and Runswick et al. (1987), the three carriers contain internal repeats which are related to each other. Several residues are highly preserved in the three repeats of the three carriers. As indicated in Fig. 3, the C-terminal extremity of UCP contains a sequence highly homologous to a sequence which has recently been identified as a cytosolic-sided ADP binding site in bovine AAC (Dalbon et al., 1988). This observation led us to propose this region as the nucleotide binding site of UCP and also to suggest a new organization of UCP in the membrane (Casteilla et al., 1988a). Finally, the alignment of the three proteins is a convincing argument to the existence of an ancestral gene for these three carriers.

THE GENE OF UCP

Rat UCP genomic sequences have been cloned in lambda vector. One clone which spans the whole transcription unit (8.5 Kb) plus 4.5 Kb in the 5'-upstream region was sequenced (Bouillaud et al., 1988b).

The overall organization of UCP gene is shown in Fig. 4. UCP gene contains 6 exons and 5 introns. The transcription site (position +1) was determined using S1 nuclease analysis and primer extension experiments. A TATA box is found in position - 26 at the 3' end of the gene and two consensus polyadenylation sites (AATAAA) separated by 367 nucleotides. The first site is confirmed by our previous cDNA sequence (Bouillaud et al., 1986). The second site explains the existence of two mRNAs in rodents and was confirmed by S1 nuclease analysis of mature mRNAs. It is know that UCP has only one gene (Bouillaud et al., 1985, 1988a and b). No data on PCP gene have been presently reported. Recently, several authors have proposed the existence of at least 2 genes for human AAC (Battini et al., 1987; Houldsworth & Attardi 1988; Neckelman et al., 1987).

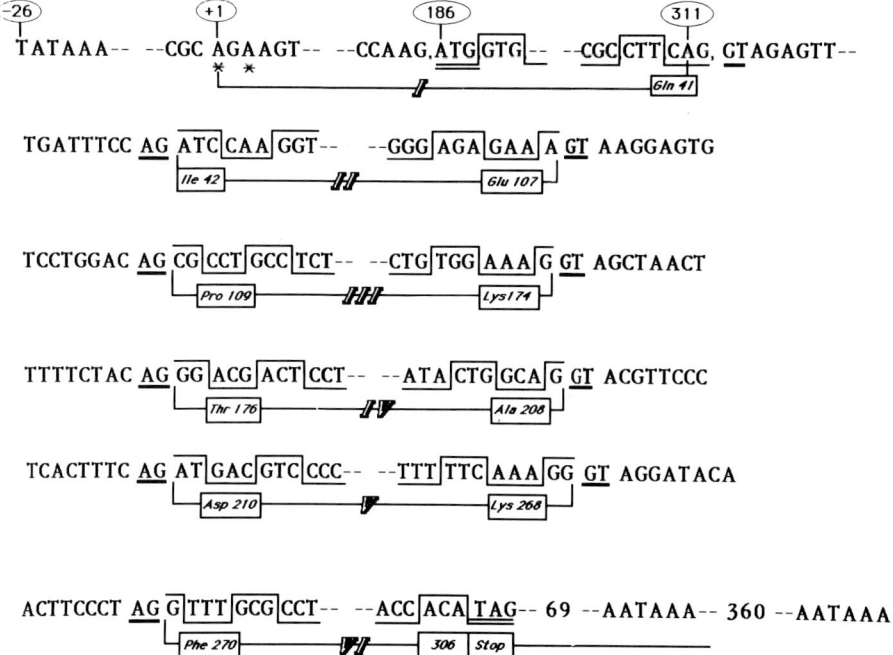

Fig. 4. Organization of rat UCP gene. UCP gene has 6 exons and 5 introns. The position +1 refers to start of transcription.

EXONS PARTITION AND EVOLUTIONARY RELATIONSHIPS

Triplicated structure: the presence of 6 exons suggests an arrangement in which each repeat corresponds to 2 exons: such an arrangement is in excellent agreement with limits of exons in UCP gene (see Figs. 3 and 4). It is thus attractive to propose that complete triplication occurred in organism in which genes were already organized in exons and introns.

Limits of exons inside each repeat: the limits between exons 3 and 4 and exons 5 and 6 both correspond to the last glycine residue which is included in the repeated G- - - - - -KG consensus sequence possibly implicated in nucleotide binding (Figs. 3 and 4). Conservation of these sequences could be due to (i) evolutionary event, (ii) amino-acid requirement for nucleotide binding and (iii) nucleotidic sequence requirement for mRNA splicing.

The determination of UCP gene sequence is very recent and detailed investigations will be necessary to a better analysis of evolutionary relationship between UCP, AAC and PCP.

CONCLUSION

The recombinant DNA technology has been used to study the uncoupling protein of brown adipose tissue mitochondria. cDNA and genomic probes were isolated and could be used to study UCP mRNA, UCP sequence and UCP gene in rodents, bovines and man. Clearly, the uncoupling protein, the ADP/ATP carrier and the phosphate carrier share structural properties and have a common ancestor. Two important questions have to be studied: the identification of UCP domain involved in proton translocation, the mechanism of control of UCP expression. Further experiments based on molecular biology could certainly help to make progress.

ACKNOWLEDGEMENTS

Our research is supported by CNRS, DRET, MRES, INSERM and INRA.

REFERENCES

Aquila H, Link TA, Klingenberg M. (1985) The uncoupling protein from brown fat mitochondria is related to the mitochondrial ADP/ATP carrier. Analysis of sequence homologies and of folding of the protein in the membrane. EMBO J 4:2369-2376

Aquila H, Link TA, Klingenberg M (1987) Solute carriers involved in energy transfer of mitochondria form a homologous protein family. FEBS Lett 212:1-9

Aquila H, Misra D, Eulitz M, Klingenberg M (1982) Complete amino-acid sequence of the ADP/ATP carrier from beef heart mitochondria. Hoppe-Seyler's Z Physiol Chem 363:345-349

Battini R, Ferrari S, Kaczmarek L, Calabretta B, Sing-Tsueng C, Baserga R (1987) Molecular cloning of a cDNA from a human ADP/ATP carrier which is growth-regulated. J Biol Chem 262:4355-4359

Bouillaud F, Raimbault S, Ricquier D (1988b) Complete sequence of the gene of rat uncoupling protein (in preparation)

Bouillaud F, Ricquier D, Thibault J, Weissenbach J (1985) Molecular approach to thermogenesis in brown adipose tissue: cDNA cloning of the mitochondrial uncoupling protein. Proc Natl Acad Sci (USA) 82:445-448

Bouillaud F, Villarroya F, Hentz E, Raimbault S, Cassard AM, Ricquier D (1988a) Detection of brown adipose tissue uncoupling protein mRNA in adult patients by a human genomic probe. Clin Sci 74 (in press)

Bouillaud F, Weissenbach J, Ricquier D (1986) Complete cDNA-derived amino-acid sequence of rat brown fat uncoupling protein. J Biol Chem 261:1487-1490

Casteilla L, Bouillaud F, Forest C, Ricquier D (1988a) Molecular cloning of bovine and ovine brown fat uncoupling protein. Structural and evolutionary relationship to bovine adenine nucleotide translocator and mitochondrial phosphate carrier (submitted)

Casteilla L, Champigny O, Bouillaud F, Robelin J, Ricquier D (1988b) Sequential changes in the expression of mitochondrial protein mRNA during the development of brown adipose tissue in bovine and ovine species. Biochem J (in press)

Dalbon P, Brandolin G, Boulay F, Höppe J, Vignais PV (1988) Mapping of the nucleotide binding sites in the ADP/ATP carrier of beef heart mitochondria by photolabeling with 2-azido [-^{32}P] adenosine diphosphate. Biochemistry 27:5141-5149

Freeman KB, Chien SM, Litchfield D, Patel HV (1983) Synthesis in vitro of rat brown adipose tissue 32,000 M_r protein. FEBS Lett 158:325-330

Houldsworth J, Attardi G (1988) Two distinct genes for ADP/ATP translocase are expressed at the mRNA level in adult human liver. Proc Natl Acad Sci (USA) 85:377-381

Jacobsson A, Stadler U, Glotzer MA, Kozak LP (1985) Mitochondrial uncoupling from mouse brown fat. Molecular cloning, genetic mapping & mRNA expression. J Biol Chem 260:16250-16254

Kolbe HVJ, Wohlrab H (1985) Sequence of N-terminal formic acid fragment and location of the N-ethylmaleimide-binding site of the phosphate transport protein from beef heart mitochondria. J Biol Chem 29:15899-15906

Mory G, Bouillaud F, Combes-George M, Ricquier D (1984) Noradrenaline controls the concentration of the uncoupling protein in brown adipose tissue. FEBS Lett 166:393-396

Neckelmann N, Kang L, Wade RP, Shuster R, Wallace DL (1987) cDNA sequence of a human skeletal muscle ADP/ATP translocator: lack of a leader peptide, divergence from a fibroblast translocator cDNA, and co-evolution with mitochondrial DNA genes. Proc Natl Acad Sci (USA) 84:7580-7584

Nicholls DG, Cunningham S, Rial E (1986) The bioenergetic mechanisms of brown adipose tissue. In: Trayhurn P, Nicholls DG (eds) Brown Adipose Tissue. Edward Arnold, London, p 52

Nicholls DG, Locke R (1984) Thermogenic mechanisms in brown fat. Physiol Rev 64:1-64

Rasmussen UB, Wohlrab H (1986) Bovine cardiac ADP/ATP carrier: two distinct mRNAs and on unusually short 3'-non coding sequence. Biochem Biophys Res Commun 138:850-857

Ricquier D, Bouillaud F, Toumelin P, Mory G, Bazin R, Arch J, Pénicaud L (1986) Expression of uncoupling protein mRNA in thermogenic or weakly thermogenic brown adipose tissue. J Biol Chem 261:13905-13910

Ricquier D, Mory G (1984) Factors affecting brown adipose tissue activity in animals and man. Clin Endocrinol Metab 13:503-519

Ricquier D, Mory G, Bouillaud F, Thibault J, Weissenbach J (1984) Rapid increase mitochondrial uncoupling protein and its mRNA in stimulated brown adipose tissue. Use of a cDNA probe. FEBS Lett 178:240-244

Ricquier D, Thibault J, Bouillaud F, Kuster Y (1983) Molecular approach to thermogenesis in brown adipose tissue. Cell-free translation of mRNA and characterization of the mitochondrial uncoupling protein. J Biol Chem 258:6675-6677

Ridley RG, Patel HV, Gerber GE, Morton RC, Freeman KB (1986) Complete nucleotide sequence and derived amino-acid sequence of cDNA encoding the mitochondrial uncoupling protein of rat brown adipose tissue: lack of mitochondrial targeting pre-sequence. Nucl Ac Res 14:4025-4035

Runswick RJ, Powell SJ, Nyren P, Walker JE (1987) Sequence of the bovine mitochondrial phosphate carrier protein: structural relationship to ADP/ATP translocase and the brown fat mitochondrial uncoupling protein. EMBO J 6:1367-1373

On the Mechanism of Transport by the Uncoupling Protein from Brown Adipose Tissue Mitochondria

Eduardo Rial and David G. Nicholls

Department of Biochemistry, Medical Sciences Institute, University of Dundee, Dundee DD1 4HN, Scotland (UK)

The uncoupling protein enables brown adipose tissue mitochondria to uncouple respiration from ATP synthesis under thermogenic conditions by catalyzing re-entry of protons extruded by the respiratory chain. Two ligands, purine nucleotides and fatty acids, interact specifically with the protein to modulate its transport activity (for review see Rial & Nicholls, 1987).

Although big advances in the knowledge of the biochemistry and molecular genetics of the uncoupling protein have been made over the past ten years, including the elucidation of the primary structure and predictions of secondary structure (Aquila et al., 1985, 1987; Runswick et al., 1987) there are still very few details on the molecular mechanism of transport.

Here we review data obtained over the past fifteen years on the transport properties of the uncoupling protein, and re-interpret experiments that looked contradictory at the time. With this information we will propose a model for the transport mechanism and for its regulation by fatty acids and purine nucleotides.

Correspondence to E. Rial, Centro Investigaciones Biológicas, CSIC, Velázquez 144, 28006 Madrid, Spain

A. Azzi et al. (Eds.)
Anion Carriers of Mitochondrial Membranes
© Springer-Verlag Berlin Heidelberg 1989

LIGANDS OF THE UNCOUPLING PROTEIN

The primary sequence of the uncoupling protein and the predictions of its secondary structure show a close homology with other mitochondrial transport systems, and therefore one would predict similarities in the transport mechanism (Aquila *et al.*, 1985, 1987; Runswick *et al.*, 1987). However, the uncoupling protein shows itself as a very complex system, since it interacts, under physiological conditions, with three types of ligands: the transported species, purine nucleotides (natural inhibitors), and free fatty acids (physiological activators) (for review see Rial & Nicholls, 1987).

The transported species

Dissipation of the proton electrochemical gradient by brown adipose tissue mitochondria, with the purpose of heat generation, implies that the uncoupling protein is catalyzing either the entry of protons or the efflux of hydroxyl ions. The nature of the charge being transferred across the membrane is a matter of great importance in order to solve the relation between this protein and other mitochondrial anion carriers.

Experiments carried out in the early seventies showed that, at neutral pH, brown fat mitochondria were highly permeable to halide ions and nitrate (Nicholls & Lindberg, 1973). Subsequent work demonstrated that the heat dissipatory pathway and the anion transporting pathway were inhibited by the same range of purine nucleotides. Two crucial experiments enlightened the relation between these two transport activities. First, it was shown that mitochondrial swelling was inhibited when mitochondria were suspended in KCl in the presence of nigericin, and the inhibition could be overcome by the addition of a protonophore (Nicholls & Lindberg, 1973). These results were confirmed when it was shown that respiring mitochondria, pre-loaded with NaCl, remained coupled while chloride was being extruded from the matrix (Nicholls, 1974).

The interpretation of these two experiments is that hydroxyl and chloride ions are competing for the transport pathway, or in other words, that the uncoupling protein transports both protons (hydroxyl ions) and halide ions. This led to the proposal that under physiological conditions the uncoupling protein was an anion carrier, catalyzing the efflux of hydroxyl ions (Nicholls & Heaton, 1978; Nicholls, 1979).

Inhibitory ligands: purine nucleotides

The study of the permeability properties of brown fat mitochondria was prompted by the observation that these mitochondria showed no respiratory control when they were prepared under conditions suitable for mitochondria from other sources (Rafael *et al.*, 1969). If free fatty acids were removed from the incubation medium and purine nucleotides were added to it, brown fat mitochondria behaved as those from liver or heart, while their permeability to halide ions was abolished (Nicholls & Lindberg, 1973). Purine nucleotides were shown to block the dissipatory pathway by binding to a site in the outer face of the inner mitochondrial membrane, and the use of the analog $[^{32}P]$-8-azido-ATP allowed the identification of a 32,000 M_r polypeptide (Heaton *et al.*, 1978). In contrast to the adenine nucleotide translocator, a closely related transporter (Aquila *et al.*, 1985, 1987; Runswick *et al.*, 1987), a wide range of nucleotides can bind to the uncoupling protein and block the channel: almost any purine nucleoside di- or triphosphate acts as inhibitor.

Activating ligands: non-esterified fatty acids

As it was mentioned above, one of the requirements to attain respiratory control is the removal of free fatty acids from the incubation. Already in the early sixties, fatty acids were known as uncouplers of mitochondrial respiration, and therefore when their effects were first observed on brown adipose tissue mitochondria, these were considered as unspecific. Over the past few years mounting evidence has been presented demonstrating that they interact specifically with the uncoupling protein to increase its proton conductance, and in fact free acids, liberated during the noradrenaline-induced lipolysis, have been proposed as the acute regulators of the uncoupling protein *in vivo* (Locke *et al.*, 1982a, 1982b; Rial *et al.*, 1983; Cunningham *et al.*, 1986).

CONFLICTING DATA

In a system as complex as the one presented above, it is necessary to describe the interaction between all ligands in order to be able to outline a mechanism for ion transport and its regulation. As progress has been made, a number of observations seem not to fit with the simple idea of an anion carrier with inhibitor and activator sites. We will now present these experiments, and then propose a working model consistent with these observations.

One of the earliest puzzling data refers to the role of free fatty acids in the control of the transport pathway. It was clear from the first experiments that while the removal of fatty acids was essential to attain respiratory control, it had no influence on the permeation of halide ions (Nicholls & Heaton, 1978). This seriously argued against a specific fatty acid uncoupling in brown fat mitochondria, even when they were much more sensitive than mitochondria from other sources. Evidence for specific fatty acid interaction with the uncoupling protein has been obtained from two main lines of work.

First, experiments of passive swelling demonstrated that fatty-acid induced proton permeability could be blocked by purine nucleotides (Rial et al., 1983). The second line of evidence comes from the comparison of cold-adapted and warm-adapted animals lacking the uncoupling protein, and it shows that fatty acid uncoupling is restricted to cells or mitochondria from thermogenically active animals (Locke et al., 1982a; Rial & Nicholls, 1984; Cunningham et al., 1986).

The blockage of the fatty acid effect by purine nucleotides posed a new problem: since the cytosolic concentration of the predominant purine nucleotide, Mg.ATP, is in the millimolar range, and its calculated K_d is around 10 μM (Rial et al., 1983), one would expect that the binding site at the uncoupling protein would always be saturated (see below). How do fatty acids exert their uncoupling action? The answer to this problem relies on the difference between the driving force during passive swelling experiments (around 20 mV), and the one during controlled respiration (around 200 mV). In the presence of nucleotide, brown fat mitochondria show a "non-ohmic" membrane conductance to protons, this conductance being low until the proton electrochemical gradient rises above 200 mV and then rapidly increases (Nicholls, 1977). Fatty acids exert their action by lowering the potential at which the conductance increases to a value which is insufficient to induce respiratory control (Rial et al., 1983).

The idea, presented above, of a nucleotide binding site that is always saturated in vivo due to its high affinity, is another question under debate. There is a large disparity between the values for the dissociation constant (K_d) when calculated using equilibrium binding techniques and those deduced from experiments that investigate the inhibitory ability of nucleotides. For example, the K_d values obtained for ATP in binding studies are, in the presence of Mg^{2+}, around 10 μM (Rial et al., 1983), while the concentration required for half maximal inhibition (K_i) of the proton conductance is 1 mM (Cannon et al., 1973; LaNoue et al., 1982; Locke et al., 1982b). This discrepancy has also been observed in the inhibition of transport during passive swelling studies, although the difference are smaller (Kopecky et al., 1984, 1987); Rial & Nicholls, unpublished observations).

An alternative hypothesis for the regulation of the uncoupling protein in vivo is centered around the possibility of modulating the degree of saturation of the nucleotide binding site (LaNoue et al., 1982, 1986). For such proposal, the existence of a mechanism to decrease the cytosolic pool of purine nucleotides upon stimulation of the cell by noradrenaline is required, and this still remains hypothetical. A decrease in ATP levels, which has been shown to occur (LaNoue et al., 1982, 1986), may not be sufficient since the ADP produced also inhibits the uncoupling protein (Nicholls, 1976). However, a recent report (Klingenberg, 1988) points out that due to the lowered affinity for the nucleotides as the pH raises above 7.2, a change in the ratio ATP/ADP may influence the activity of the protein.

More recently, new developments in the study of the uncoupling protein have re-opened the question of the differences observed between the transport of chloride and protons. Klingenberg and Winkler (1985) established a procedure for the reconstitution of the uncoupling protein into liposomes and found that it could catalyze a GTP-sensitive influx and efflux of protons, although it could not be enhanced by fatty acids. But, more importantly, this preparation was unable of catalyzing Cl⁻ transport. This observation was considered to be in line with a proposal made by Kopecky et al. (1984) of the existence of two distinct proteins: one responsible for the Cl⁻ and the other for the fatty acid-sensitive H⁺-translocating activity, and it was argued that this reconstitution protocol dealt only with the proton transport system involved in energy dissipation, i.e. the uncoupling protein (Klingenberg & Winkler, 1985).

Covalent modification of the uncoupling protein has shown apparently conflicting results but still they may help in the elucidation of the proton-chloride problem. Modification by chymotrypsin revealed that the two transport activities were being inactivated simultaneously (Fernández et al., 1987). This, together with the data from competition experiments, make difficult to sustain a two-protein system. On the other hand, chemical modification experiments have shown differential effects on the two transport activities.

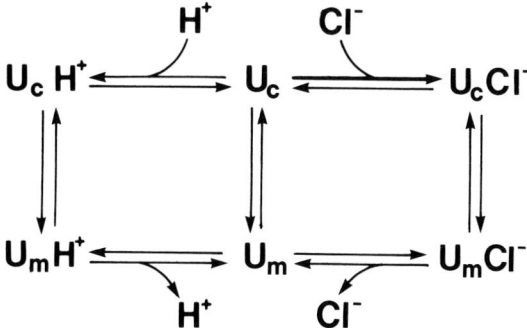

Fig. 1. Scheme representing the catalytic transport cycle for the uncoupling protein, showing the competition between Cl⁻ and H⁺ for the carrier. "U_c", free carrier with the ligand binding sites facing the cytosolic side of the inner mitochondrial membrane; "U_m", free carrier with the binding sites facing the matrix side. Competition is for the free carrier and not for a binding site, since Cl⁻ and H⁺ possess different binding sites.

This has been interpreted in terms of the existence of two regions within the uncoupling protein, each one implicated in the translocation of a different ligand (Rial & Nicholls, 1986; Kopecky et al., 1987). The reconstitution results have to be explained now as a partial inactivation of the uncoupling protein that results in the loss of the chloride transport activity and the sensitivity towards fatty acids.

A MODEL FOR THE UNCOUPLING PROTEIN MECHANISM

The basic principles behind the model that we are going to present are common to other membrane transport systems and some of them have already been proposed for the uncoupling protein (Klingenberg, 1985; Aquila *et al.*, 1987). First, the low transport index of the protein indicates that we are not facing an ion channel but rather a carrier (Nicholls & Heaton, 1978; Klingenberg & Winkler, 1985).

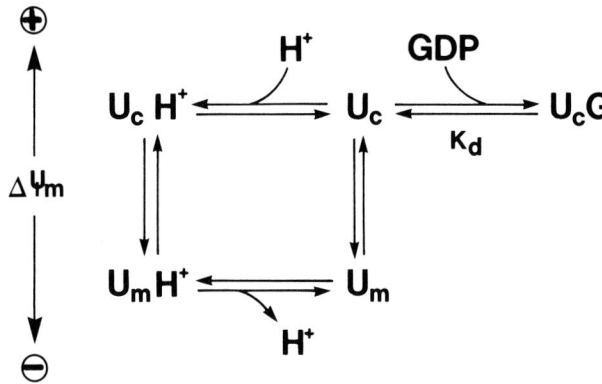

Fig. 2. Scheme representing the competitive inhibition of H^+ transport due to the binding of GDP to the uncoupling protein. GDP binds to the cytosolic side of the protein ("U_c") and blocks the carrier in this conformation. Affinity for protons is enhanced by an increased membrane potential or by the presence of free fatty acids. Under those circumstances, the apparent "K_d" for GDP binding is increased.

Klingenberg and coworkers have proposed that the translocating cycle in the uncoupling protein would be similar to those suggested for other metabolite carriers (Klingenberg, 1985; Aquila *et al.*, 1987). According to this model, the uncoupling protein would bind the ion to be transported either from the matrix side of the membrane ("U_m") or from the cytosolic side ("U_c"), and release it in the opposite side (Figs. 1 & 2). This transport process is always an electrical uniport. Binding of purine nucleotides to the uncoupling protein would block the carrier in its cytosolic conformation since they are not translocated into the matrix (Fig. 2) (Nicholls, 1976).

The most important concept introduced with this model is the idea of a competitive interaction not only between halide ions and protons, but also between each of these two species and purine nucleotides. As we said earlier, there would be two distinct binding sites, one for H^+ and another one for anions, showing different properties, one of which would be that fatty acids only affect proton transport. This proposal for a competitive interaction between ions and nucleotides would be based on the discrepancies between K_d and K_i values, explaining why, under conditions of transport, the apparent K_d for the nucleotide is higher than when measured under non-transporting conditions. Also, the higher the driving force the more effective will the competition with the nucleotide be. This would explain why inhibition of proton transport during passive swelling requires lower concentrations than those for control of respiration.

From the analysis of the experiments on the extrusion of chloride under respiring conditions (Nicholls, 1974), one could predict that the affinity for chloride ions would be higher than that for protons: as we said before, in the absence of nucleotides NaCl-loaded mitochondria remained coupled while chloride was being transported out of the matrix, and protons only re-entered into the matrix once all the Cl^- had been extruded. We have also found (Rial & Nicholls, unpublished observations) that K_i for inhibition of swelling are consistently higher than for inhibition of chloride transport, reflecting again this relatively higher affinity.

Under physiological conditions the halide permeability is not relevant, hence purine nucleotides and fatty acids will only affect the rate of proton re-entry. According to this model, nucleotides and protons cannot be bound to the uncoupling protein and therefore fatty acids will activate heat production by favoring the binding and translocation of protons. In this context it becomes clear that any variation in the total pool of adenine nucleotides or in the ratio ADP/ATP may help to modulate the uncoupling protein activity.

REFERENCES

Aquila H, Link TA, Klingenberg M (1985) The uncoupling protein from brown fat is related to the mitochondrial ADP/ATP carrier. Analysis of sequence homologies and of folding of the protein in the membrane. EMBO J 4:2369-2376

Aquila H, Link TA, Klingenberg M (1987) Solute carriers involved in energy transfer of mitochondria form a homologous protein family. FEBS Lett 212:1-9

Cannon B, Nicholls DG, Lindberg O (1973) Purine nucleotides and fatty acids in energy coupling in mitochondria from brown adipose tissue. In: Azzone GF, Ernster L, Papa S, Quagliariello E, Siliprandi N (eds) Mechanisms in Bioenergetics. Academic Press, New York, pp 357-363

Cunningham SA, Wiesinger H, Nicholls DG (1986) Quantification of fatty acid activation of the uncoupling protein in brown adipocytes and mitochondria from the guinea pig. Eur J Biochem 157:415-420

Fernández M, Nicholls DG, Rial E (1987) The uncoupling protein from brown-adipose-tissue mitochondria: chymotrypsin-induced structural and functional modifications. Eur J Biochem 164:675-680

Heaton GM, Wagenvoord RJ, Kemp A, Nicholls DG (1978) Brown adipose tissue mitochondria: photoaffinity labelling of the regulatory site for energy dissipation. Eur J Biochem 82:515-521

Klingenberg M (1985) Principles of carrier catalysis elucidated by comparing two similar membrane translocators from mitochondria, the ADP/ATP carrier and the uncoupling protein. Ann New York Acad Sci 456:279-288

Klingenberg M (1988) Nucleotide binding to uncoupling protein: Mechanism of control by protonation. Biochemistry 27:781-791

Klingenberg M, Winkler E (1985) The reconstituted isolated uncoupling protein is a membrane potential driven H^+ translocator. EMBO J 4:3087-3092

Kopecky J, Guerrieri F, Jezek P, Drahota Z, Houstek J (1984) Molecular mechanism of uncoupling in brown adipose tissue mitochondria: the nonidentity of proton and chloride conducting pathways. FEBS Lett 170:186-190

Kopecky J, Jezek P, Drahota Z, Houstek J (1987) Control of uncoupling protein in brown-fat mitochondria by purine nucleotides: chemical modification by diazobenzenesulfonate. Eur J Biochem 164:687-694

LaNoue KF, Koch CD, Meditz RB (1982) Mechanism of action of norepinephrine in hamster brown adipocytes. J Biol Chem 257:13740-13748

LaNoue KF, Strzelecki T, Strzelecka D, Koch C (1986) Regulation of the uncoupling protein in brown adipose tissue. J Biol Chem 261:298-305

Locke RM, Rial E, Nicholls DG (1982a) The acute regulation of proton conductance in cell and mitochondria from the brown fat of cold- and warm-adapted guinea-pigs. Eur J Biochem 129:381-387

Locke RM, Rial E, Scott ID, Nicholls DG (1982b) Fatty acids as acute regulators of the proton conductance of hamster brown fat mitochondria. Eur J Biochem 129:373-380

Nicholls DG (1974) Hamster brown adipose tissue mitochondria: the chloride permeability of the inner membrane under respiring conditions, the influence of purine nucleotides. Eur J Biochem 49:585-593

Nicholls DG (1976) Hamster brown adipose tissue mitochondria: purine nucleotide control of the ion conductance of the inner membrane, the nature of the nucleotide site. Eur J Biochem 62:223-228

Nicholls DG (1977) The effective proton conductance of the inner membrane of mitochondria from brown adipose tissue: dependence on proton electrochemical gradient. Eur J Biochem 77:349-356

Nicholls DG (1979) Brown adipose tissue mitochondria. Biochim Biophys Acta 549:1-29

Nicholls DG, Lindberg O (1973) Brown adipose tissue mitochondria: the influence of albumin and nucleotides on passive ion permeabilities. Eur J Biochem 37:523-530

Nicholls DG, Heaton G (1978) Anion uniport across the inner membrane of brown adipose tissue mitochondria. In: Azzone GF, Avron M, Metcalfe JC, Quagliariello E, Siliprandi N (eds) The Proton and Calcium Pumps. Elsevier, Amsterdam, pp 309-318

Rafael J, Ludolph HJ, Hohorst HJ (1969) Mitochondria from brown adipose tissue: uncoupling of oxidative phosphorylation by long chain fatty acids and recoupling by guanosine triphosphate. Hoppe-Seyler's Z Physiol Chem 350:1121-1131

Rial E, Nicholls DG (1984) The mitochondrial uncoupling protein from guinea-pig brown adipose tissue: synchronous increase in structural and functional parameters during cold-adaptation. Biochem J 222:685-693

Rial E, Nicholls DG (1986) Chemical modification of the brown-fat-mitochondrial uncoupling protein with tetranitromethane and N-ethylmaleimide: a cysteine residue is implicated in the nucleotide regulation of anion permeability. Eur J Biochem 161:689-694

Rial E, Nicholls DG (1987) The uncoupling protein from brown adipose tissue mitochondria. Cell Biol Rev 11:75-104

Rial E, Poustie A, Nicholls DG (1983) Brown adipose tissue mitochondria: the regulation of the 32,000 M_r uncoupling protein by fatty acids and purine nucleotides. Eur J Biochem 137:197-203

Runswick MJ, Powell SJ, Nyren P, Walker JE (1987) Sequence of the bovine mitochondrial phosphate carrier: structural relationship to ADP/ATP translocase and the brown fat mitochondrial uncoupling protein. EMBO J 6:1367-1373

Regulation of the Amount and Activity of the Uncoupling Protein Thermogenin in Brown Adipose Tissue

BARBARA CANNON AND JAN NEDERGAARD

The Wenner-Gren Institute, The Arrhenius Laboratories, University of Stockholm, S-106 91 Stockholm, Sweden

The presence of the brown-fat specific, uncoupling protein thermogenin on a meeting program devoted to mitochondrial anion carriers may at first sight seem surprising. This is because, in the role normally assumed for it (depicted on Fig. 1), it does not transport anions over the mitochondrial membrane. Rather, thermogenin is envisaged to work as a transporter of protons (or proton equivalents) over the inner membrane of mitochondria from brown adipose tissue, thus facilitating uncoupled respiration and the dissipation of chemical energy in the form of heat (for detailed reviews, see Nicholls & Locke, 1984; Cannon & Nedergaard, 1985).

There are, however, several good reasons for the inclusion of this protein in the group of mitochondrial anion carriers.

One reason is functional. Even though thermogenin is thought to function as a proton conducting pathway, it seems that it endows the mitochondria in which it is found with a very high Cl^- permeability (Nicholls & Lindberg, 1973). It is, however, not fully clear that Cl^- transport is a function of thermogenin itself; at least in one reconstitution study with isolated thermogenin, Cl^- permeability was not reconstituted (Klingenberg & Winkler, 1985), but whether this is a property of thermogenin or of the reconstitution system is not known; regulatory features would tend to indicate that Cl^- transport in isolated brown-fat mitochondria is thermogenin-mediated (Nicholls et al., 1974). Whether Cl^- transport is a physiological function of thermogenin is not known; if

A. Azzi et al. (Eds.)
Anion Carriers of Mitochondrial Membranes
© Springer-Verlag Berlin Heidelberg 1989

270

it is, it may be active in mitochondrial volume control (see below). Further, it has been suggested that the uncoupling is due, not to the inward passage of protons through the mitochondrial membrane, but rather due to an outward passage of OH- (Nicholls, 1976). If this is the case, the Cl⁻ transport can be considered to be an experimental manifestation of the uncoupling action of thermogenin, and the uncoupling and anion transporting properties of thermogenin would then be identical.

The second (and perhaps the major) reason for the inclusion of thermogenin amongst the mitochondrial anion transporters is structural. After the amino acid sequence of thermogenin had been elucidated (Aquila et al., 1985; Bouillaud et al., 1986), it became clear that thermogenin is structurally very similar to the adenine nucleotide translocase, and also, although to a smaller extent, to the mitochondrial phosphate carrier (Aquila et al., 1987; Runswick et al., 1987). As thermogenin is only found in mammals, whereas the adenine nucleotide translocase and the phosphate carrier are ubiquitous amongst mitochondria, thermogenin must be considered as a late modification of these carriers.

Thus, as a member of the group of mitochondrial anion carriers, thermogenin has a natural place in this symposium. However, thermogenin demonstrates features not so far studied in other mitochondrial anion carriers: its intricate patterns of regulation. Thus, concerning most other mitochondrial anion carriers, their activity is generally believed not to be controlled, i.e. the carriers always work with their full capacity.

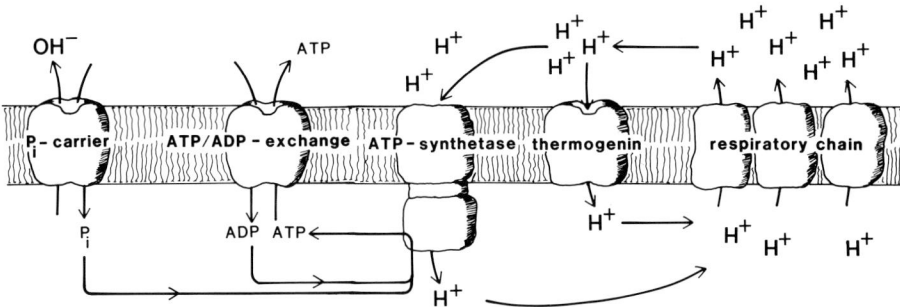

Fig. 1. Thermogenin in its functional and structural environment in the mitochondrial membrane. Thermogenin belongs to the same molecular family as the phosphate (P_i) carrier and the adenine nucleotide translocase (ATP/ADP exchange). It has a molecular weight of 32,000 dalton and it binds purine nucleotides such as GDP (generally the one used experimentally) and ATP (which is probably the nucleotide bound within the cell). It allows for the passage of proton equivalents over the mitochondrial membrane, leading to uncoupling and through this, to heat production. In the presence of GDP or ATP, its activity is inhibited. The figure is based on circuits suggested by Nicholls (1976).

The resulting intra- and extra-mitochondrial concentrations of the transported species are not directly regulated *via* the carrier but by other metabolic processes. Further, the amount of an anion carrier in the mitochondria of a given tissue is generally considered to be constitutive (with few known exceptions, such as the citrate carrier discussed in the present symposium (Kaplan *et al.*, 1989). Finally, although not fully ubiquitous, most mitochondrial carriers are to be found, to a smaller or larger extent, in mitochondria from many different tissues.

These rather passive regulations of the amount and activity of most anion carriers can be contrasted with the very elaborate systems for regulation of thermogenin. Currently we can recognize two major levels of regulation of thermogenin, and within each of these, two sublevels. Thus, the activity of the thermogenin molecule already present in a mitochondrion is influenced both by an acute regulation, and apparently also by an "initiating" or "cancelling" regulation, corresponding to the physiological situations where thermogenesis is initiated or cancelled. Further, there is an adaptive regulation of the level of expression of the thermogenin gene, corresponding to the physiological need of the animal. Finally, there must be a regulation of the "opening" of the gene, occurring during differentiation of the tissue. The thermogenin gene shows absolute tissue specificity: it is only expressed in brown fat cells, not even in otherwise so similar cells such as white-fat cells (Cannon *et al.*, 1982; Jacobsson *et al.*, 1985).

With these intricate regulatory features, thermogenin stands alone amongst the "anion carriers" and it is so far the only carrier whose physiological regulation can be and has been studied in detail. What is known about these points of regulation are reviewed below in some detail.

REGULATION OF THERMOGENIN ACTIVITY

Acute regulation

For more than a decade it has been an enigma in brown fat research that the physiological regulation of thermogenin activity is not fully understood. Briefly, the problem to be encountered is that the cytosolic concentrations of the purine nucleotides (especially ATP and ADP) are such that the activity of thermogenin would be expected to be permanently inhibited, *i.e.* no heat production should occur (Cannon *et al.*, 1973). Although other hypotheses have been proposed, the suggestions which have gained most interest have been those that imply that the inhibitory effects of the cytosolic nucleotides are counteracted by some physiological activator. Two such activators have been suggested: the free fatty acids released from the triglyceride droplets within the cells when the cells are adrenergically stimulated (Prusiner *et al.*,

1968; Cannon *et al.*, 1973; Locke & Nicholls, 1981), and the acyl-CoA derivatives of these fatty acids (Cannon *et al.*, 1977).

Features to be expected of the physiological activator would be an ability to (re)introduce proton permeability in an *in vitro* system and that this induced proton permeability would be influenced by purine nucleotides.

In Fig. 2, the ability of both palmitate and palmitoyl-CoA to decrease the mitochondrial membrane potential is demonstrated. Under the experimental conditions used here, palmitoyl-CoA was somewhat more efficient than palmitate as a depolarising agent.

In Fig. 3 the ability of a purine nucleotide (here GDP) to affect the de-energization induced by either palmitate or palmitoyl-CoA is demonstrated. It is concluded that both proposed agents are in possession of the properties pointed out above to be expected of the physiological activator. They differ in that the binding site for palmitoyl-CoA can be envisaged to be identical to that for the purine nucleotides, and thus the interaction between the purine nucleotide inhibitors and this activator can be suggested to be competitive at the same site; such interaction has been found both for binding and for activity (Cannon *et al.*, 1977; Strieleman & Shrago, 1985; Bailey *et al.*, 1989). In contrast, the free fatty activators would be expected to interact in a noncompetitive way, on another binding site on thermogenin.

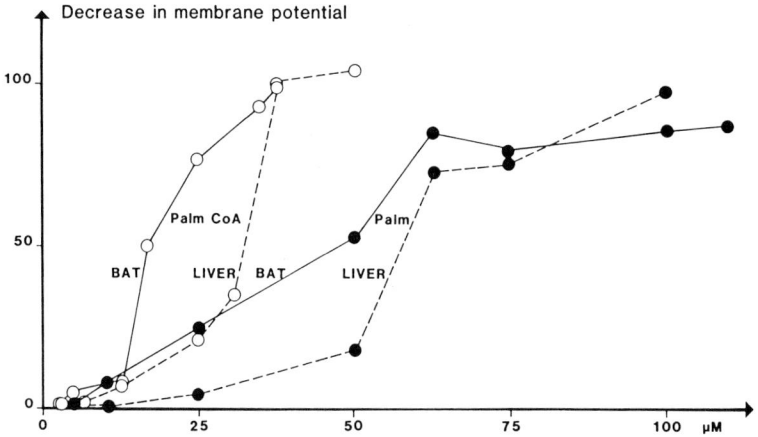

Fig. 2. A comparison of the ability of palmitate or palmitoyl-CoA to depolarise brown fat and liver mitochondria. Mitochondria were isolated and tested under identical conditions, and the membrane potential was followed with the safranin method in the presence of 0.05 % fatty acid-free bovine serum albumin. The decrease in membrane potential was expressed in arbitrary units. For further details, see Nedergaard and Cannon (1989).

Recently, based on work of Andreyev *et al.* (1988, 1989), two alternative hypotheses for the effect of free fatty acids on brown adipose tissue mitochondria could be envisaged. Andreyev *et al.* have suggested that the uncoupling effect of free fatty acids in most tissues is due to a shuttling action of the free fatty acids, with the protonated form passing freely into the mitochondria through the mitochondrial membrane, but with the fatty acid in the dissociated (anionic) form passing back to the cytosol, carried on the adenine nucleotide translocase. In the light of this theory, new interpretations of the effects of free fatty acids on brown fat mitochondria could be formulated.

Thus, one could envisage that the effect of fatty acids on brown fat mitochondria was not due to their interaction with thermogenin but rather due to their interaction with the adenine nucleotide translocase, which is present in brown fat mitochondrial membranes in levels similar to those of thermogenin (Heaton *et al.*, 1978) and similar to those found in *e.g.* liver tissue (Christiansen *et al.*, 1973). Such a suggestion would in principle be in accordance with the absence of effect of free fatty acid on proton permeability in liposomes reconstituted with isolated thermogenin (Klingenberg & Winkler, 1985), and with the fact that no binding site for free fatty acids on thermogenin has so far been observed. Also, although brown fat mitochondria are more sensitive to the uncoupling effect of fatty acids than are *e.g.* liver mitochondria

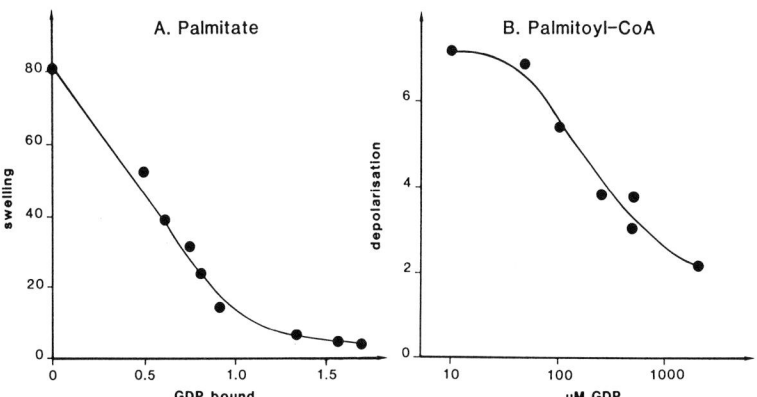

Fig. 3. The ability of GDP to influence de-energisation caused by the addition of palmitate or palmitoyl-CoA to isolated brown fat mitochondria. In A, adapted after Rial *et al.* (1985), proton permeability was followed as swelling in a K-acetate medium in the presence of valinomycin and plotted *versus* the amount of GDP bound (in nmol/mg mitochondrial protein). The experiment in B was performed as those in Fig. 2, but in the presence of different amounts of GDP (adapted from Nedergaard & Cannon, 1989).

the difference is not large (about a factor 2, see *e.g.* Fig. 2). However, no experimental evidence for the idea that the uncoupling effect of free fatty acids in brown fat mitochondria is mediated through the adenine nucleotide translocase has so far been presented.

In another interpretation of the effects of free fatty acids on brown fat mitochondria, the free fatty acids should not be considered as physiological activators, but rather as playing a functional role in the uncoupling action of thermogenin, by being the compound transporting the protons, by being carried over the membrane by thermogenin. Several observations concerning brown fat mitochondria could be re-interpretated in view of these possibilities, but again, there is presently no experimental evidence which allows for a conclusion in this matter. The main argument against this interpretation is that brown fat mitochondria are uncoupled even in the presence of fatty-acid-free albumin.

Nearly acute

In certain physiological states, an increase in [^{3}H]GDP-binding to isolated brown fat mitochondria can be observed, despite the fact that an increase in the area of the peak at 32,000 dalton on SDS-polyacrylamide gel electrophoresis (which corresponds to thermogenin) cannot be found. It would thus seem that a change in the thermogenin molecule itself, leading to exposure of previously occluded GDP-binding sites, could occur. Desautels *et al.* (1978) suggested that this would be due to what has been termed an "unmasking" phenomenon. In a series of physiological states, such a masking/unmasking phenomenon has been suggested. These states include such conditions of general interest as genetically obese mutants, cold acclimation, re-acclimation to warm, hibernation, and the effect of so-called cafeteria-feeding.

With the development of quantitative immunological assay methods for thermogenin as such (Cannon *et al.*, 1982; Lean *et al.*, 1983) it became possible to directly follow the amount of thermogenin as protein and thus to corroborate or contradict the suggested cases of unmasking. In most of the cases where an unmasking (or a masking) phenomenon was initially suggested in order to explain the data, direct immunological quantification of the amount of thermogenin has demonstrated that there are also differences in amount of thermogenin which directly parallel those originally observed for the GDP-binding capacity of the mitochondria (Nedergaard *et al.*, 1984; Nedergaard & Cannon, 1985), and an unmasking process thus does not have to be invoked to explain the changes in GDP-binding associated with these chronic states.

However, one important situation has remained: the increase in GDP-binding reported to occur after a short (1 h) cold stress. Such an increase can be observed even in the presence of the protein synthesis inhibitor cycloheximide (Desautels & Himms-Hagen, 1979), and it can thus not be due to the synthesis of new thermogenin. Further, it is apparently not due to some of the GDP-binding sites being "masked" by *e.g.* palmitoyl-CoA (which can compete with GDP for the binding site (Cannon *et al.*,

1977)), as an albumin wash, which has earlier been shown to uncover extra binding sites (Sundin & Cannon, 1980), cannot *in vitro* mimic the "physiological" unmasking (Gribskov *et al.*, 1986).

Until recently, it was only possible to study this phenomenon *in vivo*, but it is now possible to "unmask" GDP-binding sites *in vitro* (Nedergaard & Cannon, 1987). This is accomplished by pretreatment of the isolated mitochondria with isoosmotic KCl, a treatment earlier demonstrated to induce mitochondrial swelling *in vitro*. When isolated by traditional means, brown fat mitochondria are condensed, probably due to loss of osmotic support from the mitochondrial matrix through the concerted action of thermogenin (allowing Cl^- efflux) and (probably) a specialised K^+-transporting mechanism (DiResta *et al.*, 1986; Bailey *et al.*, 1989). Apparently, the masking phenomenon is caused by a physical inability of the GDP to gain access to the binding site on thermogenin in the condensed state of the mitochondria; when the mitochondria are (re)-swelled, due to the KCl pretreatment, GDP can gain access to all sites, and the "unmasking" is observed. It would seem that activity parameters of thermogenin are also influenced by this condensation-reswelling process (Nedergaard & Cannon, 1989).

The most interesting feature of this phenomenon is, however, the indication that it is more than an effect of technical procedures. Thus, there is evidence that the mitochondria in the tissue show rapid changes in size, corresponding to the transitions from an inactive to an active state of the tissue (Vallin, 1970; Desautels & Himms-Hagen, 1980). Further, this physiologically induced condensation/expansion is "remembered" during routine preparation procedures, so that mitochondria isolated from animals in different physiological conditions but tested under identical conditions still demonstrate different densities of GDP--binding sites. Thus, processes must occur in vivo, leading to profound alterations in the internal millieu of the mitochondria. The physiological significance of these processes may be related to the fact that the mitochondrial matrix has to be in an expanded (swollen) state for respiration to be maximal (Nicholls *et al.*, 1972), but the nature of the processes involved is still unknown.

REGULATION OF EXPRESSION

Control of level of expression of the thermogenin gene

Physiologically, the amount of thermogenin is the limiting factor for the amount of nonshivering thermogenesis an animal can evolve (Cannon *et al.*, 1981), and the regulation of the synthesis of thermogenin is thus of interest, not only as a problem for mitochondriogenesis, but also in physiological contexts. Although a long series of investigations have earlier approached the regulation problem in different physiological and pharmacological states, the ability to follow the level of thermogenin mRNA has revolutionized these studies.

It should, however, be stated that the level of thermogenin mRNA is, of course, not the only determinant for the amount of thermogenin found in the mitochondria. So far it would seem that a good correlation exists between thermogenin mRNA levels and the ensuing amount of thermogenin, but exceptions to this rule may exist. To this could be added that there must be a limit to the amount of functional thermogenin which can be inserted into existing mitochondria, and the synthesis of thermogenin has thus to be synchronised with the synthesis of new mitochondria, and, to a certain extent, with that of new cells.

When a small mammal is exposed to cold, there is a very rapid increase in the amount of thermogenin mRNA found in the tissue (Bouillaud $et\ al.$, 1984; Jacobsson $et\ al.$, 1985, 1986, 1987, 1989; Ricquier at al., 1986), and the new steady-state level is reached within hours, even though it takes days before the synthesis of new thermogenin influences the thermogenic capacity sufficiently to reach a new level (Jacobsson $et\ al.$, 1989). The cold-induced increase in thermogenin mRNA has lent itself to pharmacological investigations. It can be mimicked by norepinephrine injections (Bouillaud $et\ al.$, 1984; Jacobsson $et\ al.$, 1986, 1987), indicating that it is caused by the increased firing, during cold exposure, of the sympathetic nerves found in the tissue (Niijima $et\ al.$, 1984).

A simultaneous α- and β-adrenergic stimulation is necessary in order to obtain this increase in thermogenin mRNA levels in vivo (Jacobsson $et\ al.$, 1986), and thus α-1-adrenergic stimulation may be an essential component in thermogenesis. It may be pointed out that the necessity for α-adrenergic stimulation may be of a permissive character, and thus it may only be observed if animals living at thermoneutral temperatures are investigated.

At lower environmental temperatures, there is probably already an ongoing sympathetic stimulation, and the need for parallel α-adrenergic stimulation may not be evident ($i.e.$ an increase in the magnitude of the β-adrenergic stimulation may then be sufficient to increase thermogenin expression). This necessity for simultaneous α- and β-adrenergic stimulation has also been observed concerning other features of the tissue (Ma & Foster, 1984; Cunningham & Nicholls, 1987; Granneman, 1988).

An unexpected regulatory feature concerning thermogenin mRNA is observed in cold-stressed animals. In contrast to what would be expected (and what has been observed for other mRNA species studied in other systems), the increase in thermogenin mRNA level is not enhanced by a stabilization of the mRNA; rather thermogenin mRNA has a shorter half-life in the cold than in control conditions (Jacobsson *et al.*, 1987). The physiological significance of this unexpected feature is unknown.

At birth, there is a rapid increase in thermogenin mRNA (Ricquier at al., 1986; Obregon *et al.*, 1987). This postnatal expression is only observed in the euthyroid state (Obregon *et al.*, 1987), and at least in newborn rats it is fully due to the cold stress experienced by the pups when they leave the intrauterine environment and are exposed to normal environmental temperatures (Obregon *et al.*, 1989). Thus, if the pups are instead exposed even after birth to a "uterine" temperature, no postnatal increase in thermogenin mRNA is observed (Fig. 4).

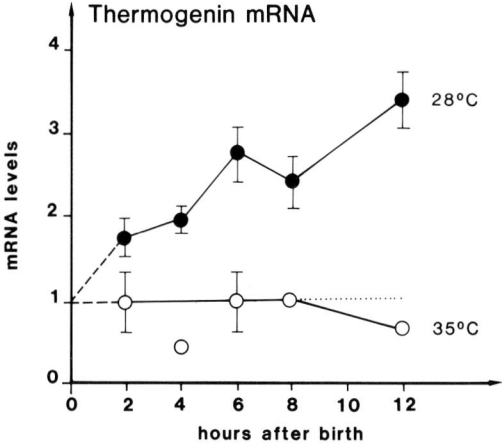

Fig. 4. Control of postnatal expression of thermogenin. Newborn rat pups were removed from the dam and were exposed to either their thermoneutral temperature 35°C or to a mild cold stress of 28°C, and the level of thermogenin mRNA in their brown adipose tissue followed. Adapted from Obregon *et al.* (1989).

As pointed out elsewhere (Nedergaard *et al.*, 1986), the control of thermogenin expression in certain types of newborns (such as guinea-pigs) cannot be under environmental control, and in these animals, special mechanisms must have developed to allow for stimulated thermogenin expression before birth.

Tissue specificity

The gene for thermogenin is localized on chromosome 8 in the mouse (Jacobsson *et al.*, 1985). It is not expressed in any other cells than brown fat cells and the "opening" of the gene for transcription must thus be under differentiational control. Thermogenin is a "late marker" of differentiation of the brown fat cell, both *in vivo* and *in vitro*. Thus, during *in utero* development of mice, thermogenin occurs much later than other mitochondrial proteins (Houštěk *et al.*, 1988; Houštěk & Kopecký 1989). In cell cultures, developing *in vitro* from undifferentiated precursors obtained from brown fat deposits, many adipocyte characteristics are readily expressed (Néchad *et al.*, 1983; Kuusela *et al.*, 1986). However, until recently it was not possible to demonstrate that such cells differentiating in culture were able to advance so far in the differentiation process that the gene for thermogenin was opened. Very recently, we have been able to observe that the gene for thermogenin may be expressed in such cells, and that the expression can be stimulated with nor-epinephrine (Kopecky *et al.*, 1988; Rehnmark *et al.*, 1989). This thus opens the possiblity also to investigate tissue specific expression of genes, even of those for mitochondrial anion carriers.

REFERENCES

Andreyev AY, Bondareva TO, Dedukhova VI, Mokhova EN, Skulachev VP, Tsofina LM, Volkov NI, Vygodina TV (1989) The ATP/ADP-antiporter is involved in the uncoupling effect of fatty acids. In: Azzi A, Nałęcz KA, Nałęcz MJ, Wojtczak L (eds) The Anion Carriers of the Mitochondrial Membranes, Springer-Verlag Heidelberg, pp 159-168

Andreyev AY, Bondareva TO, Dedukhova VI, Mokhova EN, Skulachev VP, Volkov NI (1988) Carboxyatractylate inhibits the uncoupling effect of free fatty acids. FEBS Lett 226:265-269

Aquila H, Link TA, Klingenberg M (1985) The uncoupling protein from brown fat mitochondria is related to the mitochondrial ADP/ATP carrier. Analysis of sequence homologies and of folding of the protein in the membrane. EMBO J 4:2369-2376

Aquila H, Link TA, Klingenberg M (1987) Solute carriers involved in energy transfer of mitochondria form a homologous protein family. FEBS Lett 212:1-9

Bailey C, Nedergaard J, Cannon B (1989) Interaction between palmitoyl-CoA and GDP in control of Cl⁻-permeability of brown adipose tissue mitochondria. Submitted for publication

Bailey C, Cannon B, Nedergaard J (1989) Specific induction of K⁺-permeability in brown fat mitochondria by palmitoyl-CoA. Submitted for publication

Bouillaud F, Ricquier D, Mory G, Thibault J (1984) Increased level of mRNA for the uncoupling protein in brown adipose tissue of rats during thermogenesis induced by cold exposure or norepinephrine infusion. J Biol Chem 259:11583-11586

Bouillaud F, Weissenbach J, Ricquier D (1986) Complete cDNA-derived amino acid sequence of rat brown adipose tissue uncoupling protein. J Biol Chem 261:1487-1491

Cannon B, Nicholls DG, Lindberg O (1973) Purine nucleotides and fatty acids in energy coupling of mitochondria from brown adipose tissue. In: Azzone GF, Ernster L, Papa S, Quagliariello E, Siliprandi E (eds). Mechanisms in Bioenergetics, Academic Press, New York, pp 357-364

Cannon B, Sundin U, Romert L (1977) Palmitoyl coenzyme A: a possible physiological regulator of nucleotide binding to brown adipose tissue mitochondria. FEBS Lett 74:43-46

Cannon B, Nedergaard J, Sundin U (1981) Thermogenesis, brown fat and thermogenin. In: Musacchia XJ, Jansky L (eds), Survival in Cold, Elsevier-North Holland, Amsterdam, pp 99-120

Cannon B, Hedin A, Nedergaard J (1982) Exclusive occurrence of thermogenin antigen in brown adipose tissue. FEBS Lett 150:129-132

Cannon B, Nedergaard J (1985) The biochemistry of an inefficient tissue: brown adipose tissue. Essays Biochem 20:110-164

Christiansen EN, Drahota Z, Duszynski J, Wojtczak L (1973) Transport of adenine nucleotides in mitochondria from the brown adipose tissue. Eur J Biochem 34:506-512

Cunningham SA, Nicholls DG (1987) Induction of functional uncoupling protein in guinea pigs infused with noradrenaline. Biochem J 245:485-491

Desautels M, Zaror-Behrens G, Himms-Hagen J (1978) Increased purine nucleotide binding, altered polypeptide composition, and thermogenesis in brown adipose tissue mitochondria of cold-acclimated rats. Can J Biochem 56:378-383

Desautels M, Himms-Hagen J (1979) Roles of noradrenaline and protein synthesis in the cold-induced increase in purine nucleotide binding by rat brown adipose tissue mitochondria. Can J Biochem 57:968-976

Desautels M, Himms-Hagen J (1980) Parallel regression of cold-induced changes in ultrastructure composition, and properties of brown adipose tissue mitochondria during recovery of rats from acclimation to cold. Can J Biochem 58:1057-1068

Diresta DJ, Kutschke KP, Hottois MD, Garlid KD (1986) K^+-H^+ exchange and volume homeostasis in brown adipose tissue mitochondria. Am J Physiol 251:R787-R793

Granneman JG (1988) Norepinephrine infusions increase adenylate cyclase responsiveness in brown adipose tissue. J Pharmacol Exp Therap 245:1075-1080

Gribskov CL, Henningfield MF, Swick AG, Swick RW (1986) Evidence for unmasking of rat brown adipose-tissue mitochondrial GDP-binding sites in response to acute cold exposure. Effects of washing with albumin on GDP-binding. Biochem J 233:743-747

Heaton GM, Wagenvoord RJ, Kemp A Jr, Nicholls DG (1978) Brown-adipose-tissue mitochondria: photoaffinity labelling of the regulatory site of energy dissipation. Eur J Biochem 82:515-521

Houštěk J, Kopecký J (1989) Regulation of uncoupling protein and formation of thermogenic mitochondria. In: Azzi A, Nałęcz KA, Nałęcz MJ, Wojtczak L (eds) (1989) The Anion Carriers of the Mitochondrial Membranes. Springer-Verlag Heidelberg pp 283-291

Houštěk J, Kopecký J, Rychter Z, Soukup T (1988) Uncoupling protein in embryonic brown adipose tissue - existence of nonthermogenin and thermogenic mitochondria. Biochim Biophys Acta 935:19-25

Jacobsson A, Stadler U, Glotzer MA, Kozak LP (1985) Mitochondrial uncoupling protein from mouse brown fat: Molecular cloning, genetic mapping, and mRNA expression, J Biol Chem 260:16250-16254

Jacobsson A, Nedergaard J, Cannon B (1986) α- and β-adrenergic control of thermogenin mRNA expression in brown adipose tissue. Biosci Rep 6:621-631

Jacobsson A, Cannon B, Nedergaard J (1987) Physiological activation of brown adipose tissue destabilizes thermogenin mRNA. FEBS Lett 224:353-356

Jacobsson A, Mühleisen M, Nedergaard J, Cannon B (1989) Physiological regulation of gene expression in brown adipose tissue. Submitted for publication

Kaplan RS, Mayor JA, Oliveira DL, Johnston N (1989) Recent developments in the extraction, reconstitution and purification of the mitochondrial citrate transporter from normal and diabetic rats. In: Azzi A, Nałęcz KA, Nałęcz MJ, Wojtczak L (eds) (1989) The Anion Carriers of the Mitochondrial Membranes. Springer-Verlag Heidelberg, pp 59-69

Klingenberg M, Winkler E (1985) The reconstituted isolated uncoupling protein is a membrane potential driven H^+ translocator. EMBO J 4:3087-3092

Kopecký J, Rehnmark S, Jacobsson A, Nedergaard J, Cannon B (1988) Differentiation of brown fat preadipocytes in monolayer cultures: expression of thermogenin. EBEC Rep 5:296

Kuusela P, Nedergaard J, Cannon B (1986) β-adrenergic stimulation of fatty acid release from brown fat cells differentiated in monolayer culture. Life Sci 38:589-599

Lean MEJ, Branch WJ, James WPT, Jennings G, Ashwell M (1983) Measurement of rat brown adipose tissue mitochondrial uncoupling protein by radioimmunoassay: increased concentration after cold acclimation. Biosci Rep 3:61-71

Locke RM, Nicholls DG (1981) A re-evaluation of the role of fatty acids in the physiological regulation of the proton conductance of brown adipose tissue mitochondria. FEBS Lett 135:249-252

Ma SWY, Foster DO (1984) Potentiation of *in vivo* thermogenesis in rat brown adipose tissue by stimulation of α_1-adrenoceptors is associated with increased release of cyclic AMP. Can J Physiol Pharmacol 62:943-948

Nechad M, Kuusela P, Carneheim C, Björntorp P, Nedergaard J, Cannon B (1983) Development of brown fat cells in monolayer culture. I. Morphological and biochemical distinction from white fat cells in culture. Exp Cell Res 149:105-118

Nedergaard J, Cannon B (1985) (^{3}H)GDP binding and thermogenin amount in brown adipose tissue mitochondria from cold-exposed rats. Am J Physiol 248:C365-C371

Nedergaard J, Cannon B (1987) Apparent unmasking of (^{3}H)GDP binding in rat brown fat mitochondria is due to mitochondrial swelling. Eur J Biochem 164:681-686

Nedergaard J, Raasmaja A, Cannon B (1984) Parallel increases in amount of (^{3}H)GDP binding and thermogenin antigen in brown-adipose-tissue mitochondria of cafeteria-fed rats. Biochem Biophys Res Commun 122:1328-1336

Nedergaard J, Connolly E, Cannon B (1986) Brown adipose tissue in the mammalian neonate. In: Trayhurn P, Nicholls DG (eds) Brown Adipose Tissue. Edward Arnold London, pp 152-213

Nedergaard J, Cannon B (1989) *In vitro* induction of thermogenin activity. Submitted for publication

Nedergaard J, Cannon B (1989) Effects of palmitoyl-CoA on the membrane potential of brown fat mitochondria. Submitted for publication

Nicholls DG (1976) The bioenergetics of brown adipose tissue mitochondria. FEBS Lett 61:103-110

Nicholls DG, Cannon B, Grav HJ, Lindberg O (1974) Energy dissipation in non-shivering thermogenesis. In: Ernster L, Estabrook RW, Slater EC (eds) Dynamics of Energy-Transducing Membranes. Elsevier Amsterdam, pp 529-537

Nicholls DG, Grav HJ, Lindberg O (1972) Mitochondria from hamster brown adipose tissue. Regulation of respiration *in vitro* by variations in volume of the matrix compartment. Eur J Biochem 31:526-533

Nicholls DG, Lindberg O (1973) Brown-adipose-tissue mitochondria. The influence of albumin and nucleotides on passive ion permeabilities. Eur J Biochem 37:523-530

Nicholls DG, Locke RM (1984) Thermogenic mechanisms in brown fat. Physiol Rev 64:1-64

Niijima A, Rohner-Jeanrenaud F, Jeanrenaud B (1984) Effect of cold stimulation on the efferent discharges of nerves innervating interscapular brown adipose tissue in the rat. In: Hales JRS (ed) Thermal Physiology. Raven Press, New York, pp 189-192

Obregon M-J, Jacobsson A, Kirchgessner T, Schotz MC, Cannon B, Nedergaard J (1989) Postnatal recruitment of brown adipose tissue is induced by the cold stress experienced by the pups. An analysis of mRNA levels for thermogenin and lipoprotein lipase. Submitted for publication

Obregon MJ, Pitamber R, Jacobsson A, Nedergaard J, Cannon B (1987) Euthyroid status is essential for the perinatal increase in thermogenin mRNA in brown adipose tissue of rat pups. Biochem Biophys Res Commun 148:9-14

Prusiner SB, Cannon B, Lindberg O (1968) Oxidative metabolism in cells isolated from brown adipose tissue. 1. Catecholamine and fatty acid stimulation of respiration. Eur J Biochem 6:15-22

Rehnmark S, Kopecky J, Jacobsson A, Néchad M, Nelson BD, Obregon M-J, Nedergaard J, Cannon B (1989) Brown adipocytes differentiated *in vitro* can express the gene for the uncoupling protein thermogenin. Effects of hypothyroidism and norepinephrine. Submitted for publication

Rial E, Poustie A, Nicholls, DG (1983) Brown adipose tissue mitochodria: the regulation of the 32,000 M_r uncoupling protein by fatty acids and purine nucleotides. Eur J Biochem 137:197-203

Ricquier D, Bouillaud F, Toumelin P, Mory G, Bazin R, Arch J, Penicaud L (1986) Expression of uncoupling protein mRNA in thermogenic or weakly thermogenic brown adipose tissue. Evidence for a rapid β-adrenoceptor-mediated and transcriptionally regulated step during activation of thermogenesis. J Biol Chem 261:13905-13910

Runswick MJ, Powell SJ, Nyren P, Walker JE (1987) Sequence of the bovine mitochondrial phosphate carrier protein: structural relationship to ADP/ATP translocase and the brown fat mitochondria uncoupling protein. EMBO J 6:1367-1373

Strieleman PJ, Shrago E (1985) Specific interaction of fatty acyl-CoA esters with brown adipose tissue mitochondria. Am J Physiol 248:E699-E705

Sundin U, Cannon B (1980) GDP-binding to the brown fat mitochondria of developing and cold-adapted rats. Comp Biochem Physiol 65B:463-471

Vallin I (1970) Norepinephrine response in brown adipose tissue from newborn rats. Acta Zool 51:129-139

Regulation of Uncoupling Protein and Formation of Thermogenic Mitochondria

J. HOUŠTĚK AND J. KOPECKÝ

Department of Bioenergetics, Institute of Physiology, Czechoslovak Academy of Sciences, Videnska 1083, 142 20 Prague 4, Czechoslovakia

The thermogenic function of brown adipose tissue is of principal importance for mammalian thermoregulation. To satisfy the changing demands of an organism resulting from its developmental state, dietetic conditions and thermal environment, it must be accurately balanced and regulated. It is well established that proton transport through the uncoupling protein represents the exothermic and rate-limiting step in brown adipose tissue heat production (Nicholls & Locke, 1984). Therefore, it is quite obvious that both the instant and long-term regulation of brown adipose tissue function concerns primarily the uncoupling protein. In the first case, the activation and inhibition of uncoupling protein conductance is achieved by changes in concentration of specific regulatory ligands in response to adrenergic stimulation (Nicholls & Locke, 1984; Nicholls et al., 1986) and possibly also by masking-unmasking of the uncoupling protein or its regulatory nucleotide-binding site in the membrane (Desautels et al., 1978; Gribskov et al., 1986; Nedergaard & Cannon, 1987; Peachey et al., 1988). In the second case, the regulation affects the total thermogenic potential of the tissue due to biosynthetic and degradative processes which modulate the number of mitochondria and their specific enzymic equipment, including the uncoupling protein content (Ricquier & Bouillaud, 1986).

In this report we summarize the results of our recent studies focused on acute and chronic regulation of the uncoupling protein (Kopecký et al., 1984; Houštěk et al., 1987; Kopecký et al.,1987, Houštěk et al., 1988).

A. Azzi et al. (Eds.)
Anion Carriers of Mitochondrial Membranes
© Springer-Verlag Berlin Heidelberg 1989

PROTON AND HALIDE ANION CONDUCTANCE OF THE UNCOUPLING
PROTEIN AND THEIR REGULATION BY PURINE NUCLEOTIDES

Originally it was suggested that H^+ and Cl^- are transported through uncoupling protein *via* a single channel (Nicholls, 1976; Rial *et al.*, 1983). Later on, however, in accordance with the sensitivity of H^+ conductance but insensitivity of Cl^- conductance to free fatty acids (Nicholls & Lindberg, 1973), simultaneous measurements of passive H^+ and Cl^- movements across mitochondrial membrane showed (Kopecký *et al.*, 1984) that H^+ transport is independent of anion transport and the two processes do not compete with one another. As the uncoupling protein bears only a single nucleotide binding site and only one class of high affinity GDP-binding sites is found (Heaton *et al.*, 1978; Lin & Klingenberg, 1982; Kopecký *et al.*, 1987), this indicates that two distinct, independent pathways for protons and anions exist within a single molecule of uncoupling protein. Nonidentity of pathways is further illustrated by their different sensitivity to purine nucleotides (Fig. 1).

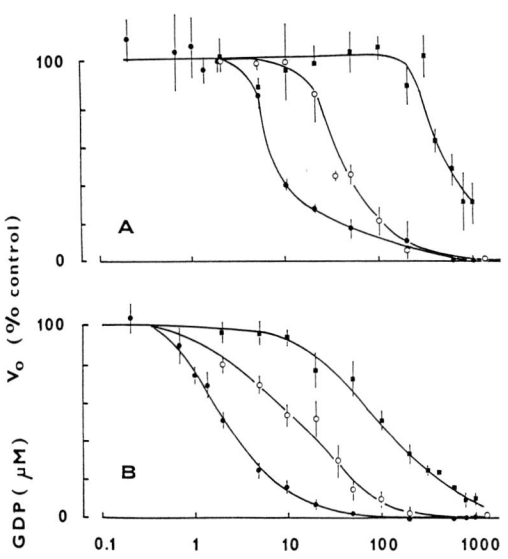

Fig. 1. Inhibition of H^+ and Cl^- transport by GDP in control (●) and DABS modified (o, 25 nmol/mg protein; ■ , 75 nmol/mg protein) mitochondria. Valinomycin-induced H^+ transport (A) and Cl^- transport (B) were measured in KCl medium as described by Kopecký *et al.* (1987).

To characterize the structural domains and functional amino acids of the uncoupling protein involved in transport and its control, we modified chemically the uncoupling protein in isolated mitochondria using the relatively nonspecific but membrane-impermeant diazobenzenesulfonate (DABS). As shown in Fig. 1, the modification markedly decreases the sensitivity of both transport activities but is without effect on the ion transport proper (Kopecký *et al.*, 1987). The effect of DABS is largely prevented by preincubation with a millimolar concentration of GDP. DABS also has no effect on the activation of H^+ transport due to removal of bound free fatty acids. It therefore appears that DABS affects directly the nucleotide-binding site.

Table 1. The effect of DABS on sensitivity of H^+ and Cl^- transport to GDP and on ^{3}H-GDP binding. K_i values were obtained from measurements shown in Fig. 1, K_d and B_{max} values from Scatchard plot analysis of ^{3}H-GDP binding. For details see Kopecký et al. (1987).

DABS	K_i		^{3}H-GDP binding	
	H^+	Cl^-	K_d	B_{max}
nmol/mg protein	μM			nmol/mg protein
0	7.7 ± 0.6	2.2 ± 0.4	2.7 ± 0.3	0.66 ± 0.05
25	40 ± 10	15 ± 6	7.9 ± 0.9	0.45 ± 0.05
75	547 ± 93	103 ± 28	13 ± 4.3	0.29 ± 0.09

The modified, although not further specified, amino acid residues are different from residues essential for ion transport and the regulatory effect of free fatty acids.

As further demonstrated in Fig. 1, even after modification with the highest concentration of DABS used a complete inhibition of transport activities is possible by high (millimolar) concentration of GDP, irrespective of the effect of DABS being irreversible. Analysis of ^{3}H-GDP binding (Table 1) shows that DABS decreases the affinity of the binding (K_d) to a greater extent than it decreases the binding capacity (B_{max}). More importantly, however, the corresponding changes in nucleotide sensitivity of transport activities are by an order of magnitude higher than the changes of binding. As only a single specific GDP binding site is detected in both control and modified mitochondria, the data were expressed as a function of the binding site occupancy. In case of Cl^- transport (Fig. 2) almost a linear relationship is found in control mitochondria while the correlation for simultaneous H^+ permeation is clearly nonlinear and biphasic, indicating that up to 50% saturation of nucleotide-binding sites the H^+ transport is not affected. After DABS treatment the nonlinear relationship of H^+ transport was further increased and it occurred, to a lesser extent, also with Cl^- transport. These results suggest that there is not a simple direct link between the binding of purine nucleotides and the inhibition of transport. Some additional intrinsic modulation of the sensitivity of transport to bound nucleotide must be involved. As different amino acid residues of the nucleotide-binding site ensure the affinity of the binding (DABS-sensitive) and the resulting gating effect (DABS-insensitive), it is likely that different conformational states of the uncoupling protein displaying different conductivity for protons and anions and/or cooperative interactions between

The physiological role of the uncoupling protein-mediated anion transport is completely unclear. The higher sensitivity of Cl⁻ transport *versus* H⁺ transport might have some physiological meaning, assuming that masking/unmasking of the nucleotide-binding sites (Desautels *et al.*, 1978; Gribskov *et al.*, 1986; Nedergaard & Cannon, 1987; Peachey *et al.*, 1988) is connected with volume changes of mitochondria (Nedergaard & Cannon, 1987) which may involve anion transport activity of the uncoupling protein. Even more obvious is, however, the potential importance of the lack of inhibition of H⁺ transport at low saturation of nucleotide-binding sites. The sharp inhibitory effect occurring close to saturation of binding appears to be advantageous for *in vivo* regulation of thermogenesis, as only a fraction of the total proton-translocating capacity of uncoupling protein is sufficient for continuous dissipation of respiration-generated proton gradient (Nicholls, 1976; Lin & Klingenberg, 1982). Thus in cooperation with changing levels of free fatty acids (Locke *et al.*, 1982; Rial *et al.*, 1983) and possibly also changes of cytosolic pH (Klingenberg, 1988), small changes in nucleotide concentration and/or ratio in cytoplasm around the basal nucleotide level (LaNoue *et al.*, 1986) would open and close the uncoupling protein proton channel. This may be sufficient to switch thermogenesis on and off.

UNCOUPLING PROTEIN IN MAMMALIAN EMBRYONIC AND NEONATE BROWN ADIPOSE TISSUE - EXISTENCE OF NONTHERMOGENIC AND THERMOGENIC MITOCHONDRIA

The perinatal period represents an original development of brown adipose tissue thermogenesis *per se*, following the "programmed" ontogenic recruitment which reflects a degree of immaturity of the newborn (Nedergaard & Cannon, 1986). The main prerequisite of thermogenically active brown adipose tissue are functional, fully competent, thermogenic mitochondria with a high oxidative capacity parallelled by sufficient capacity of the uncoupling protein-mediated proton conductivity while the phosphorylating activity is low (Nicholls, 1976; Houštěk *et al.*, 1978; Nicholls & Locke, 1984; Lin & Klingenberg, 1982). Therefore, the content of cytochrome oxidase, the uncoupling protein and F_1-ATPase should be a sensitive criterion of the functional state of the tissue.

To characterize the profile and changes in the energetics of brown adipose tissue we studied embryonic mouse and rat and newborn hamster using immunochemical quantification of the above proteins as well as immunoelectronmicroscopy which allows to assess their cellular location. In embryonic mouse and rat the brown adipose tissue could be macroscopically identified and dissected from the 14-15th day of fetal development. Already on the 15-16th day the tissue contains some, although not very

uncoupling protein molecules in the membrane are involved in ion transport and its control.

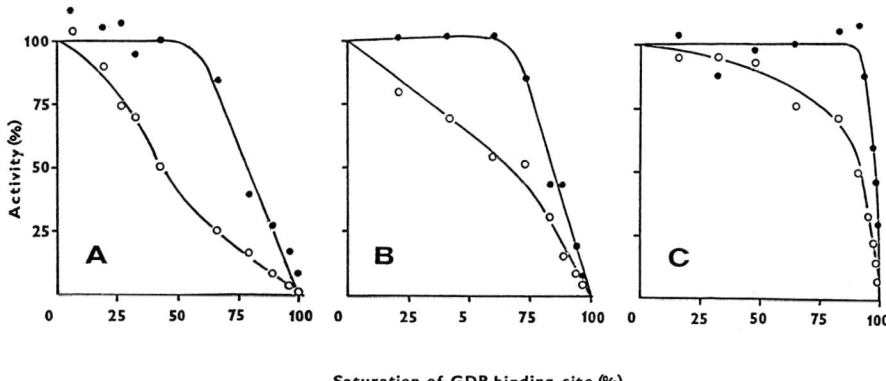

Fig. 2. Relationship between occupancy of nucleotide-binding sites and rate of H$^+$ transport (●) and Cl$^-$ transport (○) in control mitochondria (A) and mitochondria modified with (B) 25 and (C) 75 nmol DABS/mg protein, as described by Kopecký *et al.* (1987).

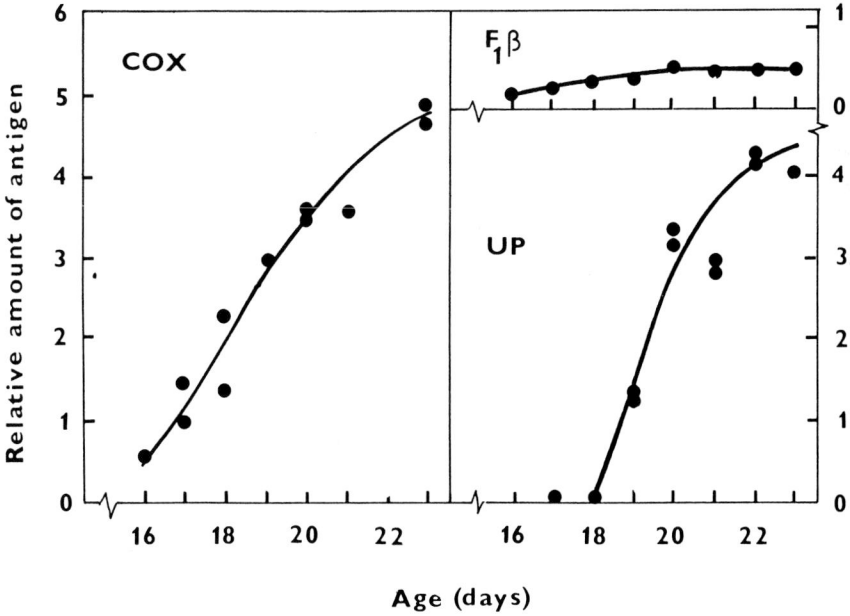

Fig. 3. Quantitative evaluation of cytochrome oxidase (COX), β-subunit of F$_1$-ATPase (F$_1\beta$) and uncoupling protein (UP) in mouse embryonic brown adipose tissue, determined by immunoblotting using rabbit antisera, as described by Houštěk *et al.* (1988).

well differentiated mitochondria. As shown in Fig. 3, cytochrome oxidase is also present and its quantity continuously increases toward birth, following an increasing number of mitochondria. F_1-ATPase is also found on the 15-16th day but reaches its maximum already on the 19th day when the uncoupling protein appears for the first time. During the subsequent days the content of the uncoupling protein rapidly increases and parallels the increase of cytochrome oxidase. Consequently, the original "immature" mitochondria with only a few and irregular cristae are continuously replaced by differentiated, highly branched mitochondria. Therefore, the original mitochondria which are present before the 19th day of pregnancy are clearly nonthermogenic as they lack the uncoupling protein but contain comparable amounts of cytochrome oxidase and ATPase, thus resembling the mitochondria of other, phosphorylating tissues.

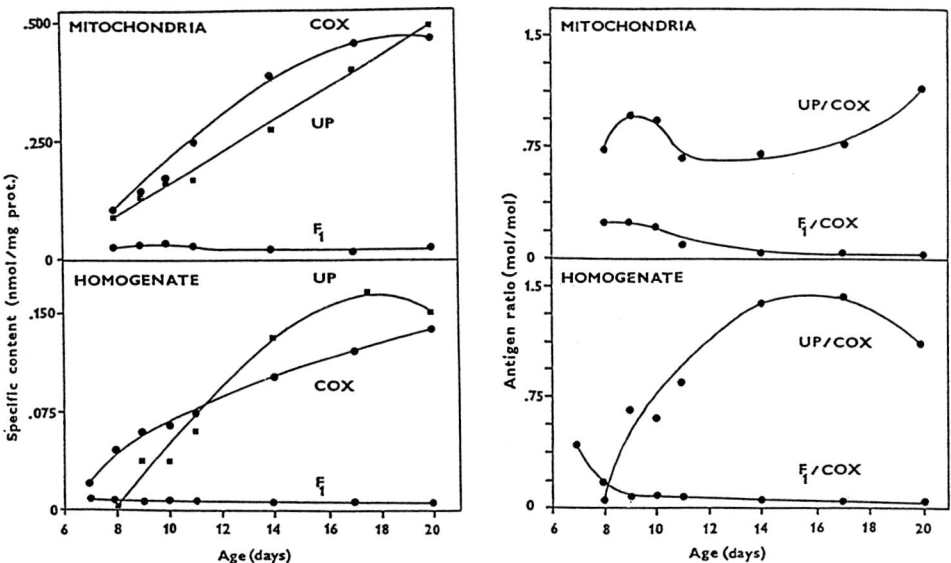

Fig. 4. The content of cytochrome oxidase (COX), F_1-ATPase (F_1) and uncoupling protein (UP) and their respective ratios in homogenates and isolated mitochondria of brown adipose tissue of newborn hamster determined by immunoblotting.

In hamster brown adipose tissue (Fig. 4) an analogous function of thermogenic, fully competent mitochondria begins to develop *post natum*, in agreement with previous measurements of ^{3}H-GDP binding (Sundin *et al.*, 1981). As indicated by the content of cytochrome oxidase, the uncoupling protein and F_1-ATPase and their respective ratios, as well as by enzyme activity measurements, before the occurrence of uncoupling protein on the 9th postnatal day F_1-ATPase is again in quantitative relation with cytochrome oxidase. Therefore, both in the case of altricical mouse and rat and immature hamster brown adipose tissue (Nedergaard & Cannon, 1986), a transformation of preexisting nonthermogenic mitochondria into thermogenic mitochondria results from abruptly induced asynchronous expression of genes of the key enzymes of mitochondrial energy conversion.

Biogenesis of brown adipose tissue mitochondria depends on stimulation of α_1- and β-adrenergic receptors (Jacobsson *et al.*, 1986) as well as on thyroid hormones (Himms-Hagen 1983; Bianco & Silva, 1987). As indicated by measurements of specific mRNA levels (Freeman & Patel, 1984; Ricquier & Bouillaud, 1986), the synthesis of coupling protein is regulated at the transcriptional level, although translational regulation is not excluded. Furthermore, highly pronounced differences in the synthesis of cytochrome oxidase and ATPase necessarily require well balanced cooperation of nuclear and mitochondrial genome and brown adipose tissue could serve here as an outstanding model.

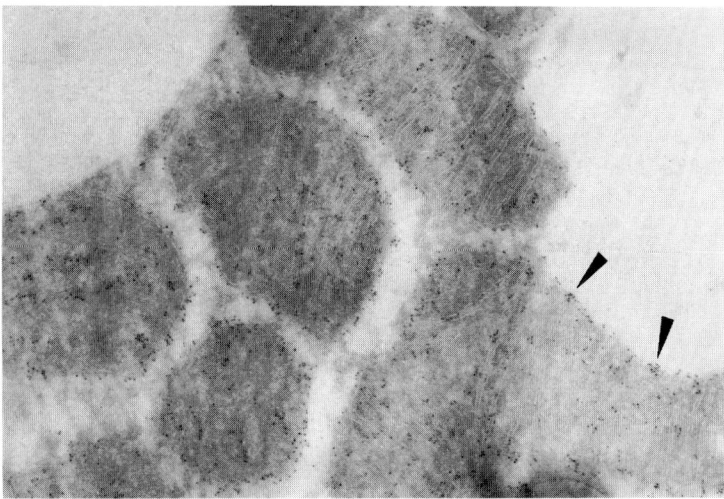

Fig. 5. Immunoelectronmicroscopy of brown adipose tissue of adult hamster. Uncoupling protein was detected by specific rabbit antibody and protein A-colloidal gold probe. From Houštěk J, Janíkova D, Bednar J, Kopecký J, Soukup T: Postnatal formation of thermogenic mitochondria in hamster brown fat (in preparation).

In contrast to immunoblotting, immunoelectronmicroscopy (Fig. 5) is sensitive enough to detect individual protein molecules in the cell. Using this technique it was possible to confirm the changes in uncoupling protein during ontogenesis as well as the higher content of F_1-ATPase in nonthermogenic brown adipose tissue mitochondria. In addition, it was found that a significant accumulation of uncoupling protein antigens in the cytoplasm shortly precedes (1-2 days) its incorporation into the mitochondrial membrane.

REFERENCES

Bianco AC, Silva JE (1987) Intracellular conversion of thyroxine to triiodothyronine is required for the optimal thermogenic function of brown adipose tissue. J Clin Invest 79:295-300

Desautels M, Zaror-Behrens G, Himms-Hagen J (1978) Increased purine nucleotide binding, altered polypeptide composition, and cold thermogenesis in brown adipose tissue mitochondria of cold-acclimated rats. Can J Biochem 56:378-383

Freeman KB, Patel HV (1984) Biosynthesis of the 32 Kdalton uncoupling protein in brown adipose tissue of developing rabbits. Can J Biochem 62:479-485

Gribskov CL, Henningfield MF, Swick AG, Swick RW (1986) Evidence for unmasking of rat brown-adipose-tissue mitochondrial GDP-binding sites in response to acute cold exposure. Biochem J 233:743-747

Heaton GM, Wagenvoord RJ, Kemp A, Nicholls DG (1978) Brown adipose tissue mitochondria: photoaffinity labeling of the regulatory site for energy dissipation. Eur J Biochem 82:515-521

Himms-Hagen J (1983) Thyroid hormones and thermogenesis. In: Girardier L, Stock MJ (eds) Mammalian Thermogenesis. Chapman and Hall, London, pp 141-177

Houštěk J, Ježek P, Kopecký J (1987) Methods for studying the structure and function of the mitochondrial uncoupling protein. In: Bertoli E, Chapman D, Cambria A, Scapagnini U (eds) Biomembrane and Receptor Mechanisms. Fidia Research Series, vol 7, Liviana Press, Padova, pp 176-191

Houštěk J, Kopecký J, Drahota Z (1978) Specific properties of brown adipose tissue mitochondrial membrane. Comp Biochem Physiol 60B:209-214

Houštěk J, Kopecký J, Rychter Z, Soukup T (1988) Uncoupling protein in embryonic brown adipose tissue - existence of nonthermogenic and thermogenic mitochondria. Biochim Biophys Acta 935:19-25

Jacobson A, Nedergaard J, Cannon B (1986) α- and β-adrenergic control of thermogenin mRNA expression in brown adipose tissue. Bioscience Reports 6:621-631

Klingenberg M (1988) Nucleotide binding to uncoupling protein. Mechanism of control by protonation. Biochemistry 27:781-791

Kopecký J, Guerrieri F, Ježek P, Drahota Z, Houštěk J (1984) Molecular mechanism of uncoupling in brown adipose tissue: the nonidentity of proton and chloride conducting pathways. FEBS Lett 170:186-190

Kopecký J, Ježek P, Drahota Z, Houštěk J (1987) Control of uncoupling protein in brown-fat mitochondria by purine nucleotides. Eur J Biochem 164:687-694

LaNoue KF, Strzelecki D, Koch C (1986) Regulation of the uncoupling protein in brown adipose tissue. J Biol Chem 261:298-305

Lin CS, Klingenberg M (1982) Characteristics of the isolated purine nucleotide binding protein from brown fat mitochondria. Biochemistry 21:2950-2956

Locke RM, Rial E, Scott ID, Nicholls DG (1982) Fatty acids as acute regulators of the proton conductance of hamster brown fat mitochondria. Eur J Biochem 129:373-380

Nedergaard J, Cannon B (1986) Brown adipose tissue in the mammalian neonate. In: Trayhurn P, Nicholls DG (eds) Brown Adipose Tissue. Edward Arnold, London, pp 152-213

Nedergaard J, Cannon B (1987) Apparent unmasking of [^{3}H]GDP binding in rat brown-fat mitochondria is due to mitochondrial swelling. Am J Biochem 164:681-686

Nicholls DG (1976) The bioenergetics of brown adipose tissue mitochondria. FEBS Lett 61:103-110

Nicholls DG, Cunningham SA, Rial E (1986) The physiological mechanisms of brown adipose tissue thermogenesis. In: Trayhurn P, Nicholls DG (eds) Brown Adipose Tissue. Edward Arnold, London, pp 52-85

Nicholls DG, Lindberg O (1973) Brown adipose tissue mitochondria: the influence of albumin and nucleotides on passive ion permeabilities. Eur J Biochem 37:523-530

Nicholls DG, Locke RM (1984) Thermogenic mechanism in brown fat. Physiol Rev 64:1-64

Peachey T, French RR, York DA (1988) Regulation of GDP binding and uncoupling protein concentration in brown fat mitochondria. Biochem J 249:451-457

Rial E, Poustie A, Nicholls DG (1983) Brown adipose tissue mitochondria: the regulation of the 32,000 M_r uncoupling protein by fatty acids and purine nucleotides. Eur J Biochem 137:197-203

Ricquier D, Bouillaud F (1986) The brown adipose tissue mitochondrial uncoupling protein. In: Trayhurn P, Nicholls DG (eds) Brown Adipose Tissue. Edward Arnold, London, pp 86-104

Sundin U, Herron D, Cannon B (1981) Brown fat thermoregulation in developing hamsters (*Mesocricetus auratus*): a GDP-binding study. Biol Neonate 39:141-149

V. Carriers and Their Cellular Environment

Biogenesis of Mitochondrial Proteins

MAXIMILIAN TROPSCHUG AND WALTER NEUPERT

Institut für Physiologische Chemie der Universität München, Goethestr. 33,
D-8000 München 2, Fed. Rep. Germany

The majority of mitochondrial proteins are encoded by nuclear genes and are imported into the organelle after being synthesized on cytoplasmic ribosomes. During the last years our knowledge of the different stages of this import process of proteins into mitochondria has increased greatly.

Most of the precursor proteins are synthesized with a transient amino-terminal extension ("presequence") which is removed in the mitochondrial matrix. On the other hand, several precursors (*e.g.* outer membrane proteins like porin or inner membrane proteins like the ADP/ATP carrier and the uncoupling protein of brown adipose tissue) are synthesized without a cleavable presequence. The import of almost all precursor proteins depends on the presence of ATP. ATP is thought to mediate the unfolding of the precursor proteins *via* cytosolic factors (unfoldases). Specific proteinaceous receptors on the mitochondrial surface recognize precursor proteins and deliver them to a general insertion protein (GIP) in the outer membrane. Further import occurs at translocation contact sites between outer and inner membrane which form a hydrophilic environment for the translocation of the precursor proteins. The insertion of precursor proteins into translocation contact sites requires a membrane potential. The completion of precursor translocation into the inner membrane or matrix is independent of the membrane potential. In the mitochondrial matrix, the amino-terminal presequences are removed by the matrix processing peptidase (MPP). MPP is a soluble protein of 57 kDa in *Neurospora crassa*. A second protein of 52 kDa, the processing enhancing protein (PEP; equivalent to the *mas1* gene product in yeast),

A. Azzi et al. (Eds.)
Anion Carriers of Mitochondrial Membranes
© Springer-Verlag Berlin Heidelberg 1989

which is largely associated with the inner membrane, markedly stimulates proteolytic processing. Several precursor proteins are processed in two steps. The second cleavage is done by MPP for some precursors or by processing activities located in the intermembrane space for several other precursors. These precursors contain an uncharged stretch of about 20 amino acids in the second part of their presequences, which is involved in the intramitochondrial sorting. This part of the presequence appears to direct the export of the precursor from the matrix across the inner membrane into the intermembrane space. We will discuss several steps of the import process of proteins into mitochondria with emphasis on mitochondrial carrier proteins like the ADP/ATP carrier or porin.

CYTOSOLIC PRECURSORS FOR MITOCHONDRIAL PROTEINS

Proteins which are imported into mitochondria are synthesized as precursors on cytoplasmic polysomes. Most precursors, but not all, are synthesized as higher molecular weight proteins with amino-terminal peptide extensions which are proteolytically removed during import. Exceptions are all proteins of the outer membrane studied to date, for example porin (Freitag et al., 1982; Mihara et al., 1982; Gasser & Schatz, 1983) or the yeast 70 kDa, 45 kDa and 14 kDa proteins (Gasser & Schatz, 1983), some proteins of the intermembrane space e.g. cytochrome c (Korb & Neupert, 1978; Zimmermann et al., 1979a; Matsuura et al., 1981) and adenylate kinase (Watanabe & Kubo, 1982) and some proteins of the inner membrane like ADP/ATP carrier (Zimmermann et al., 1979b) and the uncoupling protein from brown adipose tissue (Freeman et al., 1983; Ricquier et al., 1983; Aquila et al., 1985; Bouillaud et al., 1986; Ridley et al., 1986). All those proteins lack amino-terminal presequences. The trend that emerges is that the deeper into the mitochondrion a protein must be imported, the more likely it is to be synthesized as a larger precursor.

It has been shown by gene-fusion experiments that the cleavable presequences carry sufficient information to direct a passenger protein into the correct mitochondrial subcompartment (Hurt et al., 1984; Douglas et al., 1984; Horwich et al., 1985; Hurt et al., 1985a; Emr et al., 1986; Nguyen et al., 1986). Gene-fusions were also performed in vivo by eukaryotic cells in order to put the same protein in the cytosol and the mitochondria: examples are the yeast proteins fumarase (Wu & Tzagoloff, 1987), aminoacyl-tRNA synthetases (Gabius et al., 1983; Natsoulis et al., 1986; Chatton et al., 1988), tRNA-modifying enzymes (Hopper et al., 1982) and 2-isopropylmalate synthase (Beltzer et al., 1988) and cyclophilin, a cyclosporin A-binding protein from Neurospora crassa (Tropschug et al., 1988).

Proteins which do not contain removable pre-pieces also contain specific targeting information (see also Hase et al., 1984; 1986; Hurt et al., 1985b; Hurt & van Loon,

1986): The first 115 amino-terminal amino acids of the yeast ADP/ATP carrier contain specific targeting information (Adrian et al., 1986). Since the ADP/ATP carrier is a tripartite protein having three segments of approximately 100 amino acids which have a high degree of homology (Saraste & Walker, 1982), we proposed that similar targeting information might also exist in the other two segments. This is indeed the case: A truncated N.crassa ADP/ATP carrier molecule lacking the first 103 amino acids shows the import characteristics of the authentic precursor, including nucleoside triphosphate dependence, requirement for a protease-sensitive componenton the mitochondrial surface, two-step specific binding to the outer membrane, and membrane potential-dependent translocation into the inner membrane (Pfanner et al., 1987b). Whether each of both remaining domains contain targeting information is currently under investigation.

Porin, a protein from the outer membrane, which forms a voltage-dependent anion channel (VDAC), is also synthesized without a cleavable presequence (Freitag et al., 1982; Mihara et al., 1982; Gasser & Schatz, 1983; Mihara & Sato, 1985; Kleene et al., 1987). Structural analysis of yeast (Mihara & Sato, 1985) and N.crassa porin (Kleene et al., 1987) suggested that the first 18 amino acids of both proteins are able to form amphiphilic helices. According to von Heijne (1986), most of the mitochondrial targeting sequences studied to date form amphiphilic helices which are positively charged. In contrast, the amphiphilic helical structures at the amino-termini of N.crassa and yeast porins are negatively charged. The reason for this may be that positively-charged amphiphilic presequences that are targeted to the inner membrane or the matrix respond to the membrane potential across the inner membrane, which is inside negative. Porin which does not require a membrane potential for insertion may only require an amphiphilic structure for insertion into the outer membrane irrespective of the overall charge of this structure. Gene-fusion experiments showed that the amino-terminal region of N.crassa porin alone is not sufficient to specifically direct the protein to the outer membrane. For correct targeting also the carboxy-terminus of porin seems to be necessary (R. Kleene, M. Tropschug & W. Neupert, in preparation). An intact carboxy-terminus has also been shown to be important for correct mitochondrial import of another mitochondrial protein without a cleavable presequence, namely apocytochrome c (Stuart et al., 1987).

NUCLEOSIDE TRIPHOSPHATES ARE REQUIRED FOR MITOCHONDRIAL
PROTEIN IMPORT

It is known for a long time that an energized inner membrane is necessary to drive import of most precursor proteins into mitochondria (Schleyer et al., 1982; Pfanner & Neupert, 1985; Pfanner et al., 1988a). Exceptions include outer membrane proteins *e.g.* porin and the intermembrane space protein cytochrome c (for review see Nicholson & Neupert, 1988).

In the last two years, it was demonstrated that also nucleoside triphosphates (NTPs: ATP or GTP) are needed for the import of proteins into mitochondria (Pfanner & Neupert, 1986; Pfanner *et al.*, 1987a; Chen & Douglas, 1987a; Eilers *et al.*, 1987; Hartl *et al.*, 1987a). It has been suggested that NTPs maintain or confer an import-competent conformation in mitochondrial precursor proteins. This is supported by experiments where the proteolytic sensitivity of precursor proteins is increased in the presence of NTPs (Pfanner *et al.*, 1987a; Verner & Schatz, 1987), indicating that a less folded conformation is sustained by NTP hydrolysis and that such a conformation is necessary for import. Conformational alteration of a precursor protein (porin) can substitute for the ATP requirement (Pfanner *et al.*, 1988b). The levels of NTPs required depend primarily on the mature part of the precursor protein. For example, precursors having identical presequences but different mature polypeptides require different concentrations of NTPs for optimal import (Pfanner *et al.*, 1987a). It appears that NTPs are necessary for conferring import-competence during all steps that precede and include the interaction of the precursor with the outer membrane (Pfanner & Neupert, 1987; Eilers *et al.*, 1987). Eilers & Schatz (1986) showed that a stable tertiary structure of a protein is incompatible with import of the protein into mitochondria. Methotrexate, an inhibitor of dihydrofolatereductase (DHFR), blocked import into mitochondria of a fusion protein between a mitochondrial presequence and DHFR, most probably by imposing a defined, stable, tertiary structure. A similar result was obtained with a fusion protein between the β-subunit of F_1-ATPase and copper metallothionein (Chen & Douglas, 1987b).

SPECIFIC RECOGNITION AND MEMBRANE INSERTION OF PRECURSOR PROTEINS

Receptors

The targeting informations contained in the precursor proteins for specific mitochondrial recognition and sorting have to be decoded by components of the mitochondrial import machinery. One type of component of this import machinery are import receptors, which are of a proteinaceous nature.

Proteinaceous receptors on the outer surface of mitochondrial membranes were first demonstrated by shaving isolated mitochondria with low concentrations of proteases which do not penetrate or destroy the outer membrane (Gasser *et al.*, 1982; Riezmann *et al.*, 1983; Zwizinski *et al.*, 1984; Pfaller & Neupert, 1987; Kleene *et al.*, 1987). Following this treatment, specific binding of precursor proteins to the outer membrane was blocked and import was greatly reduced (see also below: bypass import).

Binding of precursor proteins to receptors can be stalled either by disrupting the membrane potential, low temperature, or in the special case of apocytochrome c by deuterohemin. Deuterohemin is a hemc analogue which prevents covalent attachment of heme to the precursor apocytochrome c and thus prevents subsequent translocation across the outer membrane (Hennig & Neupert, 1981). Under these conditions, apocytochrome c could still bind to mitochondria independently of import. When the inhibition by deuterohemin was reversed by adding excess amounts of hemin, apocytochrome c was subsequently imported from its receptor sites into the intermembrane space (Hennig & Neupert, 1981; Hennig et al., 1983).

How many different receptors exist on the mitochondrial surface to mediate recognition and binding of the many different precursor protein having or having not cleavable presequences? We postulated at least three different types of receptors (Pfaller et al., 1988a). This is based on different sensitivity of the receptors to trypsin and elastase treatment. Import of a number of different precursor proteins is greatly reduced by treatment of mitochondria with low concentrations of elastase (examples include ADP/ATP carrier and porin). Import of the β-subunit of F_1-ATPase, however, is not significantly affected by pretreatment of elastase (up to 10 μg/ml; Zwizinski et al., 1984; Pfaller et al., 1988a). In contrast to this, import of all these precursors is sensitive to pretreatment of mitochondria with trypsin. This suggests that at least two different types of receptors exist. Since the precursor of porin does not compete with the ADP/ATP carrier for binding to its receptor, at least three types of receptors can be postulated (one for porin, one for ADP/ATP carrier and one for F_1-β subunit). All three receptors are thought to deliver the bound precursor proteins to a general insertion protein (GIP) in the outer membrane (see below). A fourth type of receptor is the apocytochrome c receptor (Hennig & Neupert, 1981; Hennig et al., 1983). Binding of apocytochrome c to the mitochondrial surface can be blocked only by pretreatment of mitochondria with high concentrations of trypsin (> 40 μg/ml). Recent results (Nicholson et al., 1988) indicate that the enzyme cytochrome c-heme lyase, which is located on the inner side of the outer membrane (therefore being not easily protease-accessible), could be the receptor for apocytochrome c. Apocytochrome c is able to spontaneously insert into lipid bilayers in a nonspecific manner with low affinity (Rietveld et al., 1983, 1985, 1986a,b; Rietveld & Kruijff, 1984; Dumont & Richards, 1984) and could thereby expose domains to the binding protein, cytochrome c-heme lyase. Cytochrome c seems to have a unique import pathway different from all other precursor proteins (for review see Nicholson & Neupert, 1988).

General insertion protein (GIP)

Competition experiments of precursor proteins for components of the mitochondrial import machinery are possible when precursor proteins are available in sufficient chemical amounts. This was achieved either by expression in yeast (Ohta & Schatz, 1984) or *E.coli* (Eilers *et al.*, 1987). Another approach was to isolate a mitochondrial protein without a transient presequence and alter its conformation to that of its precursor form. We isolated porin from the outer mitochondrial membrane of *Neurospora crassa* and converted it to a water-soluble form (ws-porin) which behaves in many respects like the authentic biosynthetic porin precursor (Pfaller *et al.*, 1985). Ws-porin was shown to compete for the import of precursors destined for the three other mitochondrial compartments: the Fe/S protein of the bc_1-complex (intermembrane space), the ADP/ATP carrier (inner membrane), subunit 9 of F_o-ATPase (inner membrane) and subunit β of the F_1-ATPase (matrix). Competition does not occur at the level of the receptor proteins but at a common site at which precursors are inserted into the outer membrane. We suggest that distinct receptor proteins recognize precursor proteins and transfer them to a general insertion protein (GIP) in the outer membrane. Beyond GIP, the import pathways diverge, either to the outer membrane (*e.g.* porin) or to translocation contact-sites and then subsequently to the other mitochondrial compartments (Pfaller *et al.*, 1988a).

Bypass import

Proteolytic degradation of receptor sites on the mitochondrial surface strongly reduces the efficiency of mitochondrial protein import (up to 80 - 95%; see also above). The remaining residual import (bypass import) still involves basic mechanisms of protein import, including: insertion of precursors into the outer membrane, requirement for ATP and a membrane potential, and translocation through contact sites between both mitochondrial membranes. The import of a chloroplast protein (small subunit of ribulose-1,5-biphosphate carboxylase/oxygenase) into isolated mitochondria which occurs with a low rate is not inhibited by a protease-pretreatment of mitochondria, indicating that this precursor only follows the bypass pathway. The low efficiency of bypass import suggests that this unspecific import does not disturb the uniqueness of mitochondrial protein composition. We conclude that mitochondrial protein import involves a series of steps in which receptors sites appear to be responsible for the specificity of protein uptake (Pfaller *et al.*, 1988b).

Translocation contact sites

Proteins which are imported into the mitochondrial matrix or inner membrane must cross two membranes barriers to reach their final location. It has been demonstrated for a variety of proteins that import occurs at sites where inner and outer membrane come close enough to be spanned and crossed in a single event. Involvement of translocation subunit and cytochrome c_1 (Schleyer & Neupert, 1985), the Fe/S protein of bc_1 complex (Hartl et al., 1986), the ADP/ATP carrier (Pfanner & Neupert, 1987), cytochrome b_2 (Pfanner et al., 1987c; Hartl et al., 1987b) and a number of fusion proteins. Three distinct methods yielded translocation intermediates spanning both membranes: import at low temperature, pre-binding of antibodies to carboxy-terminal precursor portions (Schleyer & Neupert, 1985) and import at low levels of NTPs (Pfanner et al., 1987a,c). Precursors were thereby trapped in an intermediate position with the amino-terminal presequence in the mitochondrial matrix and other, probably carboxy-terminal, portions of the precursors outside the outer membrane. The topology of the intermediates was examined by their accessibility to the matrix-processing peptidase (see below), and to externally added proteases and antibodies (Söllner et al., 1988). Immunochemical studies (labelling the contact site intermediates with protein A-gold particles via the bound antibodies) demonstrated the identity of morphologically described (Hackenbrock, 1968) and the biochemically defined contact sites (Schwaiger et al., 1987). Contact sites appear to be stable structures which can be enriched after subfractionation of mitochondria by sonication (Schwaiger et al., 1987). Contact site intermediates could be extracted from the membranes with hydrophilic perturbants, such as urea or at alkaline pH, suggesting that mitochondrial precursor proteins are imported through a hydrophilic membrane environment (Pfanner et al., 1987c). Specific proteins in contact sites are probably involved in constituting the architecture of these sites and participate in protein translocation.

Mitochondrial processing peptidase

During or shortly following the translocation step, the amino-terminal presequences of many proteins directed to the inner membrane or matrix are removed by a specific protease which is located in the matrix (Böhni et al., 1980; Mori et al., 1980; Conboy et al., 1982; McAda & Douglas, 1982; Miura et al., 1982; Böhni et al., 1983; Schmidt et al., 1984; Miura et al., 1986). This occurs very rapidly in vivo. Processing, however, is not obligatory for import since the precursors to subunits β and IX of F_0F_1-ATPase could be imported into mitochondria when proteolytic processing was blocked by o-phenanthroline (Zwizinski & Neupert, 1983). Similarly, the precursor to the Fe/S protein of the bc$_1$ complex could be imported and accumulated in the matrix when processing was blocked (Hartl *et al.*, 1986).

The matrix-located peptidase has been purified to homogeneity from *N.crassa*. The purification was about 10,000 fold (starting with a cell extract) and yielded two bands on SDS-PAGE (PEP: 52 kDa and MPP: 57 kDa; Hawlitschek *et al.*, 1988). The matrix processing peptidase (MPP: 57 kDa) has a low intrinsic enzyme activity in the absence of the processing enhancing protein (PEP: 52 kDa) and the latter has a strong stimulating influence on the processing, being by itself completely inactive. We have cloned the cDNAs for both proteins (Hawlitschek *et al.*, 1988; H. Schneider *et al.*, in preparation). The amino acid sequence of *N.crassa* PEP shows 60 % homology to the *mas1* gene product from the yeast *S.cerevisiae*. This suggests that *mas1* encodes the yeast equivalent to *N.crassa* PEP. *N.crassa* MPP (H. Schneider *et al.*, in preparation) has a high degree of homology to the protein encoded the *mas2* gene of *S.cerevisiae* (Yaffe & Schatz, 1984; Yaffe *et al.*, 1985), which is equivalent to the *mif2* gene described by Pollock *et al.* (1988). Interestingly, the two cooperating components MPP and PEP are structurally related, suggesting that the respective genes are of common evolutionary origin (Pollock *et al.*, 1988).

REFERENCES

Adrian GS, McCammon MT, Montgomery DL, Douglas MG (1986) Sequences required for delivery and localization of the ADP/ATP translocator to the mitochondrial inner membrane. Mol Cell Biol 6:626-634

Aquila H, Link TA, Klingenberg M (1985) The uncoupling protein from brown fat mitochondria is related to the mitochondrial ADP/ATP carrier. Analysis of sequence homologies and of folding of the protein in the membrane. EMBO J 4:2369-2376

Beltzer JP, Morris SR, Kohlhaw GB (1988) Yeast *LEU4* encodes mitochondrial and nonmitochondrial forms of isopropylmalate synthase. J Biol Chem 263:368-374

Böhni P, Gasser S, Leaver C, Schatz G (1980) A matrix-localized mitochondrial protease processing cytoplasmically made precursors to mitochondrial proteins. In: Kroon Am, Saccone C (eds) The organization and expression of the mitochondrial genome. Elsevier/North-Holland, Amsterdam, pp 423-433

Böhni PC, Daum G, Schatz G (1983) Import of proteins into mitochondria. Partial purification of a matrix-localized protease involved in cleavage of mitochondrial precursor polypeptides. J Biol Chem 258:4937-4943

Bouilland F, Weissenbach J, Ricquier D (1986) Complete cDNA-derived amino acid sequence of rat brown fat uncoupling protein. J Biol Chem 261:1487-1490

Chatton B, Walter P, Ebel J-P, Lacroute F, Fasiolo F (1988) The yeast VAS1 gene encodes both mitochondrial and cytoplasmic valyl-tRNA synthetases. J Biol Chem 263:52-57

Chen W-J, Douglas MG (1987a) Phosphodiester bond cleavage outside mitochondrial matrix. Cell 49:651-658

Chen W-J, Douglas MG (1987b) The role of protein structure in the mitochondrial import pathway: Unfolding of mitochondrially bound precursors is required for membrane translocation. J Biol Chem 262:15605-15609

Conboy JG, Fenton WA, Rosenberg LE (1982) Processing of pre-ornithine transcarbamylase requires a zinc-dependent protease localized to the mitochondrial matrix. Biochem Biophys Res Commun 105:1-7

Douglas MG, Geller BL, Emr SD (1984) Intracellular targeting and import of an F_1-ATPase β-subunit-β-galactosidase hybrid protein into yeast mitochondria. Proc Natl Acad Sci (USA) 81:3983-3987

Dumont ME, Richards FM (1984) Insertion of apocytochrome c into lipid vesicles. J Biol Chem 259:4147-4156

Eilers M, Schatz G (1986) Binding of a specific ligand inhibits import of a purified precursor protein into mitochondria. Nature 322:228-232

Eilers M, Oppliger W, Schatz G (1987) Both ATP and an energized inner membrane are required to import a purified precursor protein into mitochondria. EMBO J 6:1073-1077

Emr SD, Vassarotti A, Garrett J, Geller BL, Takeda M, Douglas MG (1986) The amino terminus of the yeast F_1-ATPase β-subunit precursor functions as a mitochondrial import signal. J Cell Biol 102:523-533

Freeman KB, Chien S-M, Lichtfield D, Patel HV (1983) Synthesis in vitro of rat brown adipose tissue 32000 M_r protein. FEBS Lett 158:325-330

Freitag H, Janes M, Neupert W (1982) Biosynthesis of mitochondrial porin and insertion into the outer mitochondrial membrane of Neurospora crassa. Eur J Biochem 126:197-202

Gabius HJ, Engelhardt R, Piel N, Sterbach H, Cramer F (1983) Phenylalanyl-tRNA synthetases from yeast cytoplasm and mitochondria: The presence of a carbohydrate moiety in the mitochondrial enzyme and immunological evidence for structural relationship. Biochim Biophys Acta 743:451-454

Gasser SM, Schatz G (1983) Import of proteins into mitochondria: in vitro studies on the biogenesis of the outer membrane. J Biol Chem 258:3427-3430

Gasser SM, Daum G, Schatz G (1982) Import of proteins into mitochondria: energy-dependent uptake of precursors by isolated mitochondria. J Biol Chem 257:13034-13041

Gasser SM, Daum G, Schatz G (1982) Import of proteins into mitochondria: energy-dependent uptake of precursors by isolated mitochondria. J Biol Chem 257:13034-13041

Hackenbrock CR (1968) Chemical and physical fixation of isolated mitochondria in low-energy and high-energy states. Biochemistry 61:598-605

Hartl F-U, Schmidt B, Wachter E, Weiss H, Neupert W (1986) Transport into mitochondria and intramitochondrial sorting of the Fe/S protein of ubiquinol-cytochrome c reductase. Cell 47:939-951

Hartl F-U, Ostermann J, Pfanner N, Tropschug M, Guiard B, Neupert W (1987a) Import of cytochromes b_2 and c_1 into mitochondria is dependent on both membrane potential and nucleoside triphosphates. In: Papa S, Chance B, Ernster L (eds) Cytochrome Systems. Plenum Publishing Corporation, New York, pp 189-196

Hartl F-U, Ostermann J, Guiard B, Neupert W (1987b) Successive translocation into and out of the mitochondrial matrix: targeting of proteins to the intermembrane space by a bipartite signal peptide. Cell 51:1027-1037

Hase T, Müller U, Riezmann H, Schatz G (1984) A 70-kd protein of the yeast mitochondrial outer membrane is targeted and anchored via its extreme amino terminus. EMBO J 3:3157-3164

Hase T, Nakai M, Matsubara H (1986) The N-terminal 21 amino acids of a 70 kDa protein of the yeast mitochondrial outer membrane direct E.coli β-galactosidase into the mitochondrial matrix space in yeast cells. FEBS Lett 197:199-203

Hawlitschek G, Schneider H, Schmidt B, Tropschug M, Hartl P-U, Neupert W (1988) Mitochondrial protein import: identification of processing peptidase and of PEP, a processing enhancing protein. Cell 53:795-806

Hennig B, Neupert W (1981) Assembly of cytochrome c. Apocytochrome c is bound to specific sites on mitochondria before its conversion to holocytochrome c. Eur J Biochem 81:535-544

Hennig B, Kohler H, Neupert W (1983) Receptor sites involved in posttranslational transport of apocytochrome c into mitochondria: Specificity, affinity and number of sites. Proc Natl Acad Sci (USA) 80:4963-4967

Hopper AK, Furukawa AH, Pham HD, Martin NC (1982) Defects in modification of cytoplasmic and mitochondrial transfer RNAs are caused by single nuclear mutations. Cell 28:543-550

Horwich AL, Kalousek F, Mellman I, Rosenberg LE (1985) A leader peptide is sufficient to direct mitochondrial import of a chimeric protein. EMBO J 4:1129-1135

Hurt EC, van Loon APGM (1986) How proteins find mitochondria and intramitochondrial compartments. Trends Biochem Sci 11:204-207

Hurt EC, Pesold-Hurt B, Schatz G (1984) The amino-terminal region of an imported mitochondrial protein is sufficient to direct cytosolic dihydrofolate reductase into the mitochondrial matrix. EMBO J 3:3149-3156

Hurt EC, Pesold-Hurt B, Suda K, Oppliger W, Schatz G (1985a) The first twelve amino acids (less than half of the pre-sequence) of an imported mitochondrial protein can direct mouse cytosolic dihydrofolate reductase into the yeast mitochondrial matrix. EMBO J 4:2061-2068

Hurt EC, Müller U, Schatz G (1985b) The first twelve amino acids of a yeast mitochondrial outer membrane protein can direct a nuclear-encoded cytochrome oxidase subunit to the mitochondrial inner membrane. EMBO J 4:3509-3518

Kleene R, Pfanner N, Pfaller R, Link TA, Sebald W, Neupert W, Tropschug M (1987) Mitochondrial porin of *Neurospora crassa*: cDNA cloning, *in vitro* expression and import into mitochondria. EMBO J 6:2627-2633

Korb H, Neupert W (1978) Biogenesis of cytochrome c in *Neurospora crassa*: synthesis of apocytochrome c, transfer to mitochondria and conversion to holocytochrome c. Eur J Biochem 91:609-620

Matsuura S, Arpin M, Hannum C, Margoliash E, Sabatini DD, Morimoto T (1981) *In vitro* synthesis and posttranslational uptake of cytochrome c into isolated mitochondria: role of a specific addressing signal in the apocytochrome. Proc Natl Acad Sci (USA) 78:4368-4372

McAda PC, Douglas M (1982) A neutral metallo endoprotease involved in the processing of an F_1-ATPase subunit precursor in mitochondria. J Biol Chem 257:3177-3182

Mihara K, Sato R (1985) Molecular cloning and sequencing of cDNA for yeast porin, an outer mitochondrial membrane protein: A search for targeting signal in the primary structure. EMBO J 4:769-774

Mihara K, Blobel G, Sato R (1982) *In vitro* synthesis and integration into mitochondria of porin, a major protein of the outer mitochondrial membrane of *Saccharomyces cerevisiae*. Proc Natl Acad Sci (USA) 79:7102-7106

Miura S, Mori M, Amaya Y, Tatibana M (1982) A mitochondrial protease that cleaves the precursor of ornithine carbamoyltransferase: purification and properties. Eur J Biochem 122:641-647

Miura S, Amaya Y, Mori M (1986) A metalloprotease involved in the processing of mitochondrial precursor proteins. Biochem Biophys Res Commun 134:1151-1159

Mori M, Miura S, Tatibana M, Cohen PP (1980) Characterization of a protease apparently involved in processing of pre-ornithine transcarbamylase of rat liver. Proc Natl Acad Sci (USA) 77:7044-7048

Natsoulis G, Hilger F, Fink GR (1986) The HTS1 gene encodes both the cytoplasmic and mitochondrial histidine tRNA synthetases of *S. cerevisiae*. Cell 46:235-243

Neupert W, Schatz G (1981) How proteins are transported into mitochondria. Trends Biochem Sci 6:1-4

Nguyen M, Argan C, Lusty CJ, Shore GC (1986) Import and processing of hybrid proteins by mammalian mitochondria *in vitro*. J Biol Chem 261:800-805

Nicholson DW, Neupert W (1988) Synthesis and assembly of mitochondrial proteins. In: Das RC, Robbins PW (eds) Protein Transfer and Organelle Biogenesis. Academic Press, San Diego New York, pp 677-746

Nicholson DW, Hergersberg C, Neupert W (1988) Role of cytochrome c-heme lyase in the import of cytochrome c into mitochondria. J Biol Chem, in press

Ohta S, Schatz G (1984) A purified precursor polypeptide requires a cytosolic protein fraction for import into mitochondria. EMBO J 3:651-657

Pfaller R, Neupert W (1987) High-affinity sites involved in the import of porin into mitochondria. EMBO J 6:2635-2642

Pfaller R, Freitag H, Harmey MA, Benz R, Neupert W (1985) A water-soluble form of porin from the mitochondrial outer membrane of *Neurospora crassa*. J Biol Chem 260:8188-8193

Pfaller R, Steger HF, Rassow J, Pfanner N, Neupert W (1988a) Import pathways of precursor proteins into mitochondria: multiple receptor sites are followed by a common membrane insertion site. J Cell Biol, in press

Pfaller R, Pfanner N, Neupert W (1988b) Mitochondrial protein import: Bypass of proteinaceous surface receptors can occur with low specificity and efficiency. J Biol Chem, in press

Pfanner N, Neupert W (1985) Transport of proteins into mitochondria: a potassium diffusion potential is able to drive the import of ADP/ATP carrier. EMBO J 4:2819-2825

Pfanner N, Neupert W (1986) Transport of F_1-ATPase subunit β into mitochondria depends on both a membrane potential and nucleoside triphosphates. FEBS Lett 209:152-156

Pfanner N, Neupert W (1987) Distinct steps in the import of ADP/ATP carrier into mitochondria. J Biol Chem 262:7528-7536

Pfanner N, Tropschug M, Neupert W (1987a) Mitochondrial protein import: nucleoside triphosphates are involved in conferring import-competence to precursors. Cell 49:815-823

Pfanner N, Hoeben P, Tropschug M, Neupert W (1987b) The carboxy-terminal two-thirds of the ADP/ATP carrier polypeptide contains sufficient information to direct translocation into mitochondria. J Biol Chem 262:14851-14854

Pfanner N, Hartl F-U, Guiard B, Neupert W (1987c) Mitochondrial precursor proteins are imported through a hydrophilic membrane environment. Eur J Biochem 169:289-293

Pfanner N, Hartl F-U, Neupert W (1988a) Import of proteins into mitochondria: a multi-step process. Eur J Biochem 175:205-212

Pfanner N, Pfaller R, Kleene R, Ito M, Tropschug M, Neupert W (1988b) Role of ATP in mitochondrial protein import: Conformational alteration of a precursor protein can substitute for ATP requirement. J Biol Chem 263:4049-4051

Pollock RA, Hartl F-U, Cheng MY, Ostermann J, Horwich A, Neupert W (1988) The processing peptides of yeast mitochondria: the two cooperating components MPP and PEP are structurally related. EMBO J, in press

Ricquier D, Thibault J, Bouilland F, Kuster Y (1983) Molecular approach to thermogenesis in brown adipose tissue. J Biol Chem 258:6675-6677

Ridley RG, Patel HV, Gerber GE, Morton RC, Freeman KB (1986) Complete nucleotide and derived amino acid sequence of cDNA encoding the mitochondrial uncoupling protein of rat brown adipose tissue: lack of a mitochondrial targeting presequence. Nucleic Acids Res 14:4025-4035

Rietveld A, de Kruijff B (1984) Is the mitochondrial precursor protein apocytochrome c able to pass a lipid barrier? J Biol Chem 259:6704-6707

Rietveld A, Sijens P, Verkleij AJ, de Kruijff B (1983) Interaction of cytochrome c and its precursor apocytochrome c with various phospholipids. EMBO J 2:907-913

Rietveld A, Ponjee GAE, Schiffers P, Jordi W, Van de Coolwijk PJFM, Demel RA, Marsh D, de Kruijff B (1985) Investigations on the insertion of the mitochondrial precursor protein apocytochrome c into model membranes. Biochim Biophys Acta 818:398-409

Rietveld A, Jordi W, de Kruijff B (1986a) Studies on the lipid dependency and mechanism of the translocation of the mitochondrial precursor protein apocytochrome c across model membranes. J Biol Chem 261:3846-3856

Rietveld A, Berkhout TA, Roenhorst A, Marsh D, de Kruijff B (1986b) Preferential association of apocytochrome c with negatively charged phospholipids in mixed model membranes. Biochim Biophys Acta 858:38-46

Riezmann H, Hay R, Witte C, Nelson N, Schatz G (1983) Yeast mitochondrial outer membrane specifically binds cytoplasmically synthesized precursors of mitochondrial proteins. EMBO J 2:1113-1118

Saraste M, Walker JE (1982) Internal sequence repeats and the path of polypeptide in mitochondrial ADP/ATP translocase. FEBS Lett 144:250-254

Schleyer M, Neupert W (1985) Transport of proteins into mitochondria: translocational intermediates spanning contact sites between outer and inner membranes. Cell 43:339-350

Schleyer M, Schmidt B, Neupert W (1982) Requirement of a membrane potential for the posttranslational transfer of proteins into mitochondria. Eur J Biochem 125:109-116

Schmid B, Wachter E, Sebald W, Neupert W (1984) Processing peptidase of *Neurospora* mitochondria: Two-step cleavage of imported ATPase subunit 9. Eur J Biochem 144:581-588

Schwaiger M, Herzog V, Neupert W (1987) Characterization of translocation sites involved in the import of mitochondrial proteins. J Cell Biol 105:235-246

Sollner T, Pfanner N, Neupert W (1988) Mitochondrial protein import: differential recognition of various transport intermediates by antibodies. FEBS Lett 229:25-29

Stuart RA, Neupert W, Tropschug M (1987) Deficiency in mRNA splicing in a cytochrome c mutant of *Neurospora crassa:* importance of carboxy-terminus for import of apocytochrome c into mitochondria. EMBO J 6:2131-2137

Tropschug M, Nicholson DW, Hartl F-U, Köhler H, Pfanner N, Wachter E, Neupert W (1988) Cyclosporin A-binding protein (cyclophilin) of *Neurospora crassa:* one gene codes for both the cytosolic and mitochondrial forms. J Biol Chem 263:14433-14440

Verner K, Schatz G (1987) Import of an incompletely folded precursor protein into isolated mitochondria requires an energized inner membrane, but no added ATP. EMBO J 6:2449-2456

von Heijne G (1986) Mitochondrial targeting sequences may form amphiphilic helices. EMBO J 5:1335-1342

Watanabe K, Kubo S (1982) Mitochondrial adenylate kinase from chicken liver: Purification, characterization and its cell-free synthesis. Eur J Biochem 123:587-592

Wu M, Tzagoloff A (1987) Mitochondrial and cytoplasmic fumarases in *Saccharomyces cerevisiae* are encoded by a single nuclear gene FUM1. J Biol Chem 262:12275-12282

Yaffe MP, Ohta S, Schatz G (1985) A yeast mutant temperature-sensitive for mitochondrial assembly is deficient in a mitochondrial protease activity that cleaves imported precursor polypeptides. EMBO J 4:2069-2074

Zimmermann R, Paluch U, Neupert W (1979a) Cell-free synthesis of cytochrome c. FEBS Lett 108:141-146

Zimmermann R, Paluch U, Sprinzl M, Neupert W (1979b) Cell-free synthesis of the mitochondrial ADP/ATP carrier protein from *Neurospora crassa.* Eur J Biochem 99:247-252

Zwizinski C, Neupert W (1983) Precursor proteins are transported into mitochondria in the absence of proteolytic cleavage of the additional sequences. J Biol Chem 258:13340-13346

Zwizinski C, Schleyer M, Neupert W (1984) Proteinaceous receptors for the import of mitochondrial precursor proteins. J Biol Chem 259:7850-7856

Insensitivity of Carbamoyl-Phosphate Synthetase Towards Inhibition by Carbamoyl Phosphate Makes it Unlikely that Mitochondrial Metabolite Transport Controls Ornithine Cycle Flux

A.J. MEIJER

Laboratory of Biochemistry, University of Amsterdam, P.O. Box 20151, 1000 HD Amsterdam, The Netherlands

The first two enzymes involved in the synthesis of urea in the mammalian liver, - carbamoyl phosphate synthetase (ammonia) (CPS) and ornithine carbamoyl transferase (OCT), are located in the mitochondria whereas the other 3 enzymes (argininosuccinate synthetase, argininosuccinate lyase and arginase) are cytosolic. Because of this dual localization continuous transport of metabolites across the mitochondrial inner membrane (*e.g.* of ornithine, citrulline, glutamate, aspartate and ATP) is an essential part of the pathway of urea synthesis. Whether these transport systems contribute to the control of urea synthesis under *in vivo* conditions is an open question. In principle, flux control analysis (Kacser & Burns, 1973) should enable one to answer this question. One experimental approach would be to titrate these transport systems with specific inhibitors (Groen *et al.*, 1982b). Unfortunately, except in the case of the adenine nucleotide translocator, suitable inhibitors of these transport systems are not available. However, as will be discussed below, an alternative approach is to make use of knowledge of the *in situ* kinetic properties of the ornithine cycle enzymes.

Abbreviations: CPS, carbamoylphosphate synthase; OCT, ornithine carbamoylphosphate transferase

---

A. Azzi et al. (Eds.)
Anion Carriers of Mitochondrial Membranes
© Springer-Verlag Berlin Heidelberg 1989

METHODS

Details about the isolation of rat-liver mitochondria and rat hepatocytes, the incubation conditions and the determination of metabolites can be found elsewhere (Wanders et al., 1984; Meijer et al., 1985; Boon & Meijer, 1988).

RESULTS

Experiments with isolated liver mitochondria

A central parameter in the analysis of flux control of metabolic pathways is the flux control coefficient, defined as:

$$C^J_{E_i} = \left(\frac{dJ/J}{dE_i/E_i} \right)_{ss}$$

in which J refers to the steady state (ss) pathway flux and E_i to the enzyme under consideration (Kacser & Burns, 1973; Groen et al., 1982b). If the concentrations of initial substrate(s) and end product(s) of the pathway are either kept constant or are such that variation does not affect the pathway flux, the sum of all flux control coefficients of the enzymes in that pathway is equal to one. Kacser & Burns (1973) have also shown that there is an inverse relationship between the flux control coefficients of two adjacent enzymes in a pathway and the elasticity coefficients of both enzymes towards their common intermediate. Thus, for the pathway of citrulline synthesis in isolated liver mitochondria from ammonia, ornithine and bicarbonate (with succinate as the respiratory substrate) the following equation holds:

$$\epsilon^{CPS}_{CP} \cdot C^{J_{cit}}_{CPS} + \epsilon^{OCT}_{CP} \cdot C^{J_{cit}}_{OCT} = 0 \qquad (1)$$

in which the elasticity coefficients are defined as:

$$\epsilon^{CPS}_{CP} = \frac{\delta v_{CPS}/v_{CPS}}{\delta CP/CP} \quad \text{and} \quad \epsilon^{OCT}_{CP} = \frac{\delta v_{OCT}/v_{OCT}}{\delta CP/CP}$$

In these equations v_{CPS} and v_{OCT} are the rates of the reactions catalyzed by carbamoyl-phosphate synthetase and ornithine carbamoyl transferase, respectively (equal to the rate of citrulline production in the steady state), and CP is the intramitochondrial concentration of carbamoyl phosphate.

By titrating the pathway of citrulline synthesis with L-norvaline, a specific inhibitor of ornithine carbamoyl transferase, we have previously been able to determine the flux control coefficient of ornithine carbamoyl transferase directly; a value of 0.01 was found (Wanders et al., 1984). For calculation of the flux control coefficient of carbamoyl-phosphate synthetase the values for the elasticity coefficients (see equation (1)) must be known.

In order to determine the elasticity coefficient of ornithine carbamoyl transferase for carbamoyl phosphate the flux through the citrulline synthesizing pathway was varied by decreasing the rate of ATP production with increasing concentrations of malonate, the inhibitor of succinate dehydrogenase. As shown in Fig. 1A an almost linear relationship was found between the flux through ornithine carbamoyl transferase (measured as citrulline production) and the intramitochondrial concentration of carbamoyl phosphate. From this graph a value of 0.92 for the elasticity coefficient of ornithine carbamoyl phosphate

$$\left(\frac{\delta Jcit}{Jcit} \Big/ \frac{\delta CP}{CP}\right)$$

at zero malonate concentration can be calculated.

The elasticity coefficient of carbamoyl-phosphate synthetase for carbamoyl phosphate was determined by titration of the pathway flux with norvaline. As shown in Fig. 1B the enzyme was almost insensitive to changes in the concentration of its product; the elasticity coefficient was -0.0114.

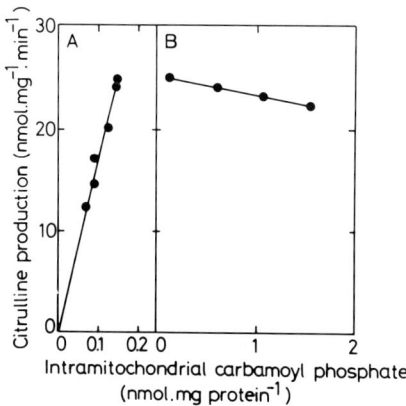

Fig. 1. Determination of the elasticity coefficients of carbamoylphosphate synthetase and of ornithine carbamoyl transferase towards carbamoyl phosphate in intact mitochondria. Liver mitochondria (2.7 mg protein/ml) from a rat fed on a high-protein diet for 2 days were incubated in a medium containing: 100 mM KCl, 50 mM Tris/HCl, 1 mM Tris-EGTA, 1 mM Tris-ATP, 20 mM Tris-succinate, 5 mM potassium phosphate, 3 mM ornithine, 16.6 mM KHCO$_3$, 1 mM aminooxyacetate, 2 μg rotenone/ ml and either increasing concentrations of malonate (0, 1, 2, 3, 4, 5 mM; A) or L-norvaline (0, 1, 2, 3 mM; B); gas phase, 95% O$_2$/5% CO$_2$ (v/v); pH 7.4; temp. 25°C. Citrulline synthesis was started by addition of 5 mM NH$_4$Cl after a 1 min preincubation period. Intramitochondrial carbamoyl phosphate was measured after 5 min of incubation. From Wanders et al. (1984).

Substituting the values for the two elasticity coefficients and the flux control coefficient of ornithine carbamoyl transferase in equation (1) yields the value for the flux control coefficient of carbamoyl-phosphate synthetase: $C_{CPS} = 0.96$.

Since the sum of the flux control coefficients of the two enzymes ($0.96 + 0.01 = 0.97$) is close to one there is, apparently, little residual control for the other steps involved in citrulline synthesis, *e.g.* transport of succinate, ornithine and NH_3 into the mitochondria and the mitochondrial production of HCO_3^- (the direct substrate of carbamoylphosphate synthetase) from CO_2 via carbonic anhydrase; participation of the latter reaction in citrulline synthesis is essential since the mitochondrial membrane is only permeable to CO_2, not to HCO_3^- (cf. Dodgson *et al.*, 1983).

Since carbonic anhydrase can be specifically inhibited by acetazolamide the flux control coefficient of this enzyme can be determined directly. Inhibition of carbonic anhydrase activity by this compound was measured by studying the effect of acetazolamide on mitochondrial swelling in an iso-osmotic solution of CO_2-saturated NH_4HCO_3 (Chappell & Crofts, 1966). Swelling occurs because CO_2, after its entry into the mitochondria, is converted to HCO_3^- *via* carbonic anhydrase. Strong inhibition was already observed at a concentration of acetazolamide of as low as 0.6 μmol/mg mitochondrial protein (Fig. 2A).

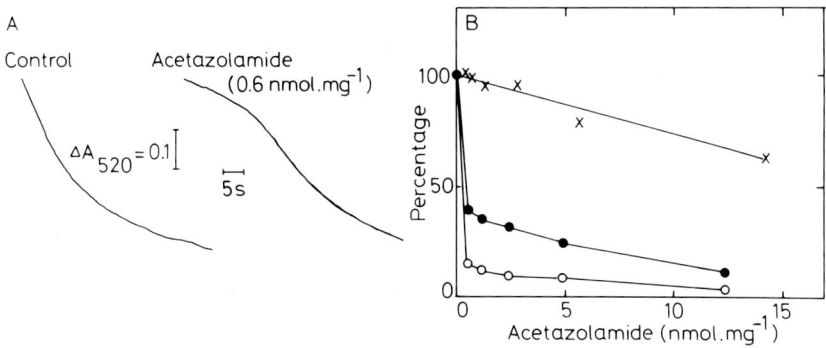

Fig. 2. Inhibition by acetazolamide of mitochondrial swelling in iso-osmotic NH_4HCO_3 and of citrulline synthesis. For experimental details, see Wanders *et al.* (1984). Swelling was monitored as the decrease in absorbance at 520 nm. In B, both the initial rate of swelling (o) and that of the second phase (●) were plotted against the amounts of acetazolamide used; (x), the rate of citrulline production. In the absence of acetazolamide, the rate of citrulline production was 31 nmol·min^{-1}·mg $protein^{-1}$ at 25°C. From Wanders *et al.* (1984).

At this concentration, however, citrulline production was hardly affected (Fig. 2B). It thus appears that the flux control coefficient of carbonic anhydrase in the pathway of citrulline synthesis in isolated rat-liver mitochondria is very low (<0.02), in agreement with the conclusion drawn above.

Experiments with isolated hepatocytes

Insensitivity of carbamoyl-phosphate synthetase towards its product can also be demonstrated in intact hepatocytes. This is shown in Fig. 3. When hepatocytes were incubated in the presence of lactate and ammonia the rate of urea synthesis, and thus flux through carbamoyl-phosphate synthetase, was already constant after 1 min (Fig. 3B) even though the intracellular concentration of carbamoyl phosphate, of which at least 90% is located in the mitochondria (Meijer et al., 1985), showed large variations in the first 10 min of incubation, reaching a peak value of 1.7 μmol/g dry weight of cells after 2 min, equivalent to about 8 mM in the mitochondrial matrix (Fig. 3A). When ornithine was also present carbamoyl phosphate dropped to undetectable low levels within 10 min (Fig. 3A) and ornithine cycle flux was greatly enhanced (Fig. 3B).

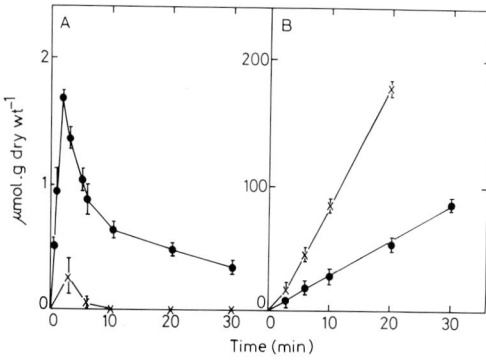

Fig. 3. Intracellular carbamoyl phosphate and urea synthesis. Hepatocytes were incubated with 10 mM lactate, 20 mM NH$_4$Cl, 0.5 mM oleate (plus 2% serum albumin) and, where indicated, 3 mM ornithine. (A) Carbamoyl phosphate; (B) urea. ($\bullet$), Ornithine absent; (x) ornithine present. Data are from 4 different hepatocyte preparations. From Meijer et al. (1985).

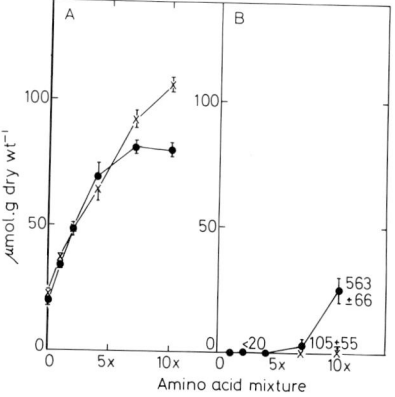

Fig. 4. Urea synthesis in the presence of physiological mixtures of amino acids. Hepatocytes were incubated for 40 min with multiple concentrations of a physiological mixture of amino acids, 0.5 mM oleate and 2% serum albumin. (x), Control; ($\bullet$), ornithine omitted from the mixtures. The numbers in (B) indicate the amount (mean $\pm$ SE) of intracellular carbamoyl phosphate (in nmol/g dry weight) after 40 min. (A), Urea production; (B), ammonia accumulation. Data are from 3 different hepatocyte preparations. From Meijer et al. (1985).

In the experiment of Fig. 4 hepatocytes were incubated under conditions closer to the physiological situation. Ammonia was not added and the cells were incubated with multiples of a physiological mixture of all 20 amino acids, each at a concentration equivalent to that normally present in the portal vein of a starved rat (for composition, see Meijer *et al.*, 1985). In order to avoid overestimation of ornithine cycle flux arginine was replaced by ornithine. As expected, urea synthesis increased with increasing amino acid concentration (Fig. 4A). However, even at high amino acid concentration, NH_3 did not accumulate and intracellular carbamoyl phosphate remained below the limit of detection (Fig. 4B). In order to study the role of ornithine in the control of urea synthesis under these conditions hepatocytes were also incubated in the absence of ornithine. Omission of this amino acid only slightly inhibited urea synthesis at the lower amino acid concentrations; at high amino acid concentration the inhibition became stronger and both NH_3 and carbamoyl phosphate accumulated (Fig. 4A,B).

Because carbamoyl phosphate did not accumulate in hepatocytes incubated with all 20 amino acids, flux control exerted by carbamoyl-phosphate synthetase must have been greater than that exerted by the other enzymes of the ornithine cycle. Indeed, low concentrations of norvaline did not affect urea synthesis under these conditions (Meijer *et al.*, 1985).

The role of carbonic anhydrase in the control of urea synthesis from amino acids was investigated in the experiment of Fig. 5, again using acetazolamide as the experimental tool. Since the activity of carbonic anhydrase could not be measured directly acetazolamide was first titrated under conditions where flux through the ornithine cycle, and thus through carbonic anhydrase, was maximal, *i.e.* in the presence of saturating concentrations of NH_3, ornithine and lactate (Fig. 5A).

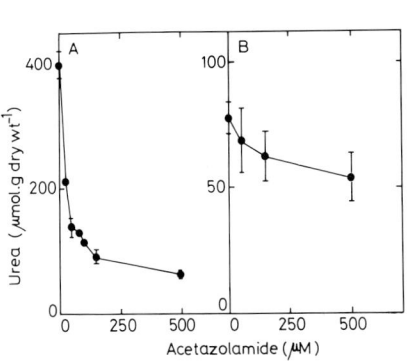 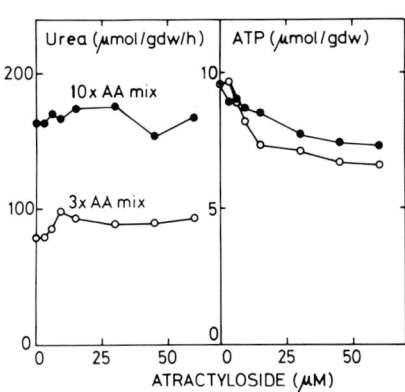

Fig. 5. Inhibition of urea synthesis by acetazolamide. Hepatocytes were incubated with 10 mM glucose, 2 mM octanoate and either 10 mM NH_4Cl, 10 mM lactate and 3 mM ornithine (A) or a complete physiological mixture of amino acids (B). Acetazolamide was added as indicated. Time, 40 min. Data are from experiments with 4 different hepatocyte preparations. From Boon and Meijer (1988).

Fig. 6. Mitochondrial ATP transport and urea synthesis from amino acids. Hepatocytes (4.3 mg dry weight/ml) were incubated with a physiological mixture of amino acids (AA mix) in which each of the amino acids was present at either three or ten times their concentration in the portal vein of a starved rat. Atractyloside was added as indicated.

Under these conditions, the percentage inhibition of urea synthesis obtained with a particular concentration of acetazolamide gives a minimal value for the percentage inhibition of carbonic anhydrase activity (Boon & Meijer, 1988). Next, acetazolamide was titrated in the presence of the amino acid mixture (Fig. 5B). The maximum value for the flux control coefficient of carbonic anhydrase was then calculated from the ratio of the percentage inhibition of urea synthesis obtained with a particular low concentration of acetazolamide in both systems. Thus, at 12.5 μM acetazolamide 26% inhibition of urea synthesis from saturating concentrations of NH_3 was observed (Fig. 5A). At the same concentration acetozalamide inhibited urea synthesis from amino acids by only 3% (Fig. 5B). This yields a maximum value of $3/26 = 0.12$ for the flux control coefficient of carbonic anhydrase in the pathway of urea synthesis from amino acids.

Transport of ATP out of the mitochondrion is required to support synthesis of argininosuccinate in the cytosol. However, when hepatocytes were incubated with physiological mixtures of amino acids atractyloside, at a concentration sufficient to decrease intracellular ATP by 25-30%, did not affect the rate of urea synthesis (Fig. 6). This indicates that mitochondrial ATP transport does not control ornithine cycle flux under these conditions.

The intrahepatic concentration of carbamoyl phosphate *in vivo*

The concentration of carbamoyl phosphate in the liver of fed rats was below the limit of detection, less than 20 nmol/g dry weight (Meijer *et al.*, 1985). Even if all carbamoyl phosphate were in the mitochondria its mitochondrial concentration would be below 0.1 mM (assuming that 1 g dry weight is equivalent to 220 μl mitochondrial matrix space); this is in agreement with recent data of Cooper *et al.* (1987). The fact that the concentration of carbamoyl phosphate is so low allows one to conclude that the activity of the ornithine cycle must exceed that of carbamoylphosphate synthetase.

DISCUSSION

Isolated carbamoyl-phosphate synthetase is hardly sensitive to product inhibition by carbamoyl phosphate ($K_i = 10^{-19}$ mM; Elliott & Tipton, 1974). Our experiments with intact liver mitochondria and with isolated hepatocytes have confirmed this.

The low intramitochondrial concentration of carbamoyl phosphate, in combination with the low elasticity coefficient of carbamoyl-phosphate synthetase towards carbamoyl phosphate and the high elasticity coefficient of the ornithine cycle enzymes towards their substrates under physiological conditions (cf. Meijer *et al.*, 1985), indicates that ureogenesis under physiological conditions cannot be controlled by the activity of the ornithine cycle nor by any other step distal to the formation of carbamoyl phosphate. In other words, the flux control coefficients of all these steps must be low. This also includes the transport of ornithine, citrulline, glutamate, aspartate and ATP across the mitochondrial membrane since a low activity of each of these systems would lead to accumulation of carbamoyl phosphate and this does not occur; in the case of the adenine nucleotide translocator this was verified directly (Fig. 6). Since the sum of the flux control coefficients of all enzymes in the pathway of urea synthesis must be one, it follows that carbamoyl-phosphate synthetase must have a flux control coefficient of close to one (about 0.9, allowing 0.1 for carbonic anhydrase), as long as ammonia is considered as the beginning of the metabolic pathway of urea synthesis, *i.e.* as long as its concentration is kept constant: only in this situation will activation of carbomoyl-phosphate synthetase lead to an increased rate of urea synthesis (Meijer *et al.*, 1985). In the liver *in vivo*, however, amino acids are the main precursor of ammonia for synthesis of carbamoyl phosphate (Walser, 1983). As shown previously, degradation of amino acids is insensitive to inhibition by ammonia (Meijer *et al.*, 1985; Boon & Meijer, 1988). Thus, a change in the activity of carbamoyl-phosphate synthetase, *e.g.* by a change in the concentration of intramitochondrial N-acetylglutamate, the essential activator of the enzyme, will alter the steady state concentration of ammonia rather than affect urea synthesis *per se*. Under these conditions ureogenic flux is almost entirely controlled by the reactions involved in the initial degradation of amino acids and/or by their transport across the plasma membrane (Groen *et al.*, 1982a; Salter *et al.*, 1986; Boon & Meijer, 1988).

CONCLUSIONS

Flux through carbamoyl-phosphate synthetase (ammonia) in intact rat-liver mitochondria is not sensitive to inhibition by its product carbamoyl phosphate. *In vivo*, under normal conditions, the intramitrochondrial concentration of carbamoyl phosphate in the liver is very low. Application of flux control analysis leads to the conclusion that urea synthesis is not controlled by steps distal to the formation of carbamoyl phosphate. This includes the transport of metabolites across the mitochondrial inner membrane.

REFERENCES

Boon L, Meijer AJ (1988) Control by pH of urea synthesis in isolated rat hepatocytes. Eur J Biochem 172:465-469

Chappell JB, Crofts AR (1966) Ion transport and reversible volume changes of isolated mitochondria. Biochim Biophys Acta 7:293-314

Cooper AJL, Nieves E, Coleman AE, Filc-DeRicco S, Gelbard AS (1987) Short-term metabolic fate of [^{13}N]ammonia in rat liver *in vivo*. J Biol Chem 262:1073-1080

Dodgson SJ, Forster RE, Schwed DA, Storey BT (1983) Contribution of matrix carbonic anhydrase to citrulline synthesis in isolated guinea pig liver mitochondria. J Biol Chem 258:7696-7701

Elliott KRF, Tipton KF (1974) Product inhibition studies on bovine liver carbamoyl phosphate sythetase. Biochem J 141: 817-824

Groen AK, Sips HJ, Vervoorn RC, Tager JM (1982a) Intracellular compartmentation and control of alanine metabolism in rat liver parenchymal cells. Eur J Biochem 122:87-93

Groen AK, van der Meer R, Westerhoff HV, Wanders RJA, Akerboom TPM, Tager JM (1982b) Control of metabolic fluxes. In: Sies H (ed) Metabolic Compartmentation. Academic Press, London, pp 9-37

Kacser H, Burns JA (1973) The control of flux. In: Davies DD (ed) Rate Control of Biological Processes. Cambridge University Press, London, pp 65-104

Meijer AJ, Lof C, Ramos IC, Verhoeven AJ (1985) Control of ureogenesis. Eur J Biochem 148:189-196

Salter M, Knowles RG, Pogson CI (1986) Quantification of the importance of individual steps in the control of aromatic amino acid metabolism. Biochem J 234:635-647

Walser M (1983) Urea cycle disorders and other hereditary hyperammonemic syndromes. In: Stanbury JB et al. (eds) Metabolic Basis of Inherited Disease. McGraw-Hill, New York, pp 402-438

Wanders RJA, Van Roermund CWT, Meijer AJ (1984) Analysis of the control of citrulline synthesis in isolated rat-liver mitochondria. Eur J Biochem 142:247-254

Mitochondrial Adenine Nucleotide Translocation During Fatty Acid Metabolism in the Intact Cell

SIBYLLE SOBOLL

Institut für Physiologische Chemie I, Universität Düsseldorf, D-4000 Düsseldorf 1, Federal Republic of Germany

Mitochondrial adenine nucleotide translocation is the functional link between ATP generation in mitochondrial oxidative phosphorylation and ATP utilization mainly occurring in the cytosolic space. Regulation of cytosolic ATP delivery is one of the main aspects in cellular bioenergetics. The role of fatty acids as regulators of cellular bioenergetics has been discussed for about 30 years since Lehninger and his coworkers reported an uncoupling effect of free fatty acids on oxidative phosphorylation in isolated mitochondria.

Whereas free fatty acids can regulate the degree of coupling of oxidative phosphorylation their coenzyme A derivatives have been shown to inhibit adenine nucleotide translocation in isolated systems (Shug *et al.*, 1971; Woldegiorgis *et al.*, 1981). However as pointed out by Wojtczak in 1976 "to prove or disprove the assumption that long chain acyl CoA plays a regulatory role of this kind in the intact cell it should be necessary to show whether or not an increase in cellular acyl CoA produces a decrease of the cytoplasmic and an increase in mitochondrial phosphorylation potentials and affects, in a positive or negative way, processes depending on ATP concentrations in the two cellular compartments".

The aim of the present study is to show relationships between changes in the subcellular contents of long chain acyl CoA, the distribution of adenine nucleotides and cellular functions by means of subfractionation studies in the intact liver and heart.

A. Azzi et al. (Eds.)
Anion Carriers of Mitochondrial Membranes
© Springer-Verlag Berlin Heidelberg 1989

MATERIAL AND METHODS

Preparation and sampling of intact organs

Livers and hearts were obtained from 200-250 g male rats either fed on stock diet or starved for 24 h. Livers were either taken *in situ* from unanaesthetised unrestrained rats by the double-hatched method (Faupel *et al.*, 1972) and freeze-clamped within 3 s or perfused with hemoglobin-free Krebs-Henseleit bicarbonate buffer in an open system (Scholz *et al.*, 1973) for 1 h and then freeze-clamped.

Hearts were prepared as working hearts (Hütter *et al.*, 1985) with a static pressure afterload of 60 mm Hg and a left atrial pressure maintained at 10 mm Hg. The perfusion medium was a modified Krebs-Henseleit solution containing 5 mM glucose, 1 mM L-lactate and 0.2 mM DL-carnitine. Hypoxia was performed by perfusing the hearts with a pO_2 of 35 mm Hg and a restricted flow of 2 ml/min in a retrograde perfusion. The experiments were terminated after 45 min (normoxia) or 90 min (hypoxia) by freeze-clamp.

Fractionation of tissue in non-aqueous solvents

For determination of mitochondrial and extramitochondrial contents of ATP and ADP, the freeze-clamped tissue was ground in liquid nitrogen and freeze-dried at 0.26 Pa at -40°C. 0.1-0.3 g of the freeze-dried tissue powder was sonicated in a mixture of heptane and CCl_4 and then fractionated on a density gradient consisting of the same media as described by Soboll *et al.* (1978). The procedure was slightly modified for livers *in vivo* or hearts. The density gradient yielded 8 fractions containing different proportions of mitochondrial and cytosolic proteins. Specific activities of marker enzymes for mitochondria (citrate synthase) and extramitochondrial compartment (phosphoglycerate kinase) and contents of ATP and ADP were determined in each fraction and mitochondrial and extramitochondrial ATP and ADP contents calculated from the distribution of marker enzymes and metabolites in the fractions by extrapolation to pure mitochondrial and extramitochondrial fractions (Elbers *et al.*, 1974). They were converted into concentrations assuming mitochondrial and extramitochondrial water contents of 0.8 and 3.8 μl/mg mitochondrial and extramitochondrial protein, respectively, for liver and 1.0 and 3.5 μl/mg for heart.

Isolation and incubation of hepatocytes

Hepatocytes were isolated from livers from fed rats as described by Berry and Friend (1969) with slight modifications (Sies *et al.*, 1977) They were incubated in 25 ml conical flasks at a concentration of 15 mg/ml in Krebs-Henseleit bicarbonate buffer containing various substrates *plus* albumin as indicated in the figures. The incubation time was 25 min at 37°C, shaking frequency 120 min⁻¹. For determination of oxygen uptake 1 ml of suspension was transferred to a thermostatically controlled cylindrical plastic chamber connected with a Clark type oxygen electrode.

Digitonin fractionation of hepatocytes

The digitonin fractionation (Zuurendonk & Tager, 1974) was used with the following modifications: 0.2 ml of cells were added to 1 ml of separation medium, mixed, and 0.7 ml of the mixture was layered on top of the silicone oil in an Eppendorf cup containing 0.12 ml of 1.88 M $HClO_4$ (bottom) and 0.5 ml of the oil. After centrifugation 0.5 ml of the supernatant was deproteinised and neutralized. The upper fraction was designated "cytosolic" whereas the bottom fraction remaining after removal of the silicone oil was designated "mitochondrial". In mitochondrial and cytosolic fractions ADP and ATP were measured (Soboll *et al.*, 1978).

RESULTS AND DISCUSSION

Adenine nucleotide translocation in liver

If adding increasing concentrations of carboxyatractyloside (10-30 μM) to isolated hepatocytes, a marked increase in the mitochondrial ATP/ADP ratios and a concomitant decrease in the cytosolic ratios are observed while oxygen uptake is decreased (Fig. 1).

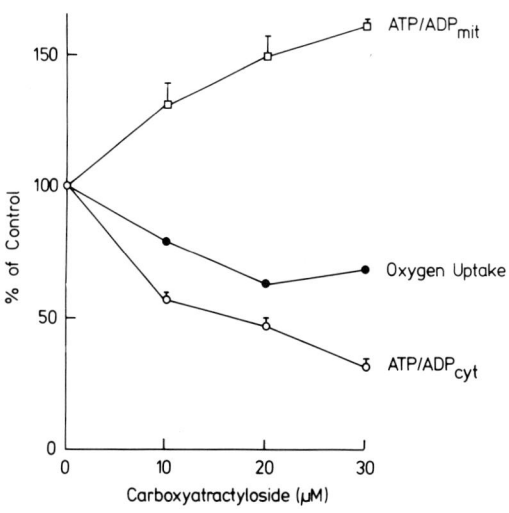

Fig. 1. Effect of carboxyatractyloside on mitochondrial and cytosolic ATP/ADP ratios in isolated hepatocytes from fed rats. Data are means ± SEM; n=4 for ATP/ADP ratios; n=2 for oxygen uptake. ATP/ADP_{mit} = 1.9 ± 0.3; ATP/ADP_{cyt} = 10.3 ± 3.5; oxygen uptake 11.0 nmol/mg protein. From Soboll *et al.* (1984).

These changes are expected under normoxic conditions when adenine nucleotide translocation is inhibited. Therefore, if long-chain acyl CoA (LCA) inhibit adenine nucleotide translocase in the intact cell, similar changes should occur.

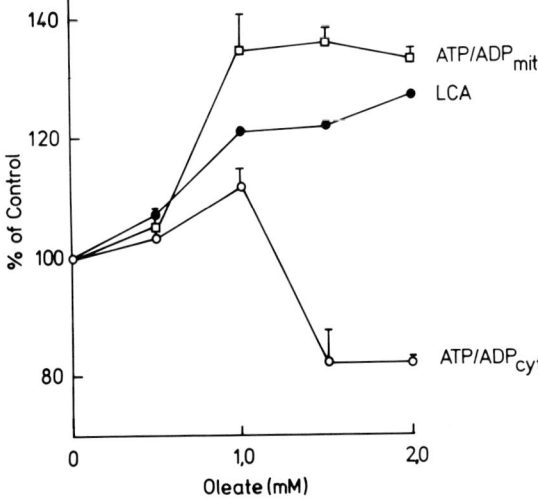

Fig. 2. Effect of oleate on mitochondrial and cytosolic ATP/ADP ratios in isolated hepatocytes from fed rats and contents of LCA. Data are mean values ± SEM; n=4. Control values were: ATP/ADP$_{mit}$ = 2.6 ± 0.8; ATP/ADP$_{cyt}$ = 13.4 ± 4.5; LCA = 138 nmol/g protein. From Soboll *et al.* (1984).

This is shown in Fig. 2. LCA were modified by enhancing the oleate concentration of the incubation medium. The changes in subcellular ATP/ADP ratios are similar to those of Fig. 1 but less pronounced. Oxygen consumption was on the contrary slightly increased from 15 to 19 nmol/mg x min at 1 mM oleate. This is explained by the uncoupling effect of non-esterified oleate.

It may be argued that the changes in mitochondrial and cytosolic ATP/ADP ratios induced by oleate are the consequence of fatty acid metabolism, *i.e.* decrease of cytosolic ATP by activation of oleate and increase of mitochondrial ATP by enhanced rate of ATP generation due to fatty acid oxidation. However no changes in mitochondrial ATP were observed in perfused liver on addition of octanoate whereas oleate caused a marked increase in the mitochondrial ATP/ADP ratio (Table 1).

In vivo a low rate of fatty acid metabolism accompanied by low plasma and tissue LCA levels are characteristic for the fed state whereas the opposite is observed in the fasted state. Thus the fasted state should be the natural metabolic condition where an inhibition of adenine nucleotide translocase could occur. In fact, we observed increased mitochondrial and decreased cytosolic ATP/ADP ratios in livers from fasted rats compared to livers from fed rats. This is seen in perfused livers as well as in livers extracted *in vivo* (Tables 1 and 2). The changes are reversible on either refeeding the animals or inducing fatty acid esterification by infusing glucose and glycerol. The latter experiment confirms that the changes in the subcellular distribution of adenine nucleotides are directly related to fatty acid metabolism.

Table 1. Mitochondrial and cytosolic ATP/ADP ratios in perfused rat livers. Results are means ± SEM for the numbers of experiments in parentheses. Substrates were infused for 20 min before the liver was freeze-clamped at 60 min of perfusion. From Soboll et al. (1984)

Animals	ATP/ADP	
	mitochondrial	cytosolic
Fed, no substrates (9)	0.18 ± 0.021	0.3 ± 1.4
Starved, no substrates (6)	0.70 ± 0.05	2.9 ± 0.2
Starved, glucose 25 mM, glycerol 1 mM (4)	0.31 ± 0.03	9.8 ± 1.9
Fed, oleate 0.5 mM, albumin 2% (4)	0.72 ± 0.07	8.5 ± 1.2
Fed, octanoate 0.5 mM, albumin 2% (2)	0.18	10.9

The physiological significance of the LCA-mediated inhibition of adenine nucleotide translocase is suggested from the following considerations: Since it has been shown that fatty acids stimulate gluconeogenesis only from precursors that require the pyruvate carboxylase step, e.g. from lactate but not from dihydroxyacetone (Brocks et al., 1980) an elevation of the mitochondrial ATP/ADP ratio would stimulate pyruvate carboxylase at the same time when pyruvate dehydrogenase is inhibited thus diverting pyruvate from oxidation to carboxylation.

Table 2. Effect of starvation and refeeding on mitochondrial and cytosolic ATP/ADP ratios in rat liver in vivo. Rats were starved for 48 h or starved for 48 and refed with pure glucose for 1 h. Data are means ± SEM for the numbers of experiments in parentheses. * $p < 0.05$; ** $p < 0.005$ for starved versus fed or refed versus starved. From Soboll et al. (1984).

Animals	ATP/ADP	
	mitochondrial	cytosolic
Fed (7)	0.85 ± 0.1	7.0 ± 0.3
Starved 48 h (7)	1.0 ± 0.04**	5.8 ± 0.4*
Refed 1 h (5)	0.60 ± 0.1*	7.0 ± 0.4*

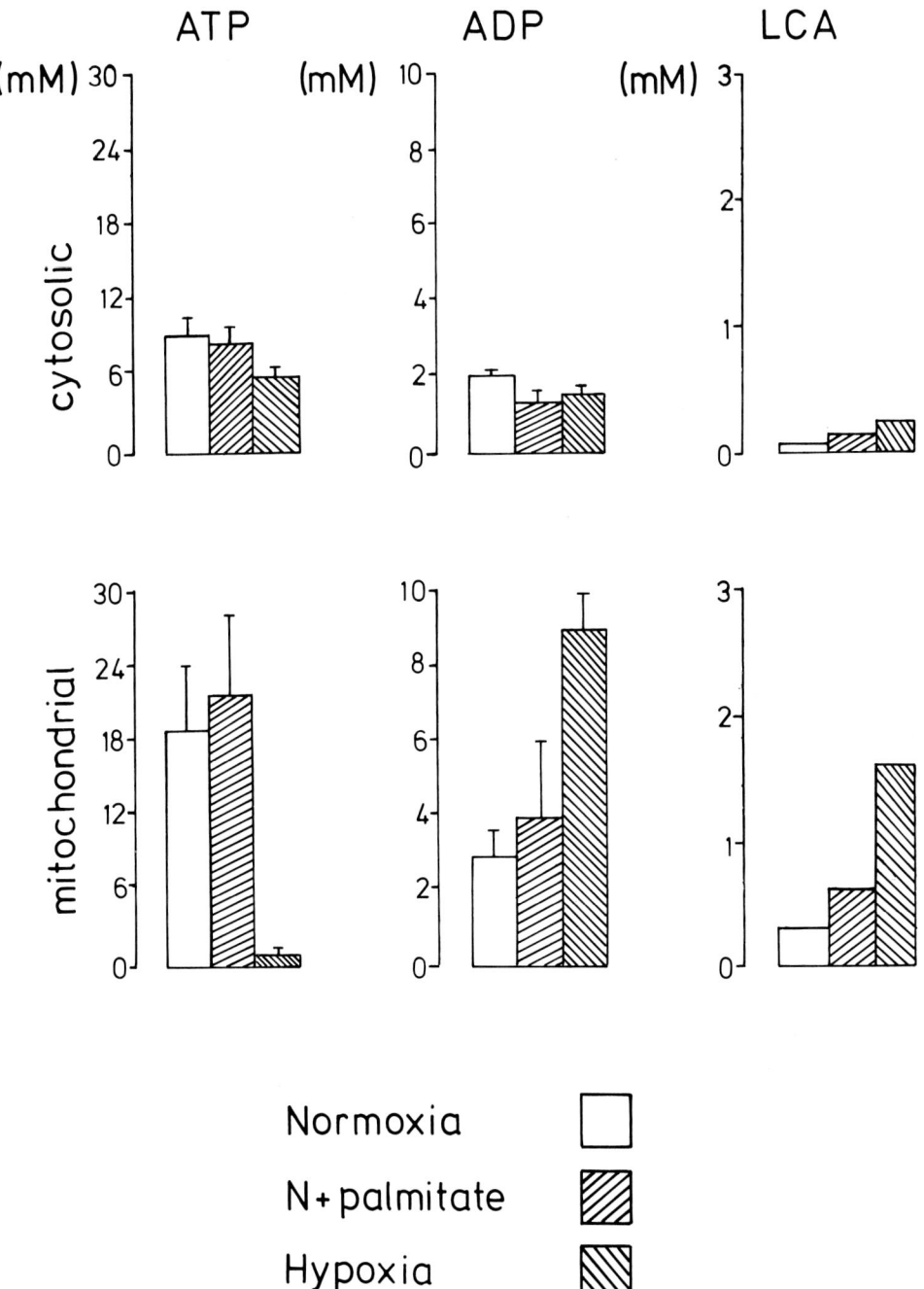

Fig. 3. Subcellular distribution of metabolites in perfused rat heart during fatty acid metabolism and during hypoxia. Fatty acid metabolism was induced by infusion of 0.5 mM palmitate to the working heart, hypoxia was performed as described in Methods (Langendorff heart). From Hütter et al. (1988).

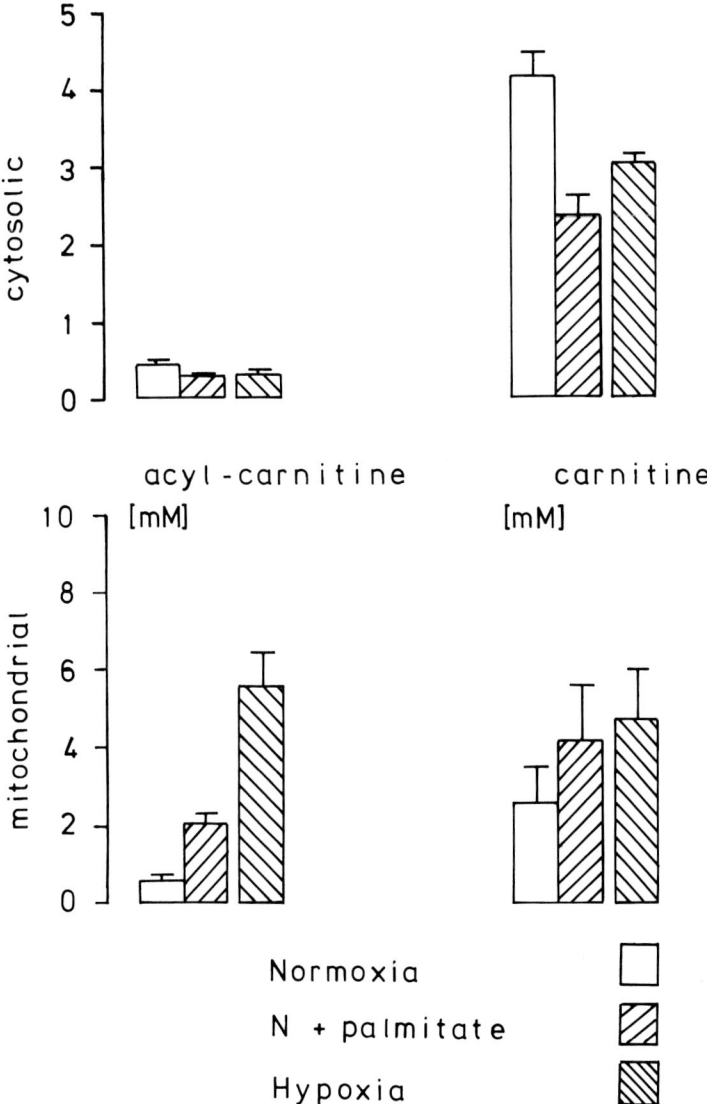

Fig. 4. Changes in subcellular distribution of acylcarnitine and carnitine during fatty acid metabolism and during hypoxia. Data are from the same experiments as in Fig. 3. From Hütter *et al.* (1988).

Adenine nucleotide translocation in heart

Normoxic condition

Addition of 0.5 mM palmitic acid increased mitochondrial LCA by a factor of 2 but did not change the mitochondrial/cytosolic distribution of ATP and ADP significantly (Fig. 3). Therefore it is concluded that in the normoxic working heart a high rate of fatty acid metabolism prevents mitochondrial LCA to reach inhibitory levels for adenine nucleotide translocation.

Hypoxia

During hypoxia a dramatic increase in mitochondrial LCA is observed accompanied by a strong decrease in the mitochondrial ATP whereas the cytosolic level is not significantly changed (Fig. 3). In line with this finding of a high cytosolic energy state in the heart is that heart rate and maximal pressure development are hardly affected (Hütter et al., 1988).

Whereas in energised mitochondria inhibition of adenine nucleotide translocation enhances mitochondrial ATP/ADP ratio and lowers the cytosolic ratio, the situation should be different in deenergised mitochondria like during hypoxia. Since adenine nucleotide transport is electrophoretic (Klingenberg, 1975), the decrease in protonmotive force following deenergisation is accompanied by a backflow of cytosolic ATP into the matrix where it is hydrolysed by H^+-ATPase to restore the protonmotive force. Transferring this situation to the hypoxic heart would lead to depletion of cytosolic and mitochondrial ATP followed by heart failure. The dramatic increase in mitochondrial LCA prevents breakdown of the heart during hypoxia by inhibiting transport of ATP back into the mitochondria. Inhibition of adenine nucleotide transport may further be supported by the lowered mitochondrial ATP level as discussed by LaNoue et al. (1981). Pande et al. (1984) have suggested that adenine nucleotide translocase protects the hypoxic heart from failure by building large amounts of cytosolic LCA thus preventing the increase of mitochondrial acylcarnitine and LCA which were believed to cause mitochondrial damage. However, our measurements show that mitochondrial LCA and acylcarnitine are dramatically enhanced during hypoxia arguing against a protective effect exerted by binding of cytosolic LCA to adenine nucleotide translocase (Fig. 4).

Taken together our results show that partial inhibition of adenine nucleotide translocation occurs in liver and heart under certain conditions and that this can be beneficial for cellular energy metabolism.

References

Brocks DG, Siess EA, Wieland OH (1980) Distinctive roles of oleate and glucagon in gluconeogenesis. Eur J Biochem 113:39-43

Elbers R, Heldt HW, Schmucker P, Soboll S, Wiese H (1974) Measurement of the ATP/ADP ratio in mitochondria and in the extramitochondrial compartment by fractionation of freeze-stopped liver tissue in non-aqueous media. Hoppe-Seyler's Z Physiol Chem 355:378-393

Faupel RP, Seitz HJ, Tarnowski W, Thiemann V, Weiss C (1972) The problem of tissue sampling from experimental animals with respect to freezing technique, anoxia, stress and narcosis. Arch Biochem Biophys 148:509-522

Hütter JF, Piper HM, Spieckermann PG (1985) An index for estimation of oxygen consumption in rat heart by hemodynamic parameters. Am J Physiol 249: H729-H734

Hütter JF, Alves C, Soboll S (1988) Effects of hypoxia and fatty acids on the distribution of metabolites in rat heart. Am J Physiol (submitted)

Klingenberg M (1975) Energetic aspect of transport of ATP and ADP through the mitochondrial membrane. In: Energy Transformation in Biological Systems (Ciba Foundation Symposium 31), Elsevier Excerpta Medica/North Holland, Amsterdam, pp 105-124

LaNoue KF, Watts JA, Koch CD (1981) Adenine nucleotide transport during cardiac ischemia. Am J Physiol 241:H663-H671

Pande SV, Goswami T, Parvin R (1984) Protective role of adenine nucleotide translocase in O_2-deficient hearts. Am J Physiol 247:H25-H34

Scholz R, Hansen W, Thurman G (1973) Interaction of mixed-function oxidation with biosynthetic processes. Eur J Biochem 38:64-72

Shug A, Lerner E, Elson C, Shrago E (1971) The inhibition of adenine nucleotide translocase activity by oleyl CoA and its reversal in rat liver mitochondria. Biochem Biophys Res Commun 43:557-563

Sies H, Akerboom TPM, Tager JM (1977) Mitochondria and cytosolic NADPH-systems and isocitrate dehydrogenase indicator metabolites during ureogenesis from ammonia in isolated hepatocytes. Eur J Biochem 72:301-307

Soboll S, Scholz R, Heldt HW (1978) Subcellular metabolite concentrations. Dependence of mitochondrial and cytosolic ATP-system on the metabolic state of perfused rat liver. Eur J Biochem 87:377-390

Soboll S, Seitz HJ, Sies H, Ziegler B, Scholz R (1984) Effect of long chain fatty acyl CoA on mitochondrial and cytosolic ATP/ADP ratios in the intact liver cell. Biochem J 220:371-376

Wojtczak L (1976) Effect of long-chain fatty acids and acyl-CoA on mitochondrial permeability, transport, and energy-coupling processes. J Bioenerg Biomembr 8:293-311

Woldegiorgis G, Shrago E, Gipp J, Yatvin M (1981) Fatty acyl coenzyme A-sensitive adenine nucleotide transport in a reconstituted liposome system. J Biol Chem 256:12297-12300

Zuurendonk PF, Tager JM (1974) Rapid separation of particulate components and soluble cytoplasm of isolated rat liver cells. Biochim Biophys Acta 333:393-399

Control of Oxidative Phosphorylation in Yeast Mitochondria: The Role of Phosphate Carrier and pH

Jean-Pierre Mazat, Eric Jean-Bart, Michel Rigoulet, Christine Reder and Bernard Guerin

IBCN-CNRS and Université Bordeaux II, 1, rue Camille Saint-Saëns F-33077 Bordeaux-Cedex. France
and
Département de Mathématiques et Informatique, Université Bordeaux I, 351 cours de la Libération, F-33405 Talence-Cedex. France

The problem of the control of oxidative phosphorylations has been the object of a great deal of discussion over the last few years. It was beautifully solved in 1982 by the application of Kacser and Burn's and Heinrich and Rapoport's theory of metabolic control to the rat liver mitochondria by Groen et al., 1982. One of the most important predictions of the theory is that control can be shared between many step within a given pathway. This is exactly one of the experimental conclusions of Groen et al.,1982 Another prediction of the theory is that the distribution of control, as measured by the control coefficients (for the official terminology see Westerhoff et al., 1984 and Burns et al., 1985) can vary according to the different steady states. This prediction has also been confirmed by Tager et al., 1983, who have studied the control coefficients dependence of different steps as a function of respiratory rate. Gellerich et al., 1983 and Doussière et al., 1984 have also shown that the control exerted by the adenine nucleotide translocator depends not only on the respiration rate, but also on the complexity of the metabolic network and on the model of ATP utilization.

All these data have been obtained studying the respiratory flux. A relevant flux for the cell is the ATP regerating flux, i.e. the rate of ATP synthesis. According to the

---

A. Azzi et al. (Eds.)
Anion Carriers of Mitochondrial Membranes
© Springer-Verlag Berlin Heidelberg 1989

chemiosmotic theory of Mitchell (1961), it is quite widely accepted that in mitochondria, the ATP synthesis flux is coupled to the respiration flux through the proton-motive force μ_H+. Nevertheless, this statement does not imply that these fluxes are identical, nor that their control is the same. One of the aims of the work, that we present here, is to compare the control of various steps on both fluxes in two different states. These two steady states are obtained by varying the inorganic phosphate concentration. Previous study in our laboratory (Rigoulet *et al.*, 1977), have shown that P_i transport respiration rate and ATP synthesis appear biphasic as a function of P_i concentration. We have studied the control of the two fluxes, respiration rate and ATP synthesis, at two different steady states induced by two different P_i concentrations: 0.5 mM and 7.7 mM. As these steady states are determined by the external phosphate concentration it appeared worthwhile to study also the control by P_i transport as a function of external P_i concentration.

MATERIALS AND METHODS

Mitochondrial preparation

Diploid *Saccharomyces cerevisiae* strain Yeast Foam cells were grown aerobically with 2% lactate as the carbon source in a complete medium containing 1% yeast extract v/v 0.1% potassium phosphate v/v 0.12% ammonium sulfate (pH 4.5). The cells were harvested in the logarithmic growth phase. Mitochondria were prepared as in Mazat *et al.*, 1986

Rates of respiration and ATP synthesis measurements

The oxygen consumption rate was measured polarographically at 27°C using a Clark electrode. The rate of ATP was measured by P_i incorporation as in Rigoulet *et al.*, (1977) O_2 consumption and ATP synthesis were always checked for linearity in order to insure that the steady state was maintained during measurements. Inhibitors were incubated 1 min before P_i addition. NADH can be used as the respiratory substrate, since there is a NADH dehydrogenase on the external side of the inner membrane in yeast mitochondria. Under these conditions, there is no control in the inflow of NADH. For most of the experiments described in this paper, both measurements were carried out simultaneously, the aliquot for labelled ATP evaluation being taken in the Clark oxygen electrode chamber. The protein concentration was measured by the biuret method using bovine serum albumin as standard.

Determination of control coefficients

The control coefficients of various steps involved in the oxidative phosphorylation were determined with specific inhibitors of these steps according to the definition:

$$C_i = \left. \frac{\dfrac{\partial \ln J}{\partial I} \quad (I=0)}{\dfrac{\partial \ln v_i}{\partial I} \quad (I=0)} \right|_{\text{steady-state}}$$

i.e. the ration of the initial slope of the whole inhibition curve J(I) over the initial slope of the inhibition curve of the isolated step $v_i(I)$ in the same conditions as in the pathway. The direct control coefficients C_i will also be considered:

$$C'_i = \left. \frac{\dfrac{\partial J}{\partial I} \quad (I=0)}{\dfrac{\partial v_i}{\partial I} \quad (I=0)} \right|_{\text{steady-state}}$$

RESULTS

Dependence of oxidative phosphorylation on P_i concentration

The dependence of respiration J_{resp} and ATP synthesis J_{ATP} on P_i concentration is shown in Fig. 1.

Experimental points are phenomenologically fitted with a sum of two hyperbolas, whereas a fit with a single hyperbola gives a much less accurate fit. The equations are:

$$J_{resp} = \frac{270\, P_i}{0.3 + P_i} + \frac{326\, P_i}{5.3 + P_i} + 265 \quad \text{and} \quad J_{ATP} = \frac{460\, P_i}{0.19 + P_i} + \frac{496\, P_i}{7.8 + P_i}$$

This equation will be used in the following, when necessary.

The existence of these two hyperbolas can be an indication of two kind of steady states involving different controlling steps. In order to check this point, we chose 0.5 mM and 7.7 mM as external P_i concentrations representative of these two different steady states.

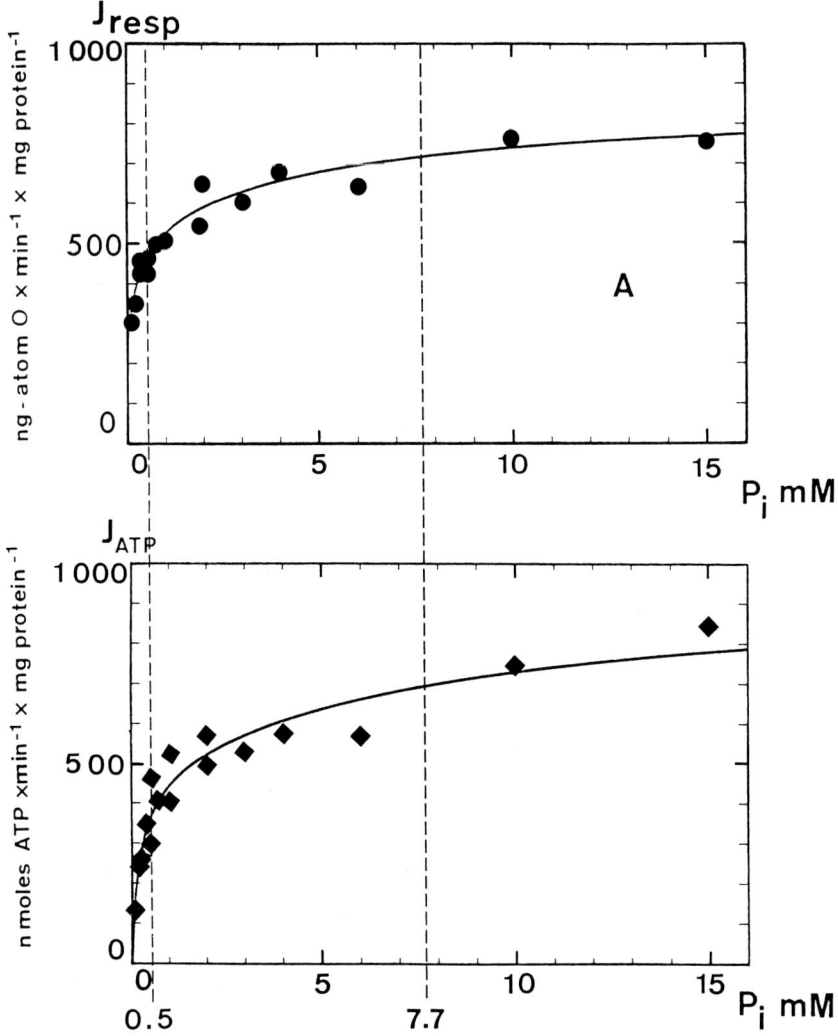

Fig. 1: Rate of O_2 consumption (A) and ATP synthesis (B) as a function of P_i concentration. 0.5 mM and 7.7 mM are the P_i concentrations of the studied steady states. The experimental conditions are described in Materials and Methods.

Control coefficients of fluxes

Table 1 summarizes the control coefficient values determined in this work for the two fluxes: O_2 consumption (J_{resp}) and ATP synthesis (J_{ATP}) at the two steady states according to the P_i concentrations $[P_o] = 0.5$ mM and $[P_i] = 7.7$ mM.

Table 1: The determination of the flux control coefficients was made using a specific inhibitor of each studied step.

	Respiration		ATP synthesis	
$[P_i]_{ext}$	0.5mM	7.7mM	0.5mM	7.7mM
Adn Translocator (CATR)	0	0	0	0
Proton leak (CCCP)	0.35	0	-0.2	0
P_i carrier (Mersalyl)	0	0.45	0	0.5
Cyt. oxidase (KCN)	0.6	0.5	0.3	0.4

The control of cytochrome c oxidase is rather high in all cases ($0.3 < C < 0.6$). It is now well established that this step is indeed a controlling one in mitochondria from various origins. With respect to the proton leak, we find the same result as Groen et al., (1982), this step plays a role when the oxidative phosphorylation rate is low.

The adenine-nucleotide carrier exerts no control except perhaps on J_{ATP} at $[P_i]$ = 7.7 mM. In these conditions, the adenine-nucleotide translocator seems to be at the threshold where it passes from a non-controlling region to a controlling one. In the threshold transition, the determination of the control coefficients is very sensitive to many external parameters including the preparation conditions of mitochondria.

Control by P_i Carrier

The control by the P_i carrier on the ATP synthesis is shown in Fig. 2. A control by the P_i carrier could appear surprising if we remember the kinetic parameters of rat liver mitochondria; a V_{max} of 3000 nmol/min per mg protein in non-respiring conditions and of 2700 nmol/min per mg protein in respiring ones (Ligeti et al., 1985). However, three properties of the P_i transport have to be taken into account:

1) At high P_i and ADP concentrations, the flux of ATP synthesis is around 700 nmol $ATP \cdot min^{-1} \cdot mg^{-1}$. This value is only 4-times less than the V_{max} of the unidirectional P_i transport in respiring rat liver mitochondria as determined by Ligeti et al., 1985

2) The V_{max} which have been measured by Ligeti et al., 1985 are the V_{max} of the unidirectional transports. However, at steady state, the relevant parameter which must be taken into account is the net flux, i.e., the difference between P_i influx and efflux. This difference is indeed lower than each unidirectional flux.

3) The P_i transport only exerts a control at high P_i concentration (Fig. 2), i.e., in a range where μ_{H^+} is decreased and pH is low (see Table 2). As the P_i transport is a co-transport H^+/P_i, a decrease in pH will limit the maximal rate of P_i carrier in the same way as in a classical two substrates reaction, i.e., the limitation by one substrate

decreases the V_{max} for the other substrate. This interpretation is confirmed by the increase of control by the

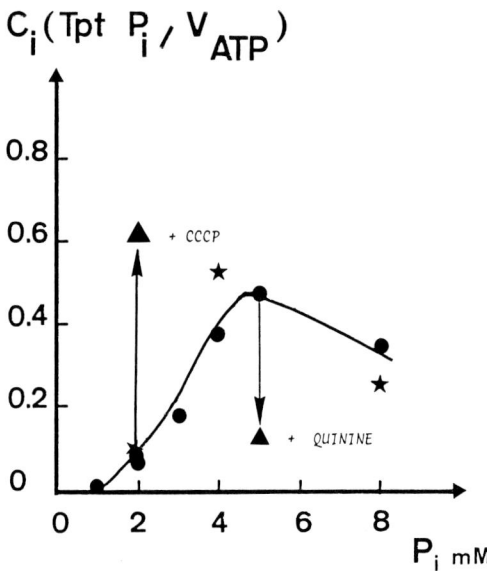

Fig. 2: Control coefficient of P_i transport on ATP synthesis flux is a function of P_i concentration (●). In order to calculate the control coefficients, we use the initial slope of the inhibition curve of J_{ATP} by mersalyl at each P_i concentration and the initial slope reference value of 1 mg protein per 15 nmol mersalyl (Arselin-de Chateaubeaudeau G., 1982) using the relationship given in Materials and Methods. (▲) are the control coefficients determined in the presence of CCCP 0.3 μM or in the presence of quinine 2.5 μmol/mg protein.

P_i carrier at low P_i concentration in the presence of an uncoupler (Fig. 2). Under these conditions, no control exists in the absence of the uncoupler. On the other hand, at the higher P_i concentration of 5 mM where the control coefficient is around 0.5, the addition of quinine which blocks the K^+/H^+ exchanger in yeast mitochondria (as in mammalian mitochondria) and maintain a high pH decreases the control coefficient. These results are summarized in Table 2.

Theoretical analysis of the control of the oxidative phosphorylation in yeast mitochondria

Reder (1988) has given a method to determine the control coefficients and the summation and connectivity relationships in any metabolic network itself and what belongs to the rate functions of each individual steps. Reder (1988) shows that several properties can be derived from the sole examination of the metabolic network structure itself. We have applied her approach to the control of the oxidative phosphorylations in yeast mitochondria.

Table 2: Determination of the control coefficient by the P_i transport of the ATP synthesis flux in various conditions.

Conditions	Control coefficient on J_{ATP} by P_i transport	Δ pH (mV)
P_i = 2 mM	0.1	-21
P_i = 2 mM + CCCP 0.3 μM	0.625	-8
P_i = 5 mM	0.53	-17
P_i = 5 mM + Quinine 2.5 μmol/mg	0.11	-30

The metabolic network, that we have considered here is represented in Fig. 3.

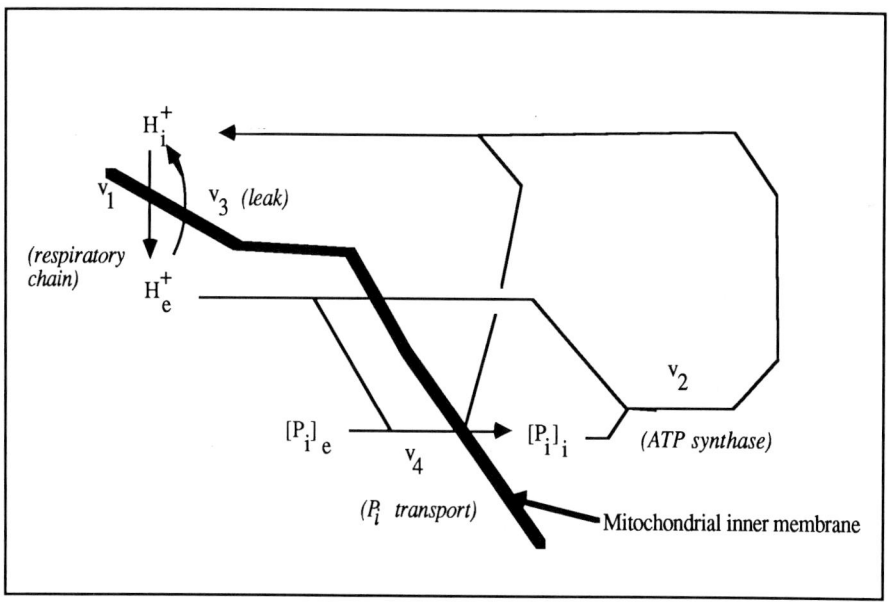

Fig. 3: Simplified model of the oxidative phosphorylations in yeast mitochondria. $v_1 = J_{resp}$ represents the extrusion of H^+, $v_2 = J6_{ATP}$ represents the synthesis of ATP coupled to the re-entry of H^+, v_3 is the H^+ leak and v_4 is the transport of P_i. The ADP/ATP translocator was omitted in first approximation because we have shown that it exerts no control in our experimental conditions.

To this oversimplified representation of the oxidative phosphorylations in yeast the theory associates the matrix (the stoichiometric matrix of the steps of the network). In our case this matrix is:

$$N = \begin{bmatrix} -n_o & n_a & 1 & 1 \\ 0 & -1 & 0 & 1 \end{bmatrix}$$

where the columns of the matrix represent the rate of the four steps, and the rows are associated with H^+_i and P_i respectively. n_o is the number of H^+ extruded by atom of oxygen consumed and n_a is the number of H^+ necessary to the synthesis of one molecule of ATP.

Applying the theory we derive a set of relationships between the direct control coefficients (see Materials and Methods).

(1) $C'_{12} - n_a C'_{14} = 0$
(2) $C'_{11} + n_o C'_{13} = 1$
(3) $-C'_{13} + C'_{14} = 0$

(4) $C'_{22} - n_a C'_{24} = 1$
(5) $C'_{22} + n_o C'_{23} = 0$
(6) $-C'_{23} + C'_{24} = 1$

Some of these relationships are verified by the experimental values of the control coefficients, we have determined, but not all. For example, (2) for $[P_i] = 7.7$ mM gives: $C'_{11} = 1$ which is not the case.

Another contradiction arises from the consideration of the previous equations in this simple model but is also apparent in fig. 1. The concentration of external phosphate is supposed to affect only the phosphate transport step. When the control coefficient of the respiratory flux or of the ATP synthesis flux is low ($[P_i] = 0.5$ mM), a variation of the external phosphate concentration might give a weak variation in both fluxes, when at high external phosphate concentration a variation of this concentration might give a big variation of both fluxes. As seen in fig. 1, the contrary is observed, *i.e.* a strong variation when the control coefficient is low and a weak variation when the control coefficient is high.

There are several possibilities to solve this contradiction.

- One is to consider a more complex (and realistic) model, that is, to add steps which could be relevant in the control coefficients determination. For instance in our model we have reduced the respiratory chain to the cytochrome c oxidase step, but other steps of this chain can be controlling in certain conditions.

- We have considered that the stoichiometry is fixed. The analysis is quite different if the stoichiometries are also functions of the steady state. We are working on this point in modelling the phosphate transport by using the approach of Pietrobon and Caplan (1985). With this approach based on Hill diagram method, the variable stoichiometry arises naturally in the rate equations.

CONCLUSIONS

These difficulties should not diminish the interest that lies in evaluating control coefficients. In the contrary, because they necessitate an internal coherence, expressed by the summation and the connectivity relationships, they could oblige the experimentator to question either the structure of the metabolic network which is proposed (additional steps or metabolites directly involved in other steps) or the rate equations of the individual steps (for instance variable stoichiometry instead of fixed stoichiometry).

ACKNOWLEDGEMENTS

This work was supported by grants from the Université Bordeaux II and from the Centre National de la Recherche Scientifique (ATP Mathématique-Biologie).

REFERENCES

Arselin-de Chateaubeaudeau G (1982) Thesis, Université Bordeaux II
Burns JA, Cornish-Bowden A, Groen AK, Heinrich R, Kacser H, Porteous JW, Rapoport SM, Rapoport TA, Stucki JW, Tager JM, Wanders, RJA, Westerhoff HV (1985) Control analysis of metabolic systems. Trends in Biochem Sci 10:16
Doussiere J, Ligeti E, Brandolin G, Vignais PV (1984) Control of oxidative phosphorylation in rat heart mitochondria. The role of the adenine nucleotide carrier. Biochim Biophys Acta 766:492-500
Gellerich FN, Bohnensack R, Kunz ZW (1983) Control of mitochondrial respiration. The contribution of the adenine nucleotide translocator depends on the ATP- and ADP-consuming enzymes. Biochim Biophys Acta 722:381-391
Groen AK, Wanders RJA, Westerhoff HV, van der Meer R, Tager JM (1982) Quantification of the contribution of various steps to the control of mitochondrial respiration. J Biol Chem 257:2754-2757
Heinrich R, Rapoport TA (1974) A linear steady state treatment of enzymatic chains. General properties, control and effector strength. Eur J Biochem 42:89-95
Kacser H, Burns JA (1973) The control of flux. In Rate Control of Biological Processes (Davies DD, ed), Cambridge University Press, pp 65-104
Ligeti E, Brandolin G, Dupont Y, Vignais PV (1985) Kinetics of P_{nci}-P_i exchange in rat liver mitochondria. Rapid filtration experiments in the millisecond time range. Biochemistry 24:4423-4428

Mazat JP, Jean-Bart E, Rigoulet M, Guerin B (1986) Control of oxidative phosphorylations in yeast mitochondria. Biochim Biophys Acta 849:7-15

Mitchell PD (1961) Nature (London) 191:144-148

Pietrobon D, Caplan SR (1985) Flow-force relationships for a six-state proton pump model: intrinsic uncoupling, kinetic equivalence of input and output forces, and domain of approximate linearity. Biochemistry 24:5764-5776

Pietrobon D (1986) A non-linear kinetic model of chemiosmotic energy coupling. Bioelectrochem and Bioenerg 15:193-209

Reder C. Mimodrame mathématique sur les systèmes biochimiques. Université Bordeaux I. Mathématiques appliquées Publication No 8608

Reder C. (1988) Metabolic control theory: a structural approach. J Theor Biol, submitted

Rigoulet M, Guerin M, Guerin B (1977) Effects of physiological manipulations on the kinetics of mitochondrial phosphate transport in *Saccharomyces cerevisiae*. Biochim Biophys Acta 471:280-295

Rigoulet M, Ezzahid Z, Guerin B (1983) Effects of tribenzylphosphate on the active phosphate transport and ATP synthesis in yeast mitochondria. Biochem Biophys Res. Commun 113:751-756

Tager JM, Wanders RJA, Groen AK, Kunz W, Bohnensack R, Kuster U, Letko, G, Bohme G, Duszynski J, Wojtczak L (1983) Control of mitochondrial respiration. FEBS Lett. 151:1-9

Westerhoff HV, Groen AK, Wanders RJA (1984) Modern theories of metabolic control and their applications. Biosci Rep 4:1-22

The Role of Pyrophosphate and the Adenine Nucleotide Transporter in the Regulation of the Intra-Mitochondrial Volume

ANDREW P. HALESTRAP AND ANNE M. DAVIDSON

Department of Biochemistry, School of Medical Sciences, University of Bristol, Bristol BS8 1TD, United Kingdom

Hepatic gluconeogenesis, urea syntheses, fatty acid oxidation and respiration are all stimulated by those hormones which cause an increase in the concentration of cyclic AMP (*e.g.* glucagon) or Ca^{2+} (*e.g.* vasopressin and α-adrenergic agonists). Activation of mitochondrial metabolism is an essential feature of the action of these hormones, and we have provided strong evidence that this is achieved through an increase in intramitochondrial matrix volume (see Halestrap et al., 1985; Halestrap, 1988). Thus the hormones cause an increase in intra-mitochondrial volume from a control value of about 1.1 μl/mg mitochondrial protein to 1.25-1.4 μl/mg protein (Quinlan et al., 1983) which, although modest, is sufficient to cause significant activation of respiration, citrulline syntheses, pyruvate carboxylation, fatty acid oxidation and glutaminase (see Halestrap et. al., 1985; Halestrap, 1988). The increase in matrix volume is brought about by influx of K^+ into the mitochondria in response to elevated cytoplasmic, and consequently mitochondrial, $[Ca^{2+}]$ (Halestrap et al., 1986). More recently we have shown that Ca^{2+} exerts its effects on K^+ permeability indirectly through an increase in mitochondrial inorganic pyrophosphate (PP_i) concentration (Davidson & Halestrap, 1987). Thus the ability of Ca^{2+} (< 1 μM) to increase the intra-mitochondrial matrix volume correlates with the observed rise in matrix PP_i. Furthermore exposure of either isolated mitochondria or hepatocytes to butyrate, which is activated to butyryl-CoA in

A. Azzi et al. (Eds.)
Anion Carriers of Mitochondrial Membranes
© Springer-Verlag Berlin Heidelberg 1989

the mitochondrial matrix with consequent production of PP_i, also increases the mitochondrial volume.

In the hepatocyte there is also a correlation between the ability of different hormones to increase the mitochondrial volume (as determined by changes in light scattering) and their effects on the PP_i content of the cell (Davidson & Halestrap, 1988). When rapid sub-cellular fractionalize was performed using a digitonin/sheer force technique it was found that more than 95% of the PP_i in the hepatocyte was intra-mitochondrial (Davidson & Halestrap, 1988). The absence of significant cytoplasmic $[PP_i]$ might be predicted since it is generally believed that those cytoplasmic biosynthetic reactions producing PP_i are driven by a very active pyrophosphatase which rapidly hydrolyses it. However the presence of a significant concentration of PP_i (0.1 mM) in mitochondria which is increased several fold by Ca^{2+} in isolated mitochondria and hormones in hepatocytes implies that mitochondrial pyrophosphatase cannot be very active. Furthermore either its activity or the synthesis of PP_i or transport of PP_i out of the mitochondria must be regulated by Ca^{2+}. In this article we will show that the mitochondrial pyrophosphatase is inhibited by low concentrations of Ca^{2+} and suggest that net transport of PP_i out of the mitochondria can only be achieved slowly by exchange with phosphate (P_i) on the adenine nucleotide translocase. We will also present data which implicate the adenine nucleotide carrier in the PP_i stimulation of K^+ entry into mitochondria.

THE SYNTHESIS AND BREAKDOWN OF PP_I IN THE MITOCHONDRIA

There are two potential sources of PP_i within the mitochondrial matrix. Firstly it could be produced from the breakdown of ATP, perhaps as a consequence of the turnover of mitochondrial phospholipids by the operation of the Ca-activated phospholipase A_2 and subsequent re-esterification of the released fatty acids (Beatrice et al., 1980). This involves activation of fatty acids to fatty acyl-CoA and consequently the production of both AMP and PP_i, but it is unlikely to occur fast enough to account for the levels of PP_i observed. Furthermore we have shown that the inhibitor of mitochondrial phospholipase A_2, tetracaine, does not influence the ability of Ca^{2+} to increase the matrix volume and $[PP_i]$ (Davidson & Halestrap, 1987). Secondly there are reports of the presence of a PP_i-synthesizing, proton translocating pyrophosphatase which is membrane bound and driven by the proton motive force (Mansurova et al., 1977; Volk & Baykov, 1984). Such an enzyme is well documented in bacteria where it plays a part in energy conservation (Baltscheffsky & Nyren, 1984), but less well studied in mammalian mitochondria.

Studies in both this and other laboratories have shown that when liver mitochondria are exposed to both butyrate and Ca^{2+} together, there is a massive increase in PP_i to values twentyfold above basal or more (Otto & Cook, 1982; Davidson & Halestrap, 1987). This is far greater than the sum of the effects of either agent on its own and may imply that Ca^{2+} is inhibiting the breakdown of the PP_i that is produced by the activation of butyrate to butyryl-CoA. There appears to be significant pyrophosphatase activity associated with the mitochondrial matrix and this has been shown to be inhibited by Ca^{2+}, but only at concentrations far in excess of those found physiologically (Irie *et al.*, 1970; Volk *et al.*, 1982, 1983).

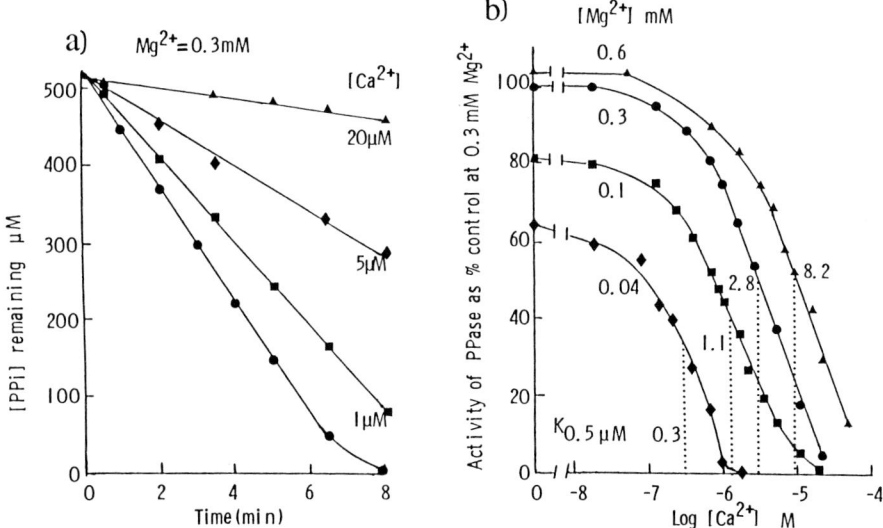

Fig. 1. Inhibition of the pyrophosphatase activity of the mitochondrial matrix by Ca^{2+} at different concentrations of Mg^{2+}. Details of the assay conditions used are given in the text. The concentrations of Ca^{2+} and Mg^{2+} given represent calculated free concentrations.

However we have studied the effects of low concentrations of Ca^{2+} on the pyrophosphatase activity of a mitochondrial matrix fraction assayed under more physiological conditions. The mitochondria were freed of microsomal, lysosomal, outer membrane and intermembrane contamination by Percoll gradient centrifugation followed by digitonin treatment (Whipps *et al.*, 1987). They were then disrupted by sonication and membranes removed by high speed centrifugation. The resulting matrix fraction was assayed for pyrophosphatase at 30° C and pH 7.2 in a buffer containing 50 mM MOPS, 50 mM KCl, 27 mM Tris, 0.5 mM PP_i, 1 mM EGTA and 1 mM N-(hydroxyethyl)ethylenediamine-triacetic acid (HEDTA) with Ca^{2+} and Mg^{2+} added at the concentration necessary to give the specified free concentrations of these cations calculated from the relevant dissociation constants. A time course of the breakdown of PP_i was obtained by stopping the reaction with $HClO_4$ and assaying the PP_i remaining as shown in Fig. 1a. It is apparent from the linearity of the loss of PP_i with time even at $[PP_i]$ below 50 μM that the Km must be less than 20 μM whilst the V_{max} of the enzyme is about 350 nmoles per mg matrix protein at 37° C. It is also clear that Ca^{2+} at micromolar concentrations can inhibit the enzyme significantly. This is shown in Fig. 1b in more detail where the effect of $[Mg^{2+}]$ on the sensitivity to $[Ca^{2+}]$ is also shown. The $K_{0.5}$ for Ca^{2+} is considerably reduced at low $[Mg^{2+}]$ but much greater at higher $[Mg^{2+}]$. At 0.3 mM Mg^{2+}, the concentration of free Mg^{2+} believed to occur in the mitochondrial matrix (Corkey et al., 1986), the $K_{0.5}$ for Ca^{2+} is about 3.5 μM. This value is somewhat higher than the $K_{0.5}$ for the activation of 2-oxoglutarate by $[Ca^{2+}]$ but not very different to the values obtained for the activation of isocitrate dehydrogenase and pyruvate dehydrogenase phosphate phosphatase (Denton & McCormack, 1985; Denton et al., 1987). Indeed in the latter case the value is also very dependent on the $[Mg^{2+}]$ (Midgley et al., 1987).

In energised mitochondria the concentration of Ca^{2+} giving half maximal increases in $[PP_i]$ is about 0.3 μM (Davidson & Halestrap, 1987), similar to the values obtained for the activation of pyruvate dehydrogenase and 2-oxoglutarate dehydrogenase (Denton & McCormack, 1985). That this value is an order of magnitude lower than the $K_{0.5}$ for Ca^{2+} on the isolated pyrophosphatase might appear inconsistent. However it must be remembered that the $[PP_i]$ within the mitochondrial matrix is about 100 μM in the absence of Ca^{2+} and the Km of the pyrophosphatase for PP_i is less than 20 μM, implying that the enzyme is saturated with substrate. Thus even a small inhibition of the enzyme would be capable of causing significant increases in mitochondrial $[PP_i]$. Indeed the increase in PP_i might be expected to be infinite unless either an increase in PP_i concentration inhibits further PP_i synthesis, or the Km for PP_i in situ is considerably greater than that measured in vitro. The former explanation seems likely since when an alternative means of generating PP_i is provided by the addition of butyrate, the matrix $[PP_i]$ can reach 3 mM or greater (Otto & Cook, 1982; Davidson & Halestrap, 1987). Another interesting implication of the saturation of the pyrophosphatase with PP_i is that we can calculate the rate of synthesis of PP_i from the Vmax of the pyrophosphate, since at equilibrium this must be equal to the rate of breakdown of PP_i. Such considerations suggest that about 5 nmoles of i are formed per min per mg mitochondrial protein, implying that a major component of the oxygen consumption in State 4 may be associated with this process.

THE TRANSPORT OF PP_I ACROSS THE MITOCHONDRIAL MEMBRANE

It is well established that PPi can exchange with adenine nucleotides on the adenine nucleotide translocase, albeit considerably less efficiently than ATP or ADP (Krämer, 1985). This phenomenon has been used to deplete mitochondria of adenine nucleotides (Asimakis & Aprille, 1980; D'Souza & Wilson, 1982) and may also explain why hormones which increase mitochondrial $[PP_i]$ produce a net uptake of adenine nucleotides into the mitochondria (see Davidson & Halestrap, 1987). However exchange of PP_i with ADP would be incapable of maintaining significant rates of transport of PP_i across the mitochondrial membrane for any time. To investigate whether an alternative transport mechanism exists we have used the technique of osmotic swelling in iso-osmotic ammonium PP_i in the presence of valinomycin. This will allow swelling to occur as PP_i enters the mitochondria whether transport is electrogenic (NH_4 enters *via* valinomycin) or protogenic (NH_3 enters). In fact we have found that optimal swelling occurs only when both these conditions are met, suggesting the movement of the anion occurs by a combination of both electrogenic and protogenic means. We believe this reflects the electrogenic nature of the exchange of PP_i^{3-} for P_i^{2-} and the protogenic movement of P_i through the phosphate carrier (see below). Results of a typical set of experiments are shown in Fig. 2.

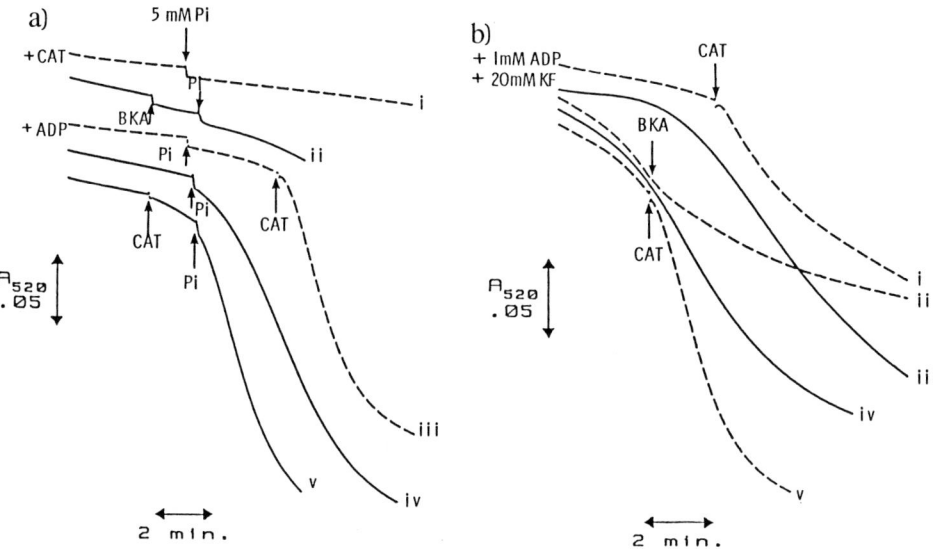

Fig. 2. The swelling of de-energized rat liver mitochondria in iso-osmotic ammonium pyrophosphate. Mitochondria (3 mg protein/ml) were added at the start of the traces to 85 mM NH_4PP_i containing 1 μg valinomycin and 1 μg/ml rotenone. In traces (aiii) and (bi) 1 mM ADP was present. In all the traces in (b) 100 mM PCMBS was present. Other additions were made as indicated: P_i, 5 mM phosphate; BKA, 10 μg/ml bongkrekate; CAT, 5 μg/ml carboxyatractyloside.

In Fig. 2a we show that little swelling occurs until phosphate is added, at which point swelling is rapid (trace iv). However pre-exposure to bongkrekate (BKA) or ADP prevents phosphate exerting its effects (traces ii & iii), but the inhibitory action of ADP can be reversed by the addition of carboxyatractyloside (CAT). If CAT is present in the medium before the addition of mitochondria it prevents P_i-stimulated swelling (trace i), whilst if added after the mitochondria it increases the effectiveness of P_i and even causes some swelling in the presence of endogenous P_i (trace v). It is know that exposure of de-energized mitochondria to PP_i rapidly depletes them of their adenine nucleotides in a carboxyatractyloside-sensitive manner (D'Souza & Wilson, 1982). Thus the difference between the effects of CAT added before or after the mitochondria may be a consequence of this depletion. The data are consistent with the adenine nucleotide translocase catalysing the exchange of PP_i for P_i in the absence of adenine nucleotides, the process being inhibited by BKA (carrier in the "m" conformation) but activated by CAT (carrier in the "c" conformation). However the "c" conformation of the carrier is unable to catalyse such exchange of PP_i for P_i when depletion of adenine nucleotides is prevented when CAT is present from the start of the experiment (trace i). This implies that intra-mitochondrial adenine nucleotides can compete with P_i for the intramitochondrial binding site, even when CAT is bound. This conclusion is not compatible with the usually accepted model for the mechanism of the carrier which does not allow the simultaneous binding of ADP or ATP and CAT (Klingenberg, 1976; Klingenberg et al., 1983). However there is evidence from ADP binding to purified inner mitochondrial membranes (Vignais et al., 1973) and disrupted mitochondria (Klingenberg et al., 1975) that CAT does not displace all the ADP bound to the carrier. This suggests that CAT binding to the cytoplasmic face of the carrier may still permit ADP binding to a side on the matrix side.

In Fig. 2b we show that entry of PP_i can occur in the absence of added P_i if 100 μM p-chloro-mercuribenzene sulphonate (PCMBS) is present to inhibit the mitochondrial P_i transporter. Under these conditions the PP_i enters initially in exchange for adenine nucleotides (an osmotically silent process) and is rapidly hydrolysed by the mitochondrial pyrophosphatase. However, since the resulting P_i cannot leave via the P_i transporter it accumulates sufficiently to exchange with the incoming PP_i on the adenine nucleotide translocase. The net result of this process is accumulation of P_i. Once again this process is prevented by ADP, the inhibition being reversed by CAT. Addition of CAT alone accelerates swelling (trace v) whilst BKA inhibits (trace ii). Addition of 20 mM KF also inhibits the onset of swelling (trace iii) since it inhibits the intra-mitochondrial pyrophosphatase and so prevents the accumulation of sufficient P_i to allow PP_i/P_i exchange to occur. The ability of the adenine nucleotide carrier to transport P_i has not been shown on the purified protein, but it has been demonstrated that P_i can deplete heart mitochondria of their adenine nucleotides by a CAT sensitive mechanism (Asimakis & Conti, 1985).

In Fig. 3 we provide evidence that the exchange of PP_i for P_i on the adenine nucleotide translocase may also be used to transport PP_i out of the mitochondria under energized, State 4 conditions. Mitochondria were incubated in the presence of substrate but in the absence of adenine nucleotides and Mg^{2+} to minimize both the formation and the breakdown of extra-mitochondrial PP_i. Methylene diphosphonate was added at 0.5 mM to further inhibit extra-mitochondrial pyrophosphatase. Under these conditions it was assumed that any PP_i found outside the mitochondria must have been transported from the inside. After 6 min (Fig. 3a) or 30 s (Fig. 3b) incubations the mitochondria were rapidly separated from the incubation medium by centrifugation through silicone oil and the intra-mitochondrial and medium [PP_i] measured.

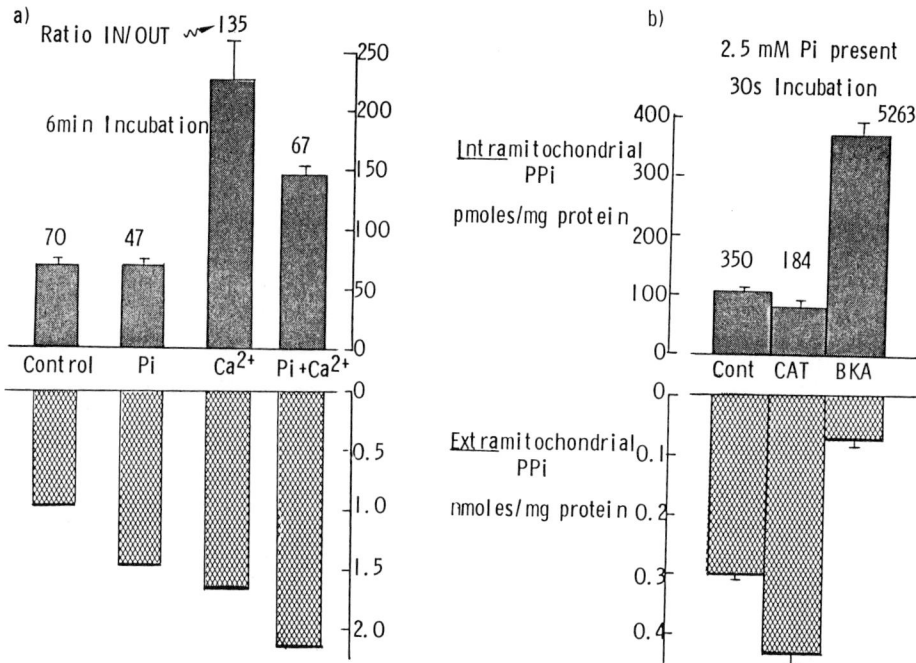

Fig. 3. Phosphate-dependent transport of PP_i out of energized rat liver mitochondria on the adenine nucleotide translocase. Intramitochondrial and extramitochondrial PP_i was measured after 30 s (b) or 6min of incubation in the presence of the additions shown. Details are given in the text. Where added P_i, CAT and BKA were at 2.5 mM, 5 μg/ml and 10 μg/ml respectively. Error bars represent the S.E.M. of at least 3 separate experiments.

In the absence of added P_i the matrix [PP_i] was higher and the extra-mitochondrial [PP_i] lower than in the presence of P_i, especially when 0.3 μM Ca^{2+} was present to inhibit intra-mitochondrial pyrophosphatase (Fig. 3a). In the presence of P_i, CAT decreased the intra-mitochondrial [PP_i] and increased the extra-mitochondrial [PP_i], whilst BKA had the opposite effect (Fig. 3b). These results are again consistent with PP_i exchange for P_i on the adenine nucleotide translocase in the "c" conformation but not the "m" conformation.

THE ROLE THE ADENINE NUCLEOTIDE TRANSLOCASE IN CATALYSING ELECTROGENIC PP_I-DEPENDENT K^+ ENTRY

The Ca^{2+}-induced swelling of energized mitochondria was found to be inhibited by the presence of extra-mitochondrial ATP but activated by the addition of atractyloside or CAT (Halestrap *et al.*, 1986; Davidson & Halestrap, 1987). This is illustrated by the data of Fig. 4. These effects were not mediated through changes in the matrix [PP$_i$] and suggested that the conformation of the adenine nucleotide translocase might influence K$^+$ permeability (Davidson & Halestrap, 1987), a suggestion that has also been made by others (Panov *et al.*, 1980; Jung & Brierley, 1981, 1984).

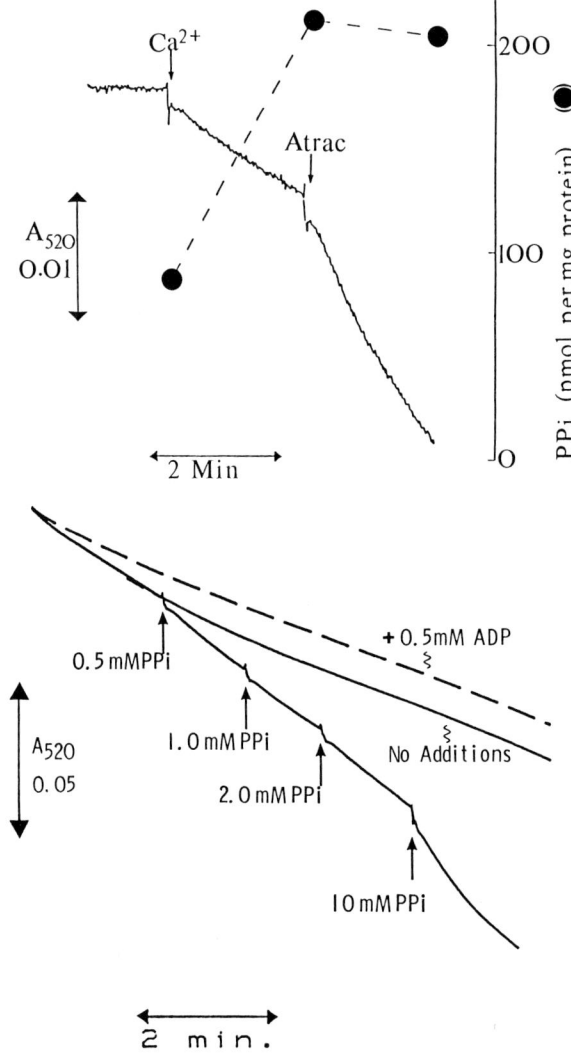

Fig. 4. The effects of atractyloside on the Ca^{2+}-mediated swelling of energized rat liver mitochondria in KCl medium. The buffer contained 2.5 mM P$_i$, 0.3 mM ATP, 2.5 mM MgCl$_2$, 0.5 mM EGTA and 5 mM succinate. Where indicated Ca^{2+} was added to give a free concentration of 0.9 μM and atractyloside at 5 μg/ml. Further details are given in Davidson & Halestrap (1987).

Fig. 5. The effects of ADP and PP$_i$ on the swelling of de-energized rat liver mitochondria in 150 mM KSCN containing 1 μg/ml rotenone, 10 mM Mops and 5mM Tris at pH 7.2. Mitochondria (2.5 mg protein/ml) were added at zero time and PP$_i$ added where shown at the concentrations indicated. The dashed trace represents a run in which 0.5 mM ADP was present in the buffer.

However we have demonstrated that addition of BKA to mitochondria causes swelling (data not shown) which is associated with the rise in [PP$_i$] already noted (Fig.

3b). Thus an increase in K^+ permeability appears to occur when the carrier is trapped in either conformation and this is further confirmed by the data of Figs. 5 and 6. In these experiments the effects of adenine nucleotides, CAT and BKA on the swelling of de-energized mitochondria in iso-osmotic KSCN was studied. Under such conditions swelling is limited only by the permeability to K^+. In Fig. 5 it is shown that the addition of 0.5 mM ADP slightly reduces the rate of swelling whilst the addition of PP_i considerably accelerates the process. The effects of these agents are studied in more detail in Figs. 6a and 6b where a split-beam spectrophotometer is used with mitochondrial suspension present in both cuvettes and additions made only to the sample cuvette. Again it can be seen that addition of ADP causes a reduction in swelling rate which is reversed by the presence of either CAT or BKA (Fig. 6a). In Fig. 6b the effects of adding 1 mM PP_i are shown in the presence and absence of 0.5 mM ADP. In both case the PP_i causes the same degree of swelling suggesting no competition between the ADP and the PP_i.

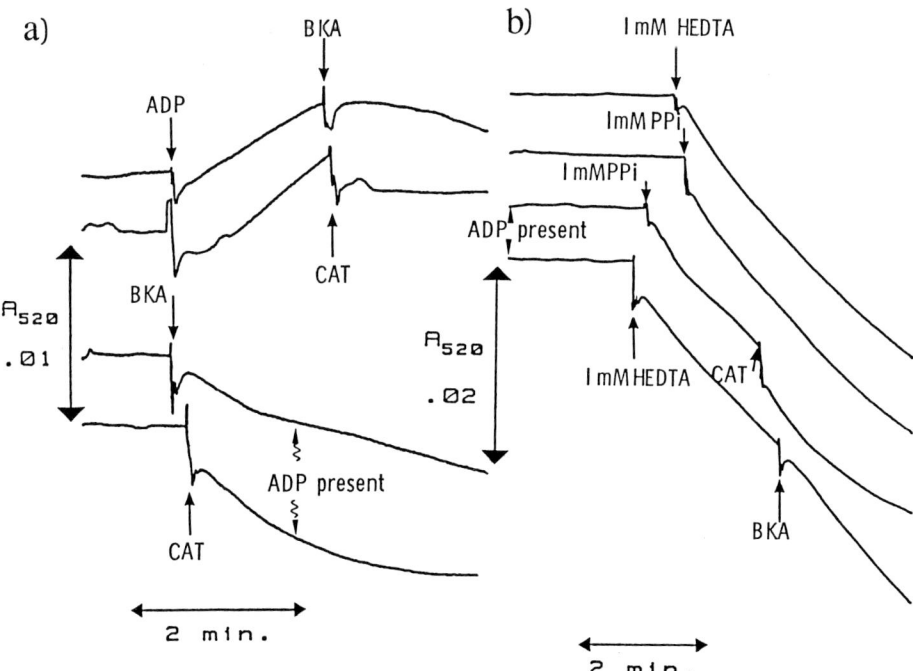

Fig. 6. The effects of ADP, PP_i, CAT, BKA and Mg^{2+}-chelation on the swelling of de-energized rat liver mitochondria in iso-osmotic KSCN. Conditions were essentially the same as in Fig. 5 but mitochondria were present in both cuvettes of a split-beam spectrophotometer and additions were made only to the sample cuvette as indicated. In the lower 2 traces of both (a) and (b) ADP was present at 0.5 mM in the buffer. Further details are given in the text.

This is not consistent with PP_i exerting its major effect through an interaction with the adenine nucleotide carrier. Rather it seemed possible that PP_i might act through reducing the extra-mitochondrial $[Mg^{2+}]$ which is known to inhibit K^+ permeability.

The K_d for $MgPP_i$ at pH 7.2 is 55 μM which is very similar to that of N-(hydroxyethyl)-ethylenediamine-triacetic acid (HEDTA) whose Mg complex has a K_d of 43 μM at pH 7.2. We therefore compared the effects of PP_i and HEDTA on the swelling of de-energized mitochondria in KSCN as shown in Fig. 6b.

It is clear that both agents had similar effects confirming that the major mode of action of PP_i under these conditions was through chelation of the extra-mitochondrial Mg^{2+}. Paradoxically it has been suggested that in energized mitochondria chelation of matrix Mg^{2+} causes their shrinkage through stimulation of the K^+/H^+ antiporter (Dordick et al., 1980). Since an increase in intra-mitochondrial $[PP_i]$ would chelate matrix Mg^{2+} and yet causes mitochondrial swelling, it is unlikely to be acting through an effect on the K^+/H^+ antiporter. Rather an alternative mechanism must be operative. Displacement of adenine nucleotides from the inner face of the adenine nucleotide translocase cannot be ruled out, although the affinity of PP_i for the carrier is considerably less than that of ADP and ATP (Krämer, 1985), which does not offer strong support for such a mechanism. An alternative explanation would be for PP_i to open another K^+ channel in the inner mitochondrial membrane. One such candidate is the 53 kD protein recently isolated from mitochondrial inner membranes by affinity chromatography on a quinine affinity column, and reported to increase the K^+ permeability of artificial liposomes (Diwan et al., 1988).

CONCLUSION

Our suggestion that hormones may exert some of their effects on intra-mitochondrial metabolism through a modest increase in mitochondrial matrix volume is now supported by a considerable number of different observations. We have demonstrated that such increases in matrix volume do occur in situ in a Ca^{2+}-dependent manner and are sufficient to activate many aspects of mitochondrial metabolism known to be stimulated by hormones. We have shown that similar increases in matrix volume can be induced by sub-micromolar $[Ca^{2+}]$ in vitro and are correlated with changes in matrix PP_i. The effects of hormones on matrix volume in situ also correlate with such changes in mitochondrial $[PP_i]$ which accounts for >90% of total cell PP_i. We have demonstrated that Ca^{2+} exerts its effects on $[PP_i]$ by inhibiting the matrix pyrophosphatase, the major route of PP_i degradation within the matrix. Transport of PP_i out of the mitochondria can occur on the adenine nucleotide translocase in exchange for P_i, but only in the absence of adenine nucleotides. In their presence a small loss occurs in exchange for the entry of ADP and this accounts for the rise in total mitochondrial adenine nucleotides following hormone treatment. How PP_i exerts its effects on the mitochondrial volume is uncertain, but appears to involve interaction with the adenine nucleotide translocase or another membrane protein causing it to act as a K^+ channel.

ACKNOWLEDGEMENT

The work described in this article was supported by the Medical Research Council.

REFERENCES

Asimakis GK, Aprille JR (1980) Net uptake of adenine nucleotides in isolated rat liver mitochondria. FEBS Lett 117:157-160

Asimakis GK, Conti VR (1985) Phosphate-induced efflux of adenine nucleotides from heart mitochondria. Am J Physiol 249:H1009 H1016

Baltscheffsky M, Nyren P (1984) The synthesis and utilisation of inorganic pyrophosphate. In: Ernster L (ed) Bioenergetics. Elsevier, Amsterdam, pp 187-206

Beatrice MC, Palmer JW, Pfeiffer DR (1980) The relationship between mitochondrial membrane permeability, membrane potential and the retention of Ca^{2+} by mitochondria. J Biol Chem 255: 8663-8671

Corkey BE, Duszynski J, Rich TL, Matschinsky B, Williamson JR (1986) Regulation of free and bound magnesium in rat hepatocytes and isolated mitochondria. J Biol Chem 261:2567-2574

Davidson AM, Halestrap AP (1987) Liver mitochondrial pyrophosphate concentration is increased by Ca^{2+} and regulates the intramitochondrial volume and adenine nucleotide content. Biochem J 246:715-723

Davidson AM, Halestrap AP (1988) Inorganic pyrophosphate is located primarily in the mitochondria of the hepatocyte and increases in parallel with the decrease in light-scattering induced by gluconeogenic hormones, butyrate and ionophore A23187. Biochem J 254:379-384

Denton RM, McCormack JG (1985) Ca^{2+} transport in mammalian mitochondria and its role in hormone action. Am J Physiol 249:E543-E554

Denton RM, McCormack JG, Midgely PJW, Rutter GA (1987) Hormonal regulation of fluxes through pyruvate dehydrogenase and the citric acid cycle in mammalian tissues. Biochem Soc Symp 54:127-143

Diwan JJ, Haley T, Sanadi DR (1988) Reconstitution of transmembrane K^+ transport with a 53 kilodalton mitochondrial protein. Biochem Biophys Res Commun 153: 224-230

Dordick RS, Brierley GP, Garlid KD (1980) On the mechanism fo A23187-induced potassium efflux in rat liver mitochondria. J Biol Chem 255:10299-10303

D'Souza MP, Wilson DF (1982) Adenine nucleotide efflux in mitochondria induced by gluconeogenic hormones, butyrate and ionophore A23187. Biochim Biophys Acta 680:28-32

Halestrap AP (1988) The regulation of the mitochondrial matrix volume in vivo and in vitro and its role in the control of mitochondrial metabolism. Biochim Biophys Acta (in press)

Halestrap AP, Quinlan PT, Armston AE, Whipps DE (1985) Mechanisms involved in the hormonal regulation of mitochondrial metabolism within intact rat hepatocytes. In: Quagliariello E et al. (eds) Achievements and Perspectives in Mitochondrial Research, vol. 1, Bioenergetics. Elsevier, Amsterdam, pp 469-480

Halestrap AP, Quinlan PT, Whipps DE, and Armston AE (1986) Regulation of the mitochondrial matrix volume *in vivo* and *in vitro*. The role of calcium. Biochem J 236:779-787

Irie M, Yabuta A, Kimura K, Shindo Y, Tomito K (1970). Distribution and properties of alkaline pyrophosphatases of rat liver. J Biochem (Tokyo) 67:47-58

Jung DW, Brierley GP (1981) On the relationship between the uncoupler-induced efflux of K^+ from heart mitochondria and the oxidation-reduction state of pyridine nucleotides. J Biol Chem 256:10490-10496

Jung DW, Brierley GP (1984) The permeability of uncoupled heart mitochondria to potassium ion. J Biol Chem 259:6904-6911

Klingenberg M (1976) The ADP-ATP carrier in mitochondrial membranes. In: Martonosi A (ed) The Enzymes of Biological Membranes, vol. 3. Plenum, New York, pp 383-483

Klingenberg M, Grebe K, Scherer B (1975) The binding of atractylate and carboxy-atractylate to mitochondria. Eur J Biochem 52:351-363

Klingenberg M, Appel M, Babel W, Aquila H (1983) The binding of bongkrekate to mitochondria. Eur J Biochem 131:647-654

Krämer (1985) Characterization of pyrophosphate exchange by the reconstituted adenine nucleotide translocator from mitochondria. Biochem Biophys Res Commun 127:129-135

Mansurova SE, Shakhov YA, Kulaev LS (1977) Synthesis of inorganic pyrophosphate by animal tissue mitochondria. FEBS Lett 74:31-34

Midgley PJW, Rutter GA, Thomas AP, Denton RM (1987) Effects of Ca^{2+} on the activity of pyruvate dehydrogenase phosphate phosphatase within toluene permeabilised mitochondria. Biochem J 241:371-377

Otto DA, Cook GA (1982) Role of Ca^{2+} in regulating the level of mitochondrial pyrophosphate. FEBS Lett 150:172-176

Panov A, Filippova S, Lyakhowich V (1980) Adenine nucleotide translocase as a site of regulation by ADP of the rat liver mitochondrial permeability to H^+ and K^+ ions. Arch Biochem Biophys 199:420-426

Quinlan PT, Thomas AP, Armston AE, Halestrap AP (1983) Measurement of intramitochondrial volume in hepatocytes without cell disruption and its elevation by hormones and valinomycin. Biochem J 214:395-404

Vignais PV, Vignais PM, Defaye G (1973) Adenosine diphosphate translocation in mitochondria: nature of the receptor site for carboxyatractyloside (gummiferin). Biochemistry 12:1508-1518

Volk SE, Baykov AA (1984) Isolation and subunit composition of membrane inorganic pyrophosphatase from rat liver mitochondria. Biochim Biophys Acta 791:198-204

Volk SE, Baykov AA, Duzhenko VS, Avaeva SH (1983) Kinetic studies on the interaction of two forms of inorganic pyrophosphatase of heart mitochondria with physiological ligands. Eur J Biochem 125:215-220

Volk SE, Baykov AA, Kostenko EB, Avaeva SH (1983) Isolation subunit structure and localization of inorganic pyrophosphatase of heart and liver mitochondria. Biochim Biophys Acta 744:127-134

Whipps DE, Armston AE, Pryor HJ, Halestrap AP (1987) Effects of glucagon and Ca^{2+} on the metabolism of phosphatidylinositol-4-phosphate and phosphatidylinositol-4,5-bisphosphate in isolated rate hepatocytes and plasma membranes. Biochem J 241:835-845

Role of the Mitochondrial Outer Membrane in Dynamic Compartmentation of Adenine Nucleotides

F.N. Gellerich, R. Bohnensack and W. Kunz

Institut für Biochemie, Medizinische Akademie Magdeburg, Leipziger Str. 44, 3090 Magdeburg, German Democratic Republic

Small molecules such as adenine nucleotides can pass the porin pores of mitochondrial outer membrane (VDAC) for exchange between mitochondria and cytosol. Estimated to range from 1.25 to 2 x 10^{-9} m (Colombini et al., 1987), the radius of the pores is sufficient to allow the passage of molecules up to a molecular weight of 6000 (Zalman et al., 1980). The effective cross-section of pores is influenced by voltage-dependent processes (Colombini et al., 1987) or colloid-osmotic effects (Zimmerberg & Parsegian, 1986). Nevertheless, the intermembrane space together with the extramitochondrial compartment is commonly believed to form a homogeneous pool for adenine nucleotides. However, it may be assumed that the transport of adenine nucleotides between the cytosol and the AdN translocator occurs by diffusion along concentration gradients, but the question arises whether or not these gradients are high enough to cause a measurable AdN compartmentation in the intermembrane space. Since concentration gradients are necessarily connected with fluxes and disappear with them, the term *dynamic compartmentation* has been proposed for diffusion-dependent inhomogeneities (Gellerich et al., 1987).

The present work was carried out to (i) answer the question if and under which conditions the mitochondrial intermembrane space forms a third compartment in addition to the well accepted AdN pools in the matrix and the extramitochondrial space; (ii) elucidate the cause of AdN compartmentation and estimate the number of

A. Azzi et al. (Eds.)
Anion Carriers of Mitochondrial Membranes
© Springer-Verlag Berlin Heidelberg 1989

pores in the outer membrane, and (iii) investigate the influence of colloid-osmotic pressure on the extent of dynamic compartmentation.

EXPERIMENTAL EVIDENCE OF THE EXISTANCE OF A DYNAMIC COMPARTMENT IN THE MITOCHONDRIAL INTERMEMBRANE SPACE

To investigate the existence of a dynamic AdN compartment in the mitochondrial intermembrane space we developed reconstituted systems consisting of (i) functionally intact mitochondria from different sources plus AdN, inorganic phosphate, substrates, and (ii) pyruvate kinase plus PEP, both competing for ADP regenerated by ATP-utilizing enzymes in varied localization. If formed extramitochondrially by hexokinase, ADP has to pass through the pores of the outer membrane on its way to the AdN translocator (hexokinase system). If formed in the intermembrane space by mitochondrial adenylate kinase or creatine kinase, ADP has to pass through the pores only on its route to pyruvate kinase (adenylate kinase or creatine kinase system). If the outer membrane presents an obstacle to ADP diffusion then, presumably, the competition between mitochondrial and extramitochondrial ADP phosphorylation is influenced by the localization of ADP-regenerating enzymes.

Using the adenylate kinase system in comparison to direct ADP addition in the extramitochondrial space lends itself to investigating whether or not ADP formation in the intermembrane space creates a dynamic ADP compartment. Fig. 1 shows the results of a typical experiment with rat liver mitochondria. Addition of ATP gave rise to a short-time active rate of respiration due to the ADP and AMP contents of the stock solution. In the absence of pyruvate kinase, addition of AMP and ADP entailed an oxygen consumption proportional to the amount of adenine nucleotides added. The ratio of oxygen consumption after further ADP and AMP additions decreased drastically with increasing pyruvate kinase activity as can be seen from the inset. After addition of sufficient amounts of pyruvate kinase, virtually all ADP added was phosphorylated by pyruvate kinase, whereas ADP formation from AMP in the intermembrane space caused a marked oxygen consumption even under these conditions. Assuming that ADP formation from added AMP in the intermembrane space is not influenced by pyruvate kinase, the ratio of oxygen consumption induced by AMP addition with and without pyruvate kinase was indicative of the amount of ADP used for oxidative phosphorylation. It was obvious that, in the presence of excessive pyruvate kinase activity, 15% of the ADP formed by adenylate kinase remained inside the outer membrane, whereas 85% diffused through the pores to the pyruvate kinase.

When, additionally, adenylate kinase was added to such a system (not shown) neither addition of ADP nor that of AMP caused stimulation of mitochondrial respiration. In this case, adenylate kinase extramitochondrially converted all added

AMP into ADP which was re-phosphorylated by pyruvate kinase (Gellerich; unpublished).

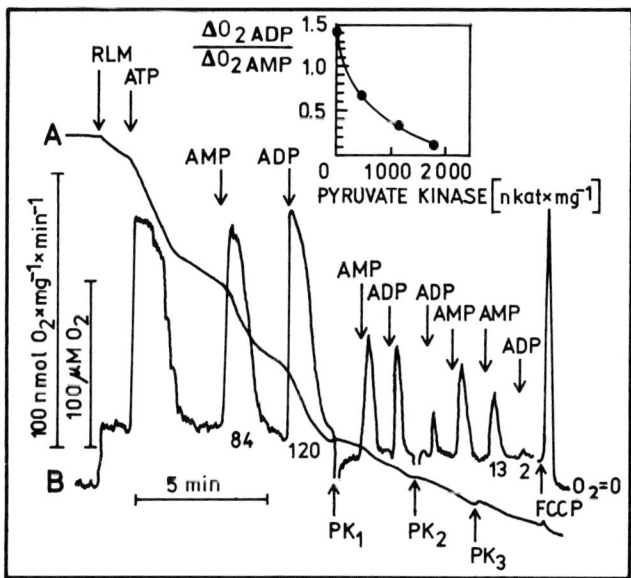

Fig. 1 Increasing differences between amounts of oxygen consumption after equivalent AMP and ADP additions produced by increasing activities of pyruvate kinase in the adenylate kinase system. Mitochondria (0.37 mg/ml) isolated from rat liver (RLM) were incubated in a medium containing: Sucrose 110 mM, Tris 10 mM, potassium phosphate 10 mM, PEP 2.5 mM, KCl 60 mM, glucose 15 mM, MgCl$_2$ 5 mM, EDTA 0.5 mM, succinate 10 mM, and rotenone 0.001 mM; pH 7.4, 25°C.
Additions: ATP 4 mM (containing 48 μM ADP and 19.9 μM AMP; ADP 73 μM (containing 19.6 μM AMP); AMP 40 μM; pyruvate kinase (PK$_1$, PK$_2$, PK$_3$) 25, 41, 103 U/mg mitochondrial protein, respectively; FCCP 0.2 μM. Inset: Pyruvate kinase activity *vs.* ΔO_2 ADP/ ΔO_2 AMP ratio calculated from the areas below the peaks. Numbers under the peaks represent specific oxygen consumption in nmol O$_2$/mg. A: Respiration trace measured with a Clark electrode, B: its first derivative d[O$_2$]/ dt measured by means of a ratemeter. Note: 16.6 nkat = 1 U.

These findings can be understood only if an effective diffusion barrier is assumed to exist between the intermembrane and the extramitochondrial spaces, resulting in a dynamic ADP compartmentation in the mitochondrial intermembrane space.

Addition of pyruvate kinase suppressed the rate of respiration down to the level resulting after complete inhibition of the AdN translocator by carboxyactractyloside. This implies that all ADP formed extramitochondrially, and even in the immediate vicinity of the pores, is freely accessible to the pyruvate kinase in the extramitochondrial compartment. It is, therefore, not reasonable to assume direct interactions between the AdN translocator and mitochondrial hexokinase as discussed in the literature (Weiler *et al.*, 1985).This conclusion is supported by the finding that mitochondrial respiration under stationary conditions as stimulated by added yeast hexokinase can be completely suppressed by addition of pyruvate kinase (Gellerich *et al.*, 1987).

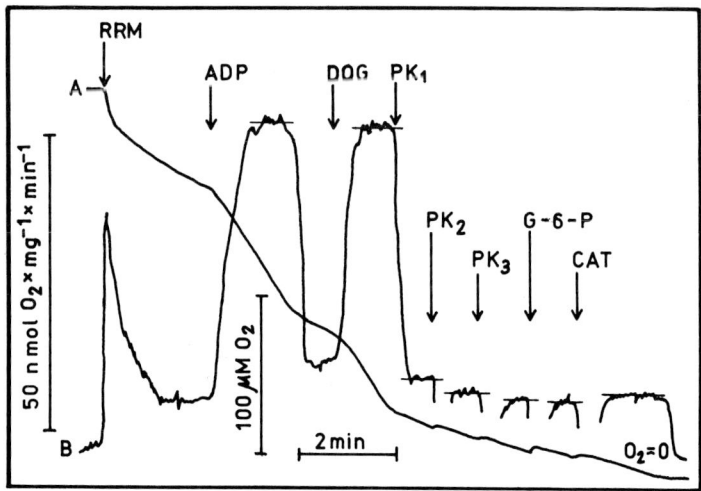

Fig. 2 Complete suppression of hexokinase-dependent stimulation of respiration by an excess of pyruvate activity. Mitochondria (0.87 mg protein/ml) isolated from rabbit reticulocytes (RRM) were incubated in a medium containing in mM: Mannitol 135, potassium phosphate 10, TRAP 1, HEPES 20, $MgCl_2$ 5, EDTA 0.5, PEP 2.5, α-ketoglutarate 10, NAD 2; pH 7.4, 25°C. Additions: ADP 0.3 mM; desoxyglucose (DOG) 10 mM; pyruvate kinase (PK_1, PK_2, PK_3) 58, 80, 85 U/mg mitochondrial protein; glucose-6-phosphate (G-6-P) 1.25 mM, and carboxyatractyloside (CAT) 11.5 nmol/mg mitochondrial protein.

This is even true of mitochondrial hexokinase bound to the outside of the outer membrane of rabbit reticulocyte mitochondria (Schlaeme *et al.*, 1981), probably directly to the pores protein (Linden *et al.*, 1982). As shown in Fig. 2 the hexokinase activity (0.13 U/mg was sufficient to adjust the active state of respiration after addition of desoxyglucose in the presence of adenine nucleotides.

WHAT ARE THE CAUSE AND THE EXTENT OF DYNAMIC AdN COMPARTMENTATION IN THE MITOCHONDRIAL INTERMEMBRANE SPACE?

To obtain quantitative data concerning the influence of the localization of ADP-regenerating enzyme relative to the outer membrane, on AdN compartmentation we analyzed heart mitochondrial respiration *vs.* addition of increasing pyruvate kinase activity in two systems: the creatine kinase and the hexokinase systems (Table I). Stationary rates of respiration were adjusted in both systems by addition of either creatine kinase or glucose plus yeast hexokinase. Then, increasing amounts of pyruvate kinase were

added additionally. It is evident that the minimal respiratory rate adjustable by excessive pyruvate kinase in the creatine kinase system was markedly elevated when compared to the hexokinase system. Assuming that this difference was attributable to a diffusion-limited exchange of adenine nucleotides through the pores of the outer membrane, we employed a relatively simple mathematical model to calculate the diffusion rate of adenine nucleotides as described previously in detail (Gellerich *et al.*, 1987). This model takes into account that in the creatine kinase system the ADP diffusion rate (v_{dADP}) through the pores must be equal to the difference between ADP production in the intermembrane space by creatine kinase (v_{dADP}) and the rate of mitochondrial phosphorylation (v_p)

$$v_{d\ ADP} = v_{CPK} - v_P . \qquad (1)$$

In the hexokinase system, ADP diffuses in the opposite direction and (v_{dADP}) is equal to the rate of mitochondrial phosphorylation (($v_{CPK} = 0$ in Eq. 1).) v_{CPK} was experimentally determined from the formation of creatine phosphate. v_p was calculated from measured respiration rates by adopting a simplified model that describes the rate of oxidative phosphorylation as a function of the extramitochondrial ATP/ADP ratio (Bohnensack, 1984). Then, assuming that an identical rate of respiration in both systems was indicative of an identical ATP/ADP ratio in the vicinity of the AdN translocator, it was possible to calculate the ATP/ADP ratios in the intermembrane compartment. A diffusion rate constant k_d was computed from the diffusion rates v_{dADP} and the different ATP/ADP ratios in both compartments, by using an equation similar to *Fick's* first law of diffusion:

$$k_d = \frac{v_{dADP}}{[ADP]_i - [ADP]_e} = \frac{v_{dATP}}{[ATP]_e - [ATP]_i} \qquad (2)$$

Using experimental data similar to those given in Table 1, it was possible to calculate a diffusion rate constant $k_d = (8.7 \pm 4.7)\ 10^5\ m^3 \times mg^{-1} \times min^{-1}$ (Gellerich *et al.*, 1987). This constant was employed together with Eq. 1 to estimate the concentration gradients between the AdN translocator and the extramitochondrial compartment for each incubation of the experiment shown in Table 1.

In the hexokinase system, the maximum flux through the pores and the maximal ADP gradient (-11.5 μM ADP) occurred at the highest rate of respiration. Since the extramitochondrial ADP concentration was higher than that in the intermembrane space (cf. Eq. 2), ΔC and v_d both have a negative sign. In the creatine kinase system,

Table 1. Direction and extent of the ADP concentration gradient between the extramitochondrial space and the AdN translocator of rat heart mitochondria *versus* localization of ADP regenerating enzyme and pyruvate kinase activity. Rat heart mitochondria (0.18 mg/ml) were incubated in a medium containing in mM: sucrose 250, HEPES 10, potassium phosphate 4, $MgCl_2$ 8, dithiothreitol 0.37, EGTA 0.3, glutamate 4, malate 2, phosphoenolpyruvate 5; 0.8 mM ATP, and either 25 mM creatine or 10 mM glucose plus yeast hexokinase. Furthermore, pyruvate kinase was added with a maximal activity as indicated. 90 s after starting the reactions by addition of ATP, stationary rates of respiration were observed and samples were quenched for enzymatic determination of adenine nucleotides, glucose 6-phosphate, creatine phosphate and pyruvate. The AdN concentration (mean ± S.D.; n = 4-6) are total values, but mainly represent extramitochondrial values, since the contribution of mitochondrial AdN (2.2 µM AdN) was low in comparison to the total one (800 µM AdN). v_{dADP} was calculated using Eqn. 1, where v_p was calculated assuming a constant leak (v_{resp} = 33.8 nmol O_2/mg) and the ADP/O ratio of 3.4$_{CADP}$, the concentration gradient of ADP between the extramitochondrial space and the AdN translocator, was estimated using Eqn. 2. (*) v_{resp}, v_{PK} and $v_{CPK/HK}$ are expressed in nmoles of O_2, pyruvate and creatine phosphate/glucose 6-phosphate, respectively, per min and mg protein.

System	PK (U/mg)	v_{resp} (*)	v_{PK} (*)	$v_{CPK/HK}$ (*)	ADPA (µM)	TP (µM)	ΔC_{ADP} (µM)	v_{dADP} (*)
CPK	0.0	147.7	0.0	746.7	151.2±10.2	726.3±25.8	+0.7	+63.3
	3.5	106.8	448.8	990.6	51.3±9.4	769.7±9.2	+6.4	+552.6
	38.7	80.1	1129.0	1284.1	20.5±3.7	869.0±17.6	+11.6	+1006.3
	193.1	53.4	1284.6	1227.3	5.4±3.6	922.0±17.6	+12.8	+1109.7
HK	0.0	200.0	0.0	n.d.	406.5±29.1	422.5±20.3	-11.5	-997.2
	4.2	78.3	424.9	645.0	43.2±5.3	876.8±29.0	-3.0	-267.0
	8.8	58.7	629.8	656.9	31.3±5.3	904.7±8.1	-1.7	-149.4
	193.1	33.8	933.0	642.9	1.8±5.9	845.6±28.5	0.0	0.0

the ADP gradient reached its maximum (12.8 μM ADP) at the lowest rate of respiration v_{resp} = 53.4 nmol O_2 x mg^{-1} x min^{-1}. Under these conditions, however, the highest diffusion rate was noted as 90% of the ADP formed by creatine kinase diffused through the pores to the pyruvate kinase. This high diffusion rate resulted from the low extramitochondrial ADP concentration (5.4 μM ADP) adjusted by excessive pyruvate kinase. Taking into account the estimable concentration gradient under these conditions, the ADP concentration in the vicinity of the AdN translocator was 12.8 μM higher than extramitochondrially (5.4 μM), high enough to ensure the indicated rate for the AdN translocator.

It is important to note that, in the presence of excessive pyruvate kinase activity, 10-15.5% of ADP formed in the intermembrane space of heart or liver mitochondria by either adenylate kinase or creatine kinase will be phosphorylated *via* oxidative phosphorylation, with the major part diffusing through the pores to the pyruvate kinase. This indicates that identical causes for dynamic compartmentation are involved in both systems. Alternatively, since creatine kinase is reversibly bound on the outside of the mitochondrial inner membrane, AdN compartmentation in the intermembrane space of creatine kinase containing mitochondria was explained by assuming that protein-protein interactions occur between mitochondrial creatine kinase and the AdN translocator (Saks *et al.*, 1980). It is, however, generally accepted that adenylate kinase is a soluble enzyme of the intermembrane space (Brdiczka *et al.*, 1968). The extent of ADP compartmentation being comparable in both systems, the necessity of a specific protein-protein interaction can be ruled out. Furthermore, heart mitochondrial creatine kinase was demonstrated to produce ADP compartmentation even under conditions in which the major part of creatine kinase was released from the inner membrane. It is therefore concluded that the diffusion limitation caused by the mitochondrial outer membrane accounts for the dynamic AdN compartmentation in the mitochondrial intermembrane space. This conclusion is consistent with results obtained by comparing experiments with mitochondria and mitoplasts from rabbit heart (Erickson-Viitanen *et al.*, 1982).

ESTIMATION OF AN APPROXIMATE NUMBER OF VDAC PORES IN THE OUTER MEMBRANE OF RAT HEART MITOCHONDRIA

The rate constant for diffusion and the resultant concentration gradients for adenine nucleotides were calculated on the basis of experimental data only. No structural premise was used for the estimations.

If, however, the limited number of pores in the outer membrane is assumed to restrict AdN diffusion between the two compartments it is possible to estimate the number of pores from the diffusion rate constant by employing Eq. 3 derived from *Fick's* first law of diffusion

356

$$k_d = \frac{D \cdot n \cdot A_{VDAC}}{\Delta l} \qquad (3)$$

where D is the diffusion coefficient, n is the number of pores, A_{VDAC} is the cross-section of a single pore, and Δl is the diffusion distance.

Assuming that (i) the AdN diffusion through the pores is comparable to that in an aqueous solution (D = 7.2×10^{-9} m^2 x min^{-1}; Kushmerick & Podolsky, 1969), (ii) the cross section of a single pore ranges from 4.9 to 12.6×10^{-18} m^2 (Colombini et al., 1987), and (iii) the diffusion distance (being equal to the thickness of the outer membrane) is 7.5×10^{-9} m (Roos et al., 1982), we estimated the number of pores to range from 0.9 to 2.3×10^{-11} pores x cm^{-2} outer membrane of rat heart mitochondria.

On the other hand, especially in heart mitochondria with their pronounced cristae structure, unstirred layer effects (Dietschy, 1978) cannot be ruled out. Such unstirred layer effects might account for part of the estimated concentration gradient and increase the estimable number of pores in the outer membrane.

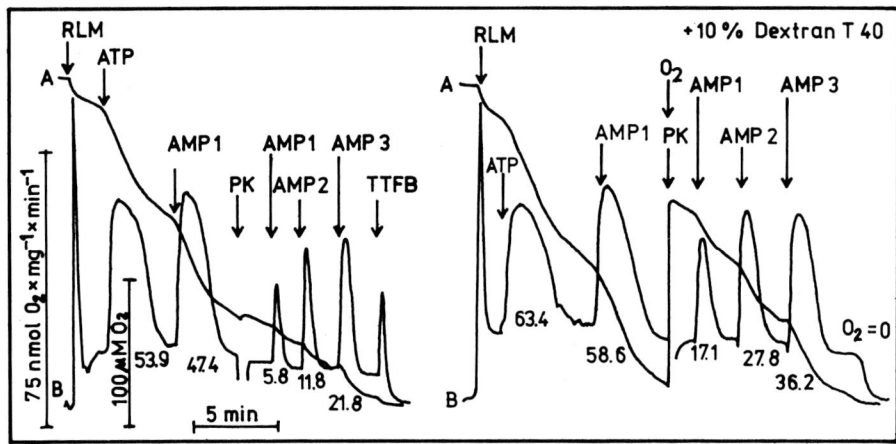

Fig. 3 Effect of addition of macromolecules on AMP-induced mitochondrial oxygen consumption in the presence of excessive pyruvate kinase activity.
Rat liver mitochondria (1.15 mg/ml) were incubated in a medium as described for Fig. 1, with (right) and without (left) addition of 10% Dextran T 40. Additions: ATP 2.4 mM (containing 0.3 mM ADP and 0.12 mM AMP); AMP$_1$, AMP$_2$, AMP$_3$ 76, 152 and 228 μM AMP; pyruvate kinase (PK) 97 U/mg mitochondrial protein; TTFB 10 μM. Numbers under the peaks represent the specific oxygen consumption in nmol O_2/mg.

EFFECT OF CHANGES IN COLLOID-OSMOTIC PRESSURE ON THE DYNAMIC COMPARTMENTATION OF ADENINE NUCLEOTIDES

Proteins of the intermembrane space exert a colloid-osmotic pressure on the inside of the outer membrane which is compensated for by cytosolic proteins of the intact cell rather than by the low-molecular-weight constituents of commonly used incubation media. As shown recently the volume of pores decreases with increasing colloid-osmotic pressure (Zimmerberg & Parsegian, 1986). Hence, addition of macromolecules to the incubation medium would presumably diminish the rate of adenine nucleotide diffusion through the pores and increase the dynamic AdN compartmentation in the intermembrane space. To test this hypothesis we applied media with and without addition of 10% Dextran T 40, resulting in an increase in colloid-osmotic pressure by 6.2 kPa from the outside of the outer membrane. Fig. 3 shows that the addition of Dextran T 40 enhanced the AMP-induced specific oxygen consumption of liver mitochondria to 123%. Also a slightly increased resting state of respiration was observed in this experiment.

Even though the cause has not yet been elucidated, the ratemeter trace was a suitable mean in determining the amounts of ADP diffusing through the pores after AMP additions with and without pyruvate kinase irrespective of differences in the efficiency of oxidative phosphorylation and/or the resting state of respiration. In the absence of Dextran, pyruvate kinase addition was seen to reduce the specific oxygen consumption from 47.4 to 5.8 nmol O_2/mg (12.3%), whereas in the presence of Dextran the remaining consumption was 17.1 nmol O_2/mg representing 29% of 58.6 nmol O_2/mg without pyruvate kinase.

Similar results were obtained with rat heart mitochondria incubated in both the adenylate kinase and the creatine kinase systems (Gellerich, unpublished results).

Obviously, the elevated colloid-osmotic pressure reduces the permeability of pores which, in turn, requires greater ADP concentration gradients. The higher ADP gradients call for greater ADP concentrations in the intermembrane space, thus favouring oxidative phosphorylation of ADP. Moreover, this increased pressure might account for decreasing the volume of intermembrane space and promoting unstirred layer phenomena which also would have a diminishing effect on AdN permeability.

CONCLUSIONS

The results presented demonstrate that diffusion of adenine nucleotides through the mitochondrial outer membrane requires rate-dependent concentration gradients of up to 12.8 μM AdN between the extramitochondrial space and the intermembrane compartment in which the detector, *viz.* the binding sites of the AdN translocator, is localized. The importance of concentration gradients to the dynamic compartmentation increases with decreasing absolute AdN concentrations, in particular that of AdP.

The limited number of pores in the mitochondrial outer membrane presumably is the main cause for the absence of a homogeneous AdN distribution among both compartments and, hence, a dynamic compartmentation.

The increasing colloid-osmotic pressure further adds to dynamic compartmentation, probably due to partial closing of the pores in the mitochondrial outer membrane, indicating the importance of dynamic AdN compartmentation to the intact cell.

One consequence of the AdN concentration gradients are the resultant gradients in phosphorylation potential across the outer membrane, being up to 3 kJ/mol greater extramitochondrially when compared to the mitochondrial intermembrane space of the creatine kinase system, a phenomenon presumably attributable to the creatine phosphate shuttle.

REFERENCES

Bohnensack R (1984) Rate law of mitochondrial respiration *versus* extramitochondrial ATP/ADP ratio. Biomed Biochim Acta 43:403-411

Brdiczka D, Pette D, Brunner G, Miller F (1968) Kompartimentierte Verteilung von Enzymen in Rattenlebermitochondrien. Eur J Biochem 5:294-304

Colombini M, Yeung CL, Tung J, König T (1987) The mitochondrial outer membrane channel, VDAC, is regulated by a synthetic polyanion. Biochim Biophys Acta 905:279-286

Dietschy JM (1978) Effect of diffusion barriers on solute uptake into biological systems. In: Srere PA, Estabrook RW (eds) Microenvironments and Metabolic Compartmentations. Academic Press New York, pp 401-432

Erickson-Viitanen S, Geiger PJ, Viitanen P, Bessman SP (1982) Compartmentation of mitochondrial creatine phosphokinase II. The importance of the outer mitochondrial membrane for mitochondrial compartmentation. J Biol Chem 257:14405-14411

Gellerich FN, Schlame M, Bohnensack R, Kunz W (1987) Dynamic compartmentation of adenine nucleotides in the mitochondrial intermembrane space of rat-heart mitochondria. Biochim Biophys Acta 890:117-126

Kushmerick MJ, Podolsky RJ (1969) Ionic mobility in muscle cells. Science 166:1297-1298

Linden M, Gellersfors P, Nelson BD (1982) Pore protein and hexokinase binding protein from outer membrane of rat liver mitochondria are identical. FEBS Lett 141:189-192

Roos M, Benz R, Brdiczka D (1982) Identification and characterization of the pore-forming protein in the outer membrane of rat liver mitochondria. Biochim Biophys Acta 686:204-214

Saks VA, Kupriyanov VV, Elizarova GV, Jacobus WE (1980) Studies on energy transport in heart cells. The importance of creatine kinase localization for the coupling of mitochondrial phosphoryl creatine production of oxidative phosphorylation. J Biol Chem 255:755-763

Schlame M, Gellerich FN, Augustin W (1981) Localization of hexokinase in mitochondria from rabbit reticulocytes and its relation to mitochondria from rabbit reticulocytes and its relation to mitochondrial ATP-formation studied by measurement of ^{32}P-fluxes. Acta Biol Med Germ 40:617-623

Weiler U, Riesinger I, Knoll G, Brdiczka D (1985) The regulation of mitochondrial bound hexokinase in the liver. Biochem Med 33:223-235

Zalmann LS, Nikaido H, Kagawa Y (1980) Mitochondrial outer membrane contains a protein producing nonspecific diffusion channels. J Biol Chem 255:1771-1774

Zimmerberg J, Parsegian A (1986) Polymer inaccessible volume changes during opening and closing of a voltage-dependent ionic channel. Nature 323:36-39

Topology of Peripheral Kinases: its Importance in Transmission of Mitochondrial Energy

DIETER BRDICZKA, VOLKER ADAMS, MATTHIAS KOTTKE AND ROLAND BENZ

Faculty of Biology, University of Constance, P.O. Box 5560, D-7750 Konstanz
and
Department of Biotechnology, University of Würzburg, Federal Republic of Germany

All transport systems for anionic mitochondrial metabolites reside in the inner mitochondrial membrane. Therefore, we are entirely accustomed to the idea that the outer membrane is freely permeable for these compounds. However, the permeability for polar metabolites through the outer membrane is restricted to a slightly anion-selective general diffusion pore protein (Colombini, 1979; Benz, 1985) which, at a voltage above 30 mV, adopts a different state, characterized by low conductance and cation selectivity (Ludwig *et al.*, 1988). The latter state of the pore was found to exclude ADP and ATP permeation in intact mitochondria (Benz *et al.*, 1988). Having accepted this fact we turned our curiosity on the question whether a membrane potential across the outer mitochondrial membrane can exist physiologically that results in regulation of anion permeability. The answer we arrived at was that the inner membrane potential might influence the outer membrane where it is in close contact with the inner membrane. The structure and function of these contact sites has been analyzed by electron microscopy in freeze fractured mitochondria. We observed a dynamic regulation of the contacts by the rate of the oxidative phosphorylation and a distance between the two membranes in the sites of 1-2 nm. The regulation of the pore lends support to the concept of a separate compartment of adenine nucleotides in the intermembrane space which would enhance the ATP translocation process because of the following reasons.

A. Azzi et al. (Eds.)
Anion Carriers of Mitochondrial Membranes
© Springer-Verlag Berlin Heidelberg 1989

The asymmetric ATP export performed by the translocator is driven electrogenically by the mitochondrial membrane potential (Klingenberg et al., 1982). Thus, a 13 times higher efflux of internal ATP could be maintained by the inner membrane potential in competitive in vitro experiments (Klingenberg et al., 1982). If we consider the existence of an ATP/ADP quotient in the cytosol of 10 in liver (Soboll et al., 1978) and muscle cells (Hebisch et al., 1986) it goes without saying that the translocation process would be at equilibrium and, therefore, would have low rates. However, it cannot be easily dismissed that under physiological conditions the flux rate in the mitochondrial ATP/ADP exchange is high in spite of a high phosphorylation potential in the cytosol.

A satisfying answer to this problem has to be in terms of a mechanism which keeps the translocator away from equilibrium. At this point the functional coupling of kinases to the translocator comes into effect. The existence of this interaction has been described for hexokinase in mitochondria from liver (Gots et al., 1974; Brdiczka et al., 1986), muscle (Viitanen et al., 1986) and brain (Inui et al., 1974) and also for creatine kinase in muscle mitochondria by several authors (Saks et al., 1987; Bessman et al., 1985; Jacobus, 1985; Wallimann et al., 1985; Gellerich et al., 1987). Provided that we agree to the functional coupling between translocator and kinases it would be a promising mechanism to dislocate the translocator from equilibrium. In this case, the direct interaction with kinases would cause an immediate transfer of the phosphorylation energy from the exported ATP to metabolites which are no substrates for the translocator.

Apart from hexokinase and creatine kinase there are quite a number of different kinases situated outside the inner membrane at the mitochondrial periphery which may be involved in this postulated mechanism. Of these kinases two groups can be defined with respect to location and to function: 1. energy-consuming kinases which bind to the outer membrane pore protein at the surface of the membrane, for example hexokinase and glycerolkinase (Fiek et al., 1984), and 2. energy-transmitting kinases which are located between the two boundary membranes, for example adenylate kinase (Brdiczka et al., 1968), nucleoside diphosphate kinase (Jacobus et al., 1977), creatine kinase (Jacobs et al., 1964).

In accordance with the above arguments we postulate that the surface-bound as well as the intermembrane kinases have access to a different ATP/ADP pool between the two boundary membranes. We shall explain in this article how the inner membrane potential may affect the outer membrane pore and limit its conductance. Furthermore, the importance of the contact sites in this type of regulation will be emphasized by description of the preferential organization of several peripheral kinases in these sites as observed by electron microscopic and biochemical investigations.

CHARACTERIZATION OF CONTACT SITES BETWEEN THE TWO
BOUNDARY MEMBRANES

The existence of contact sites between the two boundary membranes has been demonstrated by three different methods: 1. electron microscopy, 2. isolation and biochemical characterization, and 3. treatment with digitonin.

Electron microscopy, freeze fracturing

Hackenbrock in 1968 was the first to describe contact sites between the two mitochondrial boundary membranes in thin sections of rat liver mitochondria. These contacts have been assumed to be responsible for the atypical behaviour of mitochondrial membranes in freeze fractured samples which is characterized by frequent jumping of the fracture plane between the two mitochondrial limiting membranes (Van Venetie & Verkleij, 1982). Supposing that fracture plane deflections would only occur in the contact sites we suggested that the frequency of fracture plane deflections would directly correlate with the frequency of contacts. Based on these considerations we determined the frequency of fracture plane deflections in freeze fractured samples of mitochondria in different metabolic states and observed a significantly larger frequency of contacts (fracture plane deflections) in phosphorylating mitochondria compared to mitochondria in states 4 and 1 (Knoll et al., 1983; Klug et al., 1984). Uncoupling of the mitochondria by dinitrophenol caused a further reduction of the contacts. In view of these findings the contacts appeared to be dynamic structures regulated by the activity of the oxidative phosphorylation.

Electron microscopic localization of hexokinase at the mitochondrial surface

The possible involvement of the contacts in the functional coupling of peripheral kinases brought up the question of location of these enzymes at the mitochondrial surface. We analyzed the distribution of hexokinase at the surface of rat liver and brain mitochondria using immuno-gold labelling techniques (Weiler et al., 1985; Kottke et al., 1988). In both types of mitochondria the enzyme was non-randomly distributed because the gold grains were predominantly localized in areas where the inner and outer mitochondrial boundary membranes were closely apposed. Taking into account that hexokinase binds specifically to the outer membrane pore (Fiek et al., 1983; Lindén et al., 1983) these results suggested that the enzyme might preferentially bind to a contact site specific pore structure, or alternatively that pores are concentrated only in the contact regions.

Isolation of contact sites from osmotically disrupted mitochondria

Based on the above results we decided to judge hexokinase as a good marker enzyme for the contact fraction when we attempted to isolate the contact fraction by density gradient centrifugation from osmotically disrupted liver mitochondria (Ohlendieck *et al.*, 1986). By applying this method also to brain and kidney mitochondria we were able to separate a fraction in the density gradient which was distinct from the main outer and inner membrane fractions and was characterized by a high hexokinase activity (Fig. 1).

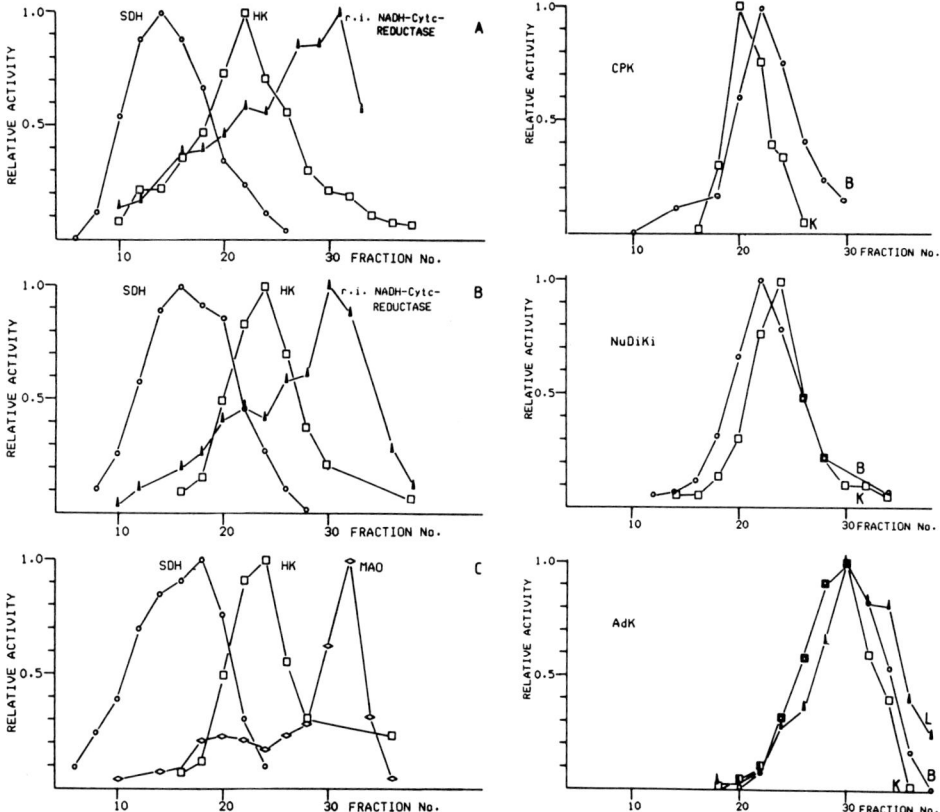

Fig. 1. Isolated mitochondria from brain (A), kidney (B), and liver (C) were disrupted by swelling (20 mM phosphate buffer), shrinking (60% sucrose) and sonication. Unbroken mitoplasts were removed by low speed centrifugation and the supernatant was loaded on a continuous sucrose density gradient varying between 1.06 and 1.22 g/ml. Left panel shows the distribution of outer membrane (rotenone insensitive NADH-cytochrome c reductase, or monoamine oxidase, MAO) and inner membrane (succinate dehydrogenase, SDH) marker enzymes and of hexokinase (HK) activity. Right panel depicts the activity profiles of adenylate kinase (AdK), creatine kinase (CPK), and nucleoside diphosphate kinase (NuDiKi) in the same density gradients separating subfractions of brain (B), kidney (K), and liver (L) mitochondria. The activity of the enzymes in every fraction (highest density No. 1) is given relative to the activity in the peak fraction of the respective enzyme.

In addition, it contained inner and outer membrane components. This presumptive contact fraction was subsequently removed from the density gradient to increase the density of hexokinase by using specific antibodies and subsequent decoration with protein-A-gold. Although this procedure labelled exclusively hexokinase, the inner and outer membrane components like hexokinase migrated to the same higher density upon recentrifugation on a second discontinuous density gradient (Fig. 2), suggesting the existence of a complex between the enzyme and the two boundary membranes.

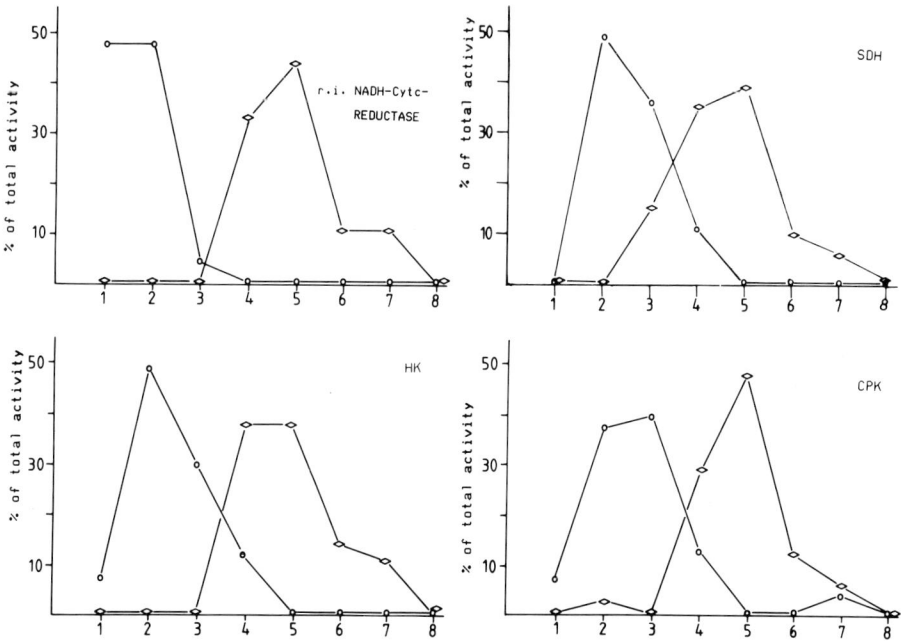

Fig. 2. The hexokinase-containing fractions from a density gradient, as shown in Fig. 1A, were incubated with specific antibodies against hexokinase I. Unbound antibodies were removed by centrifugation and the sediment was incubated with protein-A-gold. A control fraction was treated in the same way except prior decoration by specific antibodies. Both fractions were centrifuged on a density gradient containing from right to left 50%, 45%, and 40% sucrose. The activity distribution of outer and inner membrane marker enzymes and of hexokinase and creatine kinase was determined in the different gradient fractions. Circles = control gradient, abbreviations as in Fig. 1.

The structure of the isolated contact fraction was further analyzed by surface proteolysis and freeze fracture. The results of both methods agreed in that the fraction contained right-side-out outer membrane vesicles which enwrapped an inner membrane vesicle (Ohlendieck *et al.*, 1986; Kottke *et al.*, 1988). Having previously accepted the fact that hexokinase, representing peripheral energy-consuming kinases, resided specifically inside the contacts, we next investigated the distribution of energy-transmitting kinases. The mitochondrial isozymes of creatine kinase (Jocobs *et al.*, 1964), nucleoside diphosphate kinase (Brdiczka *et al.*, unpublished) and adenylate kinase (Brdiczka *et al.*, 1968) are located in the intermembrane space. By applying the same method as described above for hexokinase to kidney and brain mitochondria, we observed a concentration of the activity of creatine kinase and nucleoside diphosphate kinase in the contact site fraction, while the activity of adenylate kinase remained on top of the gradients in the fraction of soluble proteins (Fig. 1).

On the basis of these results it looked rather likely that two of the energy-transmitting kinases, namely creatine kinase and nucleoside diphosphate kinase, reside inside the contacts. The question of the physiological meaning of this specific location will be discussed later.

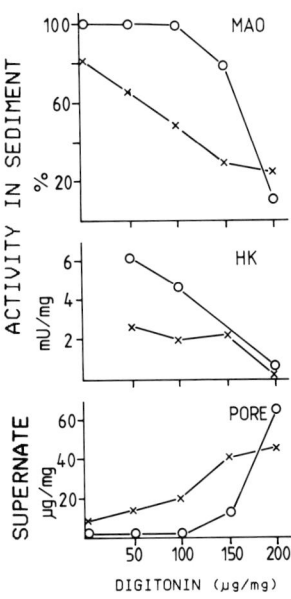

Fig. 3. Isolated liver mitochondria were incubated for 1 min with different digitonin concentrations in the presence of either 5 mM succinate, 4 mM phosphate and MgCl$_2$ and 2 mM ADP ( ) or 20% glycerol ( ). The mitochondria were subsequently centrifuged in a table-top centrifuge and activity of hexokinase (HK) and monoamine oxidase (MAO) was determined in the pellet. The outer membrane in the supernatant was sedimented by ultracentrifgation to measure the concentration of porin by immuno decoration of transblotted polyacrylamide electrophoreses.

Effect of digitonin

By comparing the outer membrane in the contact fraction with the pure outer membrane separated by density gradient centrifugation we observed differences in protein composition and cholesterol content (Ohlendieck *et al.*, 1986; Kottke *et al.*, 1988). The uneven effect of digitonin on the outer membrane may result from this heterogeneity because it is characterized by an almost specific detachment of those parts of the outer membrane beyond the contacts (Hackenbrock *et al.*, 1975; Brdiczka *et al.*, 1976). In accordance with this the activity of enzymes residing in the contact zones, like hexokinase and creatine kinase, could not be removed by digitonin treatment, whereas the activity of adenylate kinase was liberated parallel to the detachment of the outer membrane and marker enzymes like monoamine oxidase (Kottke *et al.*, 1988). Furthermore, digitonin treatment proved to be quite a good method to indicate differences in contact site frequency. When we studied the digitonin effect on mitochondria in different functional states, we observed that more outer membrane markers and pore protein remained bound to the mitoplast fraction of phosphorylating mitochondria while a significantly higher amount of pore protein and of outer membrane was removed by digitonin from uncoupled (Ohlendieck *et al.* 1986) or glycerol treated mitochondria (Fig. 3). This result was directly comparable with the observations in electron microscopy, where uncoupling and incubation with glycerol reduced the number of contact sites, while it was increasing in phosphorylating mitochondria (Knoll *et al.*, 1983). The amount of hexokinase activity which remained bound to the mitoplast fraction correlated with the concentration of porin in this fraction (Fig. 3).

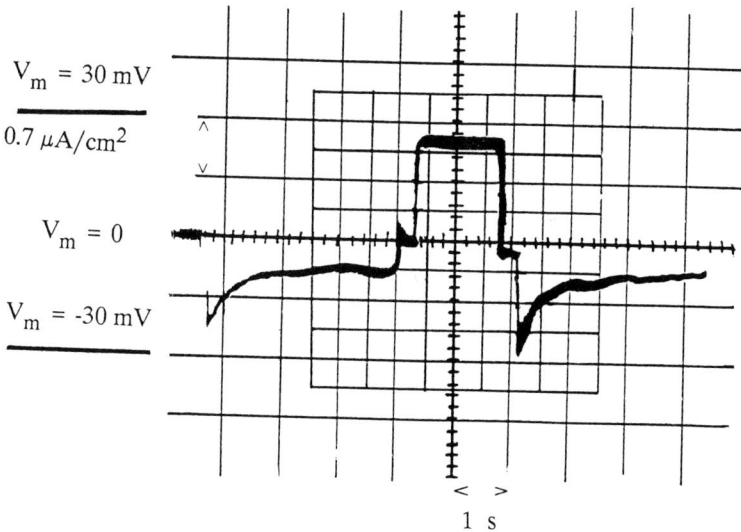

Fig. 4. Rat liver porin was reconstituted in an asolectin bilayer membrane by addition of 100 ng/ml of protein to the cis-side of the membrane. The membrane current as a function of time was recorded after application relative to the cis-side of -30 mV (lower trace) or +30 mV (upper trace). The solution on both sides of the membrane contained 1 M KCl.

REGULATION OF THE OUTER MEMBRANE PORE BY THE MEMBRANE POTENTIAL

Provided we are allowed to take for granted the difference in location of adenylate kinase and the other kinases, a satisfying explanation for the preferential location in the contact sites has to be in terms of a specific property of the outer membrane pore permeability in the contact sites. In support of this argument, several authors (Erickson-Viitanen *et al.*, 1982; Brooks & Suelter, 1987) observed a regulatory effect of the outer membrane on creatine kinase activity. To explain the regulation of the outer membrane pore it seems pertinent to think of a property which is distinct from bacterial porins, namely the voltage dependence (Benz, 1985). When reconstituted in planar bilayers the pore can switch to a lower conducting state depending on the applied voltage (Colombini, 1979; Roos *et al.*, 1982; Benz, 1985; Ludwig *et al.*, 1988). Taking into account the high ion conductivity of the pore, the existence of an electrochemical potential across the outer membrane cannot be expected. However, in the contact sites a transduction of the inner membrane potential across the outer membrane appears possible and would lead to a dynamic compartmentation at the mitochondrial surface regulated by the formation of the contact sites.

Characterization of the low conducting state of the outer membrane pore

At a voltage higher than 30 mV (Colombini *et al.*, 1979; Roos *et al.*, 1982) the pore switches to a lower conductivity and becomes cationically selective. In intact mitochondria we recently observed that the pore in the low conducting state becomes impermeable to ADP and ATP. This means that the inner membrane potential might exert control on the adenine nucleotide transport in the contacts by switching the pore to the low conducting state and thereby prevent the equilibrium of anionic metabolites with the cytosol. Supposing that low conducting pores are present in the contact sites we were then at the liberty to expect pores in the open state beyond the contacts which are characterized by high conductivity and low anion selectivity and would thus allow creatine phosphate to leave the outer mitochondrial compartment. To explain how hexokinase can get access to the mitochondrial ATP we have to consider that the membrane potential across the inner membrane is not constant and the voltage-dependent regulation could result in a frequent change between anion and cation selectivity according to the fluctuations of the inner membrane potential. At this point the asymmetry of the reduced pore conductance and cation selectivity comes into discussion. This property of the closed state was observed when the pore protein was inserted into the planar bilayer only from one side (Fig. 4).

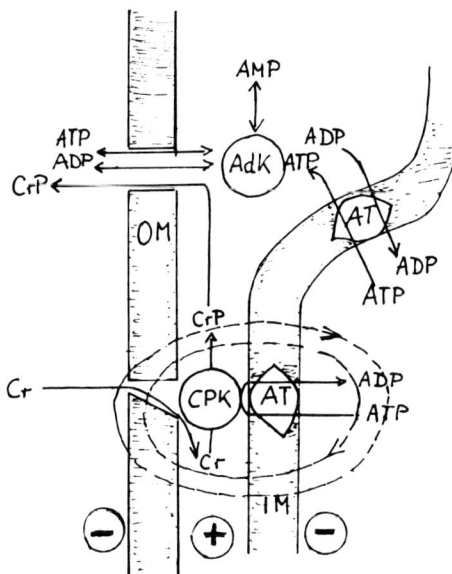

Fig. 5. Scheme showing the difference in transport properties of the pore inside and beyond the contact sites as regulated by the inner membrane potential. IM = inner membrane, OM = outer membrane.

When the potential of 30 mV was negative relative to the side where the pore was added to the bilayer, the conductance was low, while it was significantly higher and low anion selective upon application of the same voltage of the opposite polarity. It seems likely that the orientation and behaviour of the pore in the artificial membrane resembles that of the pore in the outer membrane because Tedeschi et al. (1987) by patch clamping the outer membrane in intact mitochondria observed the same dependence of the pore conductance on voltage and polarity. The observation of an asymmetric effect of the membrane potential on the pore conductance is directly transferable to a model which applies to the regulation of all kinases in the contact sites (Fig. 5). The model assumes the existence of a small, 1 nm, cleft between the two boundary membranes in the contact sites which would result in a membrane potential (negative outside) across the outer membrane created by transduction of the inner membrane potential. Consequently the pore will be low conducting and cationically-selective in the contact sites, whereas it will be high conducting and anionically-selective beyond the contacts. To base this assertion of pore regulation on some evidence in intact mitochondria, we studied the inhibition by creatine phosphate of the creatine kinase reaction starting from creatine and ATP (Fig. 6). The inhibitory effect of creatine phosphate on the creatine kinase reaction was significantly reduced in intact or digitonin-treated mitochondria compared to the free enzyme. In a comparable experiment the effect of AMP on the reaction of adenylate kinase assayed with ADP was studied (Fig. 6). In this case AMP efficiently inhibited the reaction of the enzyme al-

370

ready in intact mitochondria. This observation supported the idea of a different regulation of the outer membrane pores beyond the contacts compared to those located inside, where the membrane potential would exert control only on the latter pores and the enzymes located in the contact sites, like creatine kinase. This view is consistent with the concept of dynamic compartmentation of adenine nucleotides in the intermembrane space which has been recently postulated by Gellerich *et al.* (1987).

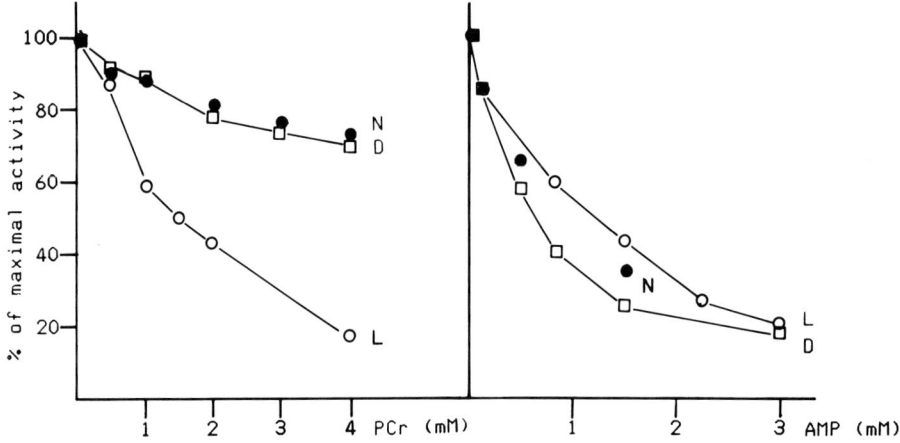

Fig. 6. The activity of creatine kinase and adenylate kinase was determined in the presence of 5 mM succinate and 2 mM phosphate and $MgCl_2$ in samples of brain mitochondria which were either intact (N), treated with digitonin (0.1 mg/mg of protein, D) or disrupted by 1% lubrol (L). The reaction of creatine kinase, measured from creatine and ATP, was inhibited by increasing concentrations of creatine phosphate and the activity of adenylate kinase, starting from ADP (in the presence of oligomycin), was inhibited by different concentrations of AMP.

CONCLUSION

The adenine nucleotide translocator performs active export of ATP without concomitant active uptake of ADP, although the ADP gradient across the inner membrane has an inside-out direction. There are two ways to meet these specifications, which come into existence inside the contact sites, first the limitation of transport through the outer membrane pore governed by the inner membrane potential, so that equilibration with the phosphorylation potential in the cytosol is prevented. The second is the direct communication of kinases with the translocator serving to displace the translocation process from equilibrium. Consistent with this model is the preferential location of several kinases inside the contact sites and the reduced permeability of the pore protein in the low conducting state for adenine nucleotides and presumably creatine phosphate.

REFERENCES

Benz R 1985) Porin from bacterial and mitochondrial outer membranes. CRC Crit Rev Biochem 19:145-190

Benz R, Wojtczak L, Bosch W, Brdiczka D (1988) Inhibition of adenine nucleotide transport through the mitochondrial porin by a synthetic polyanion. FEBS Lett 210:75-80

Bessman SP, Carpenter CL (1985) The creatine creatine-phosphate energy shuttle. Ann Rev Biochem 54:831-865

Brdiczka D, Pette D, Brunner G, Miller F. (1968) Kompartimentierte Verteilung von Enzymen in Rattenlebermitochondrien. Eur J Biochem 5:294-304

Brdiczka D, Schumacher D (1976) Iodination of peripheral mitochondrial membrane proteins in correlation to the functional states of the ADP/ATP carrier. Biochem Biophys Res Commun 73:823-832

Brdiczka D, Knoll G, Riesinger I, Weiler U, Klug G, Benz R, Krause J (1986) Micro-compartmentation at the mitochondrial surface: its function in metabolic regulation. In: Brautbar N (ed) Myocardial and Skeletal Muscle Bioenergetics. Plenum, Press New York, pp 55-69

Brooks SPJ, Suelter CH (1987) Compartmented coupling of chicken heart creatine kinase to the nucleotide translocase requires the outer membrane. Arch Biochem Biophys 257:144-153

Colombini M (1979) A candidate for the permeability pathway of the outer mitochondrial membrane. Nature 279:643-645

Erickson-Viitanen S, Viitanen P, Geiger PJ, Yang WC, Bessman SP (1982) Compartmentation of mitochondrial creatine phosphokinase. J Biol Chem 257:14395-14404

Fiek Ch, Benz R, Roos N, Brdiczka D (1982) Evidence for identity between the hexokinase-binding protein and the mitochondrial porin in the outer membrane of rat liver mitochondria. Biochim Biophys Acta 688:429-440

Gellerich FN, Schlame M, Bohnensack R, Kunz W (1987) Dynamic compartmentation of adenine nucleotides in the mitochondrial intermembrane space of rat-heart mitochondria. Biochim Biophys Acta 890:117-126

Gots RE, Bessman SP (1974) The functional compartmentation of mitochondrial hexokinase. Arch Biochem Biophys 163:7-14

Hackenbrock CR (1968) Chemical and physical fixation of isolated mitochondria in low-energy and high-energy states. Proc Natl Acad Sci (USA) 61:598-605

Hackenbrock CR, Miller KJ (1975) The distribution of anionic sites on the surface of mitochondrial membranes. J Cell Biol 65:615-630

Hebisch S, Sies H, Soboll S (1986) Function dependent changes in the subcellular distribution of high energy phosphates in fast and slow rat skeletal muscles. Pflügers Arch 406:20-24

Inui M, Ishibashi S (1979) Functioning of mitochondria-bound hexokinase in rat brain in accordance with generation of ATP inside the organelle. J Biochem 85:1151-1156

Jacobs H, Held HW, Klingenberg M (1964) High activity of creatine kinase in mitochondria from muscle and brain. Evidence for a separate mitochondrial isozyme of creatine kinase. Biochem Biophys Res Commun 16:516-521

Jacobus WE (1985) Respiratory control and the integration of heart high-energy phosphate metabolism by mitochondrial creatine kinase. Ann Rev Physiol 47:707-725

Jacobus WE, Evans JJ (1977) Nucleoside diphosphokinase of rat heart mitochondria. Dual localization in matrix and intermembrane space. J Biol Chem 252:4232-4241

Klingenberg M, Held HW (1982) The ADP/ATP translocation in mitochondria and its role in intracellular compartmentation. In: Sies H (ed) Metabolic Compartmentation. Academic Press, New York, pp 101-122

Knoll G, Brdiczka D. (1983) Changes in freeze-fracture mitochondrial membranes correlated to their energetic state. Biochim Biophys Acta 733:102-110

Klug G, Krause J, Östlund AK, Knoll G, Brdiczka D (1984) Alteration in liver mitochondrial function as a result of fasting and exhaustive exercise. Biochim Biophys Acta 764:272-282

Kottke M, Adams V, Riesinger I, Bremm G, Bosch W, Brdiczka D, Sandri G, Panfili E (1988) Mitochondrial boundary membrane contact sites in brain: Points of hexokinase and creatine kinase location and of control of Ca^{2+} transport. Biochim Biophys Acta 935:87-102

Lindén M, Gellerfors P, Nelson BD (1982) Pore protein and hexokinase-binding protein from the outer membrane of rat liver mitochondria are identical. FEBS Lett 141:189-192

Ludwig O, Benz R, Schultz IE (1988) Porin of paramecium mitochondria: Isolation, characterization and ion selectivity of the closed state. Eur J Biochem (submitted)

Ohlendieck K, Riesinger I, Adams V, Krause J, Brdiczka D (1986) Enrichment and biochemical characterization of boundary membrane contact sites in rat-liver mitochondria. Biochim Biophys Acta 860:672-689

Roos N, Benz R, Brdiczka D (1982) Identification and characterization of the pore-forming protein in the outer membrane of rat liver mitochondria. Biochim Biophys Acta 686:204-214

Saks VA, Rosenshtraukh LV, Smirnov N, Chazov EI (1987) Role of creatine phosphokinase in cellular function and metabolism. Can J Physiol Pharmacol 56:691-706

Soboll S, Scholz R, Heldt HW (1978) Subcellular metabolite concentrations. Dependence of mitochondrial and cytosolic ATP systems on the metabolic state of perfused liver. Eur J Biochem 87:377-390

Tedeschi H, Mannella CA, Bowman CL (1987) Patch clamping of outer mitochondrial membrane. J Membrane Biol 97:21-29

Van Venetie R, Verkleij AJ (1982) Possible role of non-bilayer lipids in the structure of mitochondria. Biochim Biophys Acta 692:397-405

Viitanen PV, Geiger PJ, Erickson-Viitanen S, Bessman SP (1984) Evidence for functional hexokinase compartmentation in rat skeletal muscle mitochondria. J Biol Chem 259:9679-9684

Wallimann Th, Eppenberger HM (1985) Localization and function of M-line bound creatine kinase. M-Band model and creatine phosphate shuttle. In: Shay IW (ed) Cell and Muscle Motility, Vol 6. Plenum Press, pp 239-285

Weiler U, Riesinger I, Knoll G, Brdiczka D (1985) The regulation of mitochondrial-bound hexokinases in the liver. Biochem Medicine 33:223-235

Control of Mitochondrial Energy Production in Vivo

D.J. TAYLOR

MRC Biochemical and Clinical Magnetic Resonance Unit, University of Oxford, John Radcliffe Hospital, Oxford OX3 9DU, England

The study of purified enzymes and isolated mitochondria under well-defined conditions has provided a wealth of information about cellular metabolism over the past 50 years. However, results from studies on these reconstructed systems define the possible range of activity rather than the actual activity *in vivo*. In intact tissue the metabolic networks which we divide artificially into "glycolysis" and "oxidative phosphorylation" work as a unit, responding in concert to stimuli and effectors which are constantly in flux. The relationship of the properties of macromolecules and organelles to tissue metabolism can ultimately be understood only through non-invasive studies *in vivo*. [31]Phosphorus nuclear magnetic resonance (^{31}P MRS) allows continuous measurement of the major phosphorus-containing metabolites and intracellular pH (pH_i) in many tissues (Radda & Taylor, 1985; Radda *et al.*, 1988). These metabolites are ATP, phosphocreatine (PCr), P_i, phosphodiesters (PDE) and phosphomonoesters (PME). The information on all of these variables is gathered simultaneously and without affecting the processes being investigated. This technique provides an excellent means of investigating the integration of mitochondrial activity into the overall energy metabolism of muscle. The approach in our laboratory has been to study the control and coordination of these pathways in normal and diseased human by observing with MRS the biochemical response to stress. Exercise provides an energetic demand that can be varied by intensity and duration; glycolytic, glycogenolytic and oxidative defects impose other specific metabolic stresses on cellular energetics.

A. Azzi et al. (Eds.)
Anion Carriers of Mitochondrial Membranes
© Springer-Verlag Berlin Heidelberg 1989

374

The investigations described below have been carried out over several years by many investigators in Professor G.K. Radda's laboratory in Oxford, and this is reflected in the authorship of many of the references cited in the text.

INVESTIGATION OF HUMAN SUBJECTS

Finger flexors and gastrocnemius are the two muscles most often investigated in our laboratory (Taylor *et al.*, 1983; Hands *et al.*, 1986). These tissues are easily interrogated by a surface coil and they can be exercised within the magnet bore. Finger flexors are exercised by squeezing a rubber bulb or a handgrip, gastrocnemius by pushing the foot against a pedal. The subject is positioned in a 1.9 tesla superconducting magnet in such a way that the tissue to be investigated lies next to the transmitter/receiver coil in the homogeneous portion of the magnetic field. In the case of finger flexor muscle this involves inserting the arm into the 25 cm bore of the magnet. For gastrocnemius studies a whole body magnet with a 60 cm bore is used. MRS signals are received from resting or contracting muscle and accumulated over ≥ 15 s depending on the conditions of the experiment. They are then processed into a series of spectra (Fig. 1).

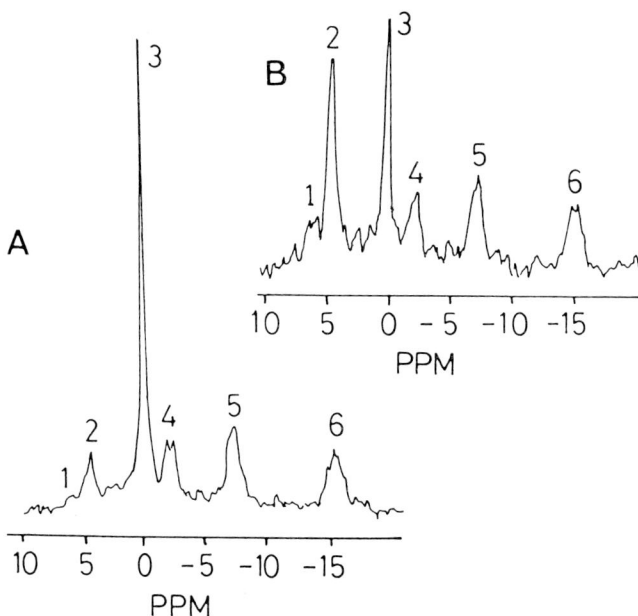

Fig. 1. ^{31}P MR spectra of normal human skeletal muscle (A) at rest and (B) during exercise. The x axis is resonant frequency expressed in parts per million relative to PCr. The y axis is signal intensity in arbitrary units. Peak assignment: (1) PME, (2) P_i, (3) PCr, (4) γ-ATP, (5) α-ATP + NAD(H), 6) β-ATP. The pH_i was (A) 7.02 and (B) 6.86.

Mitochondria make up only a small fraction of the muscle cell volume, so the MRS signals originate almost exclusively in cytoplasmic metabolites. The pH measurement is derived from the separation of the PCr and P_i peaks. This distance varies with pH because the exact position of the P_i peak depends on the relative contributions of $H_2PO_4^-$ and HPO_4^{2-}. These species resonate at different frequencies but are in fast chemical exchange. Peak areas are proportional to concentration, but because there is no internal standard the area of the ATP peak in the resting muscle is taken as 8.2 mM, and other concentrations are derived from the proportions of peak area to ATP peak area (Arnold et al., 1985). Results are given as the mean ± S.D.

NORMAL MUSCLE VERSUS MITOCHONDRIAL MYOPATHY

^{31}P MR spectra from normal muscle at rest (Fig. 1) show little variation between normal individuals. For example, results from finger flexors of 18 subjects were: pH_i 7.03±0.02; PCr, 39.3±1.7 mM; P_i, 4.1±0.8 mM. The concentration of free (metabolically active) ADP was 6 ± 3 when derived from the creatine kinase equilibrium expression (Veech et al., 1979):

$$PCr + ADP + H^+ \rightleftharpoons ATP + creatine$$

In patients with defective oxidative phosphorylation an abnormality can usually be detected even at rest (Arnold et al., 1985). In 15 patients with mitochondrial myopathy of several different types in whom we have investigated finger flexor muscle, 11 had PCr concentrations more than 2 S.D. lower than the normal and in 12 ADP was more than 2 S.D. higher. We were able to measure the ATP concentration in two of these patients and it was normal. Four patients had abnormally elevated P_i. Thus, a general feature of impaired oxidative phosphorylation was a low phosphorylation potential ($[ATP]/[ADP] \times [P_i]$) at rest. Intracellular pH was not low in any of these patients even though most had high blood lactates at rest. In three patients the pH_i was abnormally high (≥ 7.10). A spectrum from a patient with an electron transport defect showing these abnormalities is illustrated in Fig. 2.

Mitochondrial defects can affect tissues in addition to muscle, and not all muscles are equally affected. Three of the patients who had abnormalities in resting finger flexors had normal gastrocnemius at rest. Even with exercise, results are often much more easily detectable in finger flexors than in gastrocnemius. We have studied two patients with electron transport disorders who have shown abnormal ^{31}P MR spectra of brain as well as in muscle (Hayes et al., 1985; Edwards et al., 1985). One patient with central nervous system disorder (brain atrophy, possible calcification near the basal ganglia, seizures, exercise-induced vomiting) had low phosphorylation potential in brain as well as in muscle. Detailed biochemical studies on a muscle biopsy sample showed that the lesion was in the region of NADH-coenzyme Q reductase. Muscle from the second patient oxidised pyruvate and malate at only 20% the normal rate. After strenuous exercise, his brain pH was found to be low at 6.88.

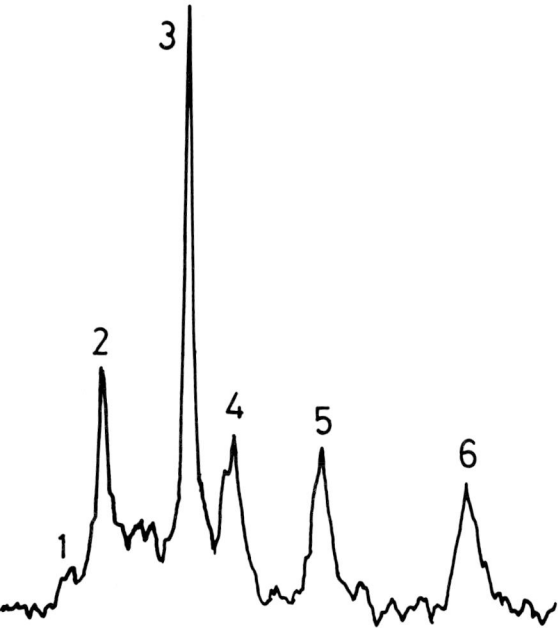

Fig. 2. ^{31}P MR spectrum of the resting skeletal muscle of a patient with coenzyme Q reductase deficiency. Note the large P_i and low PCr signals compared to normal muscle (Fig. 1A). The pH_i (7.10) was significantly elevated. Peak assignments are the same as in Fig. 1.

When the patients with mitochondrial myopathy exercised, PCr was depleted more quickly than usual. This was not unexpected because the ability to produce ATP is often severely restricted in these disorders, but we were surprised to find that the decrease in intracellular pH was compared to the change in PCr. These patients produce excessive amounts of lactate as shown by the high concentrations in the blood during and after exercise. The MRS results suggest that their muscles must have adapted to handling a large acid load and are able to eliminate hydrogen ions faster than normal muscle. The pH recovery data is consistent with this. A rapid recovery of pH_i to resting values can be seen in the recovery period after exercise (Arnold et al., 1985). Low pH_i is accompanied by slow rates of oxidative phosphorylation (Arnold et al., 1984), and these patients cannot afford to decrease the rate of energy production.

Several lines of experimental evidence obtained by MRS suggest that mitochondria provide the energy for recovery after exercise. (1) In normal subjects,

PCr repletion after exercise does not begin unless circulation is intact (Taylor *et al.*, 1983). (2) Patients with McArdle's syndrom (myophosphorylase deficiency) cannot produce ATP from breakdown of glycogen, the major source of glycosyl units for glycolysis in muscle, yet these patients do not have slow recovery after exercise. (3) Patients with severe defects in oxidative phosphorylation have half recovery times for PCr of several minutes compared to <1 min in normals (Arnold *et al.*, 1985).

Varying the biochemical state of the muscle at the end of exercise can give information about the factors which influence mitochondrial ATP synthesis. This can be measured by the PCr repletion rate because the creatine kinase reaction (see above) is near equilibrium. PCr repletion is slowed when ATP is depleted (Taylor *et al.*, 1986). In these studies, normal subjects were exercised to exhaustion. In some cases a decrease in [ATP] of 30-50% occurred, and this ATP depletion was accompanied by a very slow PCr recovery (half time 5.3 min for depleted muscle, 1.1 min when there was no depletion). If the P_i is trapped in sugar phosphates which are only slowly metabolised, as can happen in phosphofructokinase deficiency, then ATP synthesis is limited by the rate of release of P_i (Argov *et al.*, 1987). As mentioned above, a relationship between PCr recovery and pH has been reported (Arnold *et al.*, 1984). When [H$^+$] is high, PCr repletion is slow. H$^+$ is one of the reactants in the creatine kinase reaction, so changing the pH also alters [ADP]. We have found that there is a hyperbolic relationship between [ADP] at the end of excrise and PCr repletion rate during the initial recovery phase. When results from different degrees and duration of exercise are used, a K_m of 27 μM for ADP for mitochondrial ATP production and a maximum rate of ATP synthesis of 43 mM·min^{-1} are found (Fig. 3).

Most of the values shown in Fig. 3 were derived from studies of normal subjects, but those with the highest ADP concentrations (and the fastest PCr recoveries) are from patients with phosphorylase deficiency. The lack of lactic acid production in this deficiency results in a high pH_i during exercise and, consequently, higher ADP concentration are achieved than in normal muscle. Muscle defective in oxidative production of ATP also reaches high [ADP] during exercise due to an excessive decrease in PCr with only a moderate decrease in pH. However, in this type of muscle, the electron transport chain is unable to respond to the high [ADP], and a sensitive index of mitochondrial competence is the ADP half recovery time. This $T_{1/2}$ is very rapid (about 15 s) for normal muscle. In patients with defects in glycogenolysis and glycolysis, oxidative processes are intact and recovery is not impaired. The highest ADP concentrations in the muscles of these patients are found shortly after exercise begins. It decreases as blood flow to the muscle increases and mitochondrial shift from near state 4 toward state 3.

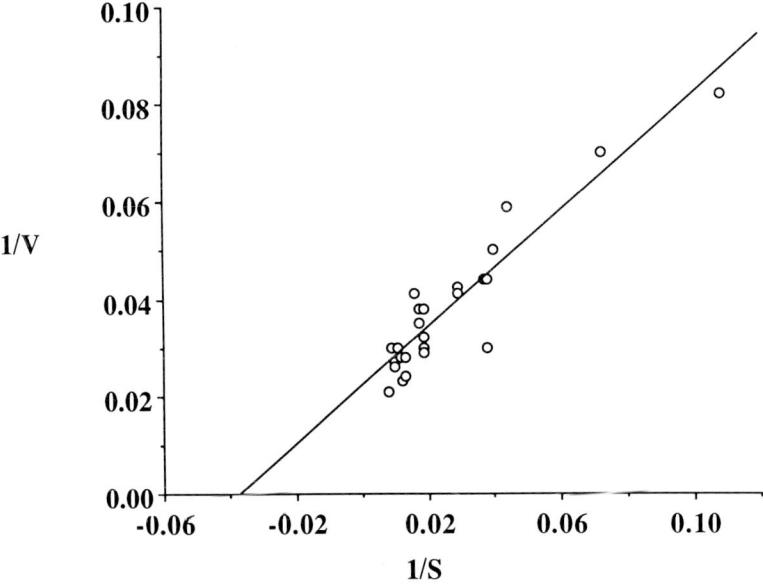

Fig. 3. The relationship between PCr resynthesis rate (mM·min^{-1}) and [ADP] (μM). Results are from initial rates of PCr repletion following exercise of different intensities and duration and from the calculated [ADP] at the end of the exercise. Each point represents data from a different subject.

MALATE-ASPARTATE SHUTTLE DEFECT

A patient with an unusual metabolic defect has given us some further insight into the relationship between cytoplasmic and mitochondrial function *in vivo* (Hayes *et al.*, 1987). The patient, a 27 year old man, presented with symptoms and signs indicative of a glycolytic or glycogenolytic disorder. He complained of exercise-induced muscle pain and passage of red urine. Histologically, the muscle showed some non-specific myopathic changes with a few clusters of subsarcolemmal mitochondria. The stain for phosphorylase activity was normal.

^{31}P MRS investigations were carried out on finger flexors and gastrocnemius. Results from resting muscle were essentially normal. Those from exercise studies were initially puzzling because they showed features we had noted in both glycolytic and mitochondrial involvement: (1) PME (mostly sugar phosphates in muscle) increased markedly in exercise. This is a usual finding in glycolytic deficiencies like phosphofructokinase, in which fructose 6-phosphate cannot be metabolized (Edwards *et al.*, 1982

Argov *et al.*, 1987). We have not observed these large changes in PME in mitochondrial disease. (2) During exercise the pH_i always remained in the range found in control subjects and was not elevated. During an ischaemic exercise, the pH_i of forearm muscle decreased to 6.43. Thus the muscle was able to produce lactate, showing that the glycolytic pathway was intact. Blood lactate was normal at rest, but bicycle exercise provoked a rise to 3.0 mM, significantly higher than the normal increase to 0.83 ± 0.41 mM. (3) PCr decreased rapidly during exercise. This is consistent with a block in either the glycolytic or oxidative pathway. (4) Recovery of PCr and P_i were both slow and both approached their resting concentrations at about the same rate. Slow recovery is associated with defects in oxidative phosphorylation, but we have found that in mitochondrial myopathy patients the recovery times of the two metabolites retain their normal relationship. That is, the P_i half recovery time, T½, is about twice as fast as the T½ for PCr (27 ± 12 s *vs* 57 ± 16 s in 20 normal subjects). We have also observed this phenomenon in isolated mitochondria, and so have attributed this "disappearance" of P_i from the spectrum to its uptake into mitochondria during recovery. Thus, it seemed possible that the defect in this patient involved the transport of P_i into the mitochondria.

Subsequently, mitochondria were isolated from a muscle biopsy taken from quadriceps. No primary defects in the respiratory chain could be demonstrated. Mitochondria were well coupled and had a normal rate of oxygen uptake with pyruvate + malate, glutamate + malate, succinate + rotenone, and ascorbate + TMPD. The MRS results suggested that the defect was interfering with both glycolysis and oxidative phosphorylation, and, indeed, rate of oxidation of exogenous NADH by the malate-aspartate shuttle system of the inner mitochondrial membrane was only about 20% of control values. The MRS data are consistent with such a defect because the slow re-oxidation of cytoplasmic NADH could impair glycolysis and slow the rate of substrate supply to the mitochondria. We therefore have concluded that the muscle problems experienced by this patient probably arise from the deficiency in malate-aspartate shuttle activity. The results suggest that re-oxidation of NADH by lactate dehydrogenase alone may not allow normal functioning of the glycolytic pathway. They also suggest that the α-glycerophosphate shuttle activity is not adequate to cope with NADH oxidation in human muscle *in vivo*.

CA^{2+}-ATPASE DEFICIENCY

Another patient, also with a rare muscle disorder, has given us the chance to observe the effects of raised intracellular Ca^{2+} on muscle energetics (Brosnan *et al.*, 1987). This patient is a man in his 40's with a 15 year history of muscle fatigue and cramps after exercise. His sister suffers from the same condition. His symptoms are partially relieved by verapamil, a Ca^{2+} antagonist. He was thoroughly investigated clinically, but no diagnosis was made. Histological and biochemical studies on muscle samples had not revealed any abnormalities. However, MRS findings, described below, lead us to assay for Ca^{2+}-ATPase in the sarcoplasmic reticulum fraction of a muscle biopsy sample, and this had only 10% of normal activity.

MR spectra from resting muscle were within normal limits. In spite of this disorder, the patient was strong and active with well-developed muscles. Yet, during aerobic or ischaemic exercise PCr decreased rapidly. The pH_i decrease was greater than normal, consistent with an increased glycolytic rate. The muscle acidified markedly during the first minute post exercise, and this suggests that glycolysis remained active after exercise ceased. PCr and P_i recovery rates were consistent with the low pH values reached during exercise, and ADP recovery was normal. Thus mitochondrial activity was apparently not inhibited, and the increased rate of PCr depletion suggests an over-utilisation of ATP during exercise.

When Ca^{2+} is released into the cytoplasm to initiate contraction, re-uptake will be very slow in the absence of sarcoplasmic Ca^{2+}-ATPase, and $[Ca^{2+}]$ must remain above the resting concentration in the cytoplasm for longer than normal. It might also reach a higher concentration than normal. The increased $[Ca^{2+}]$ would be expected to result in changes in the activity of phosphorylase which, in turn, would increase glycolytic activity. Mitochondria are thought not to be important in buffering Ca^{2+} because the K_m of the uniport mechanism for the ion is about 10 μM which is too high for the sub-μM concentrations normally found in muscle. However, in this patient it is possible that the $[Ca^{2+}]$ reaches high enough levels to lead to Ca^{2+} cycling by the mitochondria. This would result in energy dissipation and account for the excessive decrease in PCr during exercise.

CONCLUSIONS

The results presented here illustrate how the understanding of bioenergetics is enhanced by studying the interrelationships of metabolic pathways *in situ* under a variety of conditions. The technique of ^{31}P magnetic resonance spectroscopy has been used to investigate the relationship of mitochondrial ATP production to the metabolic state of the cytoplasm and to the overall pattern of muscle energy utilization and production. The results from these studies help to provide a link between our understanding of basic biochemical processes and the physical manifestations of altered intracellular control mechanism.

REFERENCES

Argov Z, Bank WJ, Maris J, Leigh JS, Chance B. (1987) Muscle energy metabolism in human phosphofructokinase deficiency as recorded by ^{31}P nuclear magnetic resonance spectroscopy. Ann Neurol 22:46-51

Arnold DL, Matthews PM, Radda GK (1984) Metabolic recovery after exercise and the assessment of mitochondrial function in human skeletal muscle by means of ^{31}P NMR. Mag Res Med 1:307-315

Arnold DL, Taylor DJ, Radda GK (1985) Investigation of human mitochondrial myopathies by phosphorus magnetic resonance spectroscopy. Ann Neurol 18:189-196

Brosnan MJ, Taylor DJ, Walton J, Radda GK (1987) A study of muscle metabolism in a rare muscle disorder. In: Abstracts of the 6th Annual Meeting, Soc Mag Res Med, New York, p 589

Edwards RHT, Dawson MJ, Wilkie DR, Gordon RE, Shaw D (1982) Clinical use of nuclear magnetic resonance in the investigation of myopathy. Lancet i:725-731

Edwards RHT, Griffiths RD, Radda GK, Taylor DJ (1985) Physiological and metabolic consequences of a defect in mitochondrial pyruvate oxidation. In: Abstracts of the 4th Annual Meeting, Soc Mag Res Med, London, pp 1221-1222

Gadian DG, Radda GK, Ross BD, Hockaday J, Bore P, Taylor D, Styles P (1981) Examination of a myopathy by phosphorus nuclear magnetic resonance. Lancet ii:774-775

Hands LJ, Bore PJ, Galloway G, Morris PJ, Radda GK (1986) Muscle metabolism in patients with peripheral vascular disease investigated by ^{31}P nuclear magnetic resonance spectroscopy. Clin Sci 71:283-290

Hayes DJ, Hilton-Jones D, Arnold DL, Galloway G, Styles P, Duncan J, Radda GK (1985) A mitochondrial encephalomyopathy: A combined ^{31}P magnetic resonance and biochemical investigation. J Neuro Sci 71:105-118

Hayes DJ, Taylor DJ, Bore PJ, Hilton-Jones D, Arnold DL, Squire MV, Gent AE, Radda GK (1987) An unusual metabolic myopathy: a malate-aspartate shuttle defect. J Neuro Sci 82:27-39

Radda GK (1986) The use of NMR spectroscopy for the understanding of disease. Science 233:640-645

Radda GK, Bore PJ, Gadian DG, Ross BD, Styles P, Taylor DJ, Morgan-Hughes JA (1982) ^{31}P NMR examination of two patients with NADH-CoQ reductase deficiency. Nature (London) 295:608-609

Radda GK, Bore PJ, Rajagopalan B (1984) Clinical aspects of ^{31}P NMR spectroscopy. Brit Med Bull 40:155-159

Radda GK, Rajagopalan B, Taylor DJ (1988) Biochemistry in vivo: An appraisal of clinical magnetic resonance spectroscopy. Magnetic Resonance Annual. Raven Press, New York (in press)

Radda GK, Taylor DJ (1985) Applications of nuclear magnetic resonance spectroscopy in pathology. In: Richter GW, Epstein MA (eds) International Review of Experimental Pathology. Academic Press, New York London, pp 1-58

Ross BD, Radda GK, Gadian DG, Rocker G, Esiri M, Falconer-Smith J (1981) Examination of a case of suspected McArdle's syndrom by ^{31}P NMR. New Eng J Med 304:1338-1342

Taylor J, Bore PJ, Styles P, Gadian DG, Radda GK (1983) Bioenergetics of intact human muscle - a ^{31}P nuclear magnetic resonance study. Mol Biol Med 1:77-94

Taylor DJ, Styles P, Matthews PM, Arnold DL, Gadian DG, Bore PJ, Radda GK (1986) Energetics of human muscle: Exercise-induced ATP depletion. Mag Res Med 3:44-54

Veech RL, Lawson JRW, Cornel NW, Krebs HA (1979) Cytosolic phosphorylation potential. J Biol Chem 254:6538-6547